# Power in a Warming World

## Earth System Governance

Frank Biermann and Oran R. Young, series editors

Oran R. Young, *Institutional Dynamics: Emergent Patterns in International Environmental Governance*

Frank Biermann and Philipp Pattberg, eds., *Global Environmental Governance Reconsidered*

Olav Schram Stokke, *Disaggregating International Regimes: A New Approach to Evaluation and Comparison*

Aarti Gupta and Michael Mason, eds., *Transparency in Global Environmental Governance*

Sikina Jinnah, *Post-Treaty Politics: Secretariat Influence in Global Environmental Governance*

Frank Biermann, *Earth System Governance: World Politics in the Anthropocene*

Walter F. Baber and Robert B. Bartlett, *Consensus in Global Environmental Governance: Deliberative Democracy in Nature's Regime*

Diarmuid Torney, *Global Climate Governance: Conflict and Cooperation between Europe, China, and India*

David Ciplet, J. Timmons Roberts and Mizan Khan, *Power in a Warming World: The New Global Politics of Climate Change and the Remaking of Environmental Inequality*

## Related books from Institutional Dimensions of Global Environmental Change: A Core Research Project of the International Human Dimensions Programme on Global Environmental Change

Oran R. Young, Leslie A. King, and Heike Schroeder, eds., Institutions and Environmental Change: Principal Findings, Applications, and Research Frontiers

Frank Biermann and Bernd Siebenhüner, eds., Managers of Global Change: The Influence of International Environmental Bureaucracies

Sebastian Oberthür and Olav Schram Stokke, eds., Managing Institutional Complexity: Regime Interplay and Global Environmental Change

# Power in a Warming World

The New Global Politics of Climate Change and the Remaking of Environmental Inequality

David Ciplet, J. Timmons Roberts, and Mizan R. Khan

The MIT Press
Cambridge, Massachusetts
London, England

© 2015 Massachusetts Institute of Technology

MIT Press books may be purchased at special quantity discounts for business or sales promotional use. For information, please email special_sales@mitpress.mit.

This book was set in Sabon LT Std by Toppan Best-set Premedia Limited. Printed on recycled paper and bound in the United States of America.

Library of Congress Cataloging-in-Publication Data is available.

ISBN: 978-0-262-02961-2
10  9  8  7  6  5  4  3  2  1

# Contents

Series Foreword   vii
Preface and Acknowledgments   ix

1  Trading a Livable World   1

2  Power Shift   23

3  Beyond the North–South Divide?   53

4  Manufacturing Consent   75

5  The Politics of Adaptation   101

6  The Staying Power of Big Fossil   133

7  Society Too Civil?   155

8  Contesting Climate Injustice   181

9  Power in a Future World   205

10  Linking Movements for Justice   235

Notes   253
References   285
Index   319

# Series Foreword

Humans now influence all biological and physical systems of the planet. Almost no species, no land area and no part of the oceans has remained unaffected by the expansion of the human species. Recent scientific findings suggest that the entire earth system now operates outside the normal state exhibited over the past 500,000 years. Yet at the same time, it is apparent that the institutions, organizations, and mechanisms by which humans govern their relationship with the natural environment and global biogeochemical systems are utterly insufficient—and poorly understood. More fundamental and applied research is needed.

Yet such research is no easy undertaking. It must span the entire globe because only integrated global solutions can ensure a sustainable co-evolution of natural and socio-economic systems. But it must also draw on local experiences and insights. Research on earth system governance must be about places in all their diversity, yet seek to integrate place-based research within a global understanding of the myriad human interactions with the earth system. Eventually, the task is to develop integrated systems of governance, from the local to the global level, that ensure the sustainable development of the coupled socio-ecological system that the Earth has become.

The series Earth System Governance is designed to address this research challenge. Books in this series will pursue this challenge from a variety of disciplinary perspectives, at different levels of governance, and with a plurality of methods. Yet all will further one common aim: analyzing current systems of earth system governance with a view to increased understanding and possible improvements and reform. Books in this series will be of interest to the academic community but will also inform practitioners and at times contribute to policy debates.

This series is related to the long-term international research effort "Earth System Governance Project," a core project of the International Human Dimensions Programme on Global Environmental Change.

Frank Biermann, *Vrije Universiteit Amsterdam*
Oran R. Young, *University of California, Santa Barbara*
Earth System Governance Series Editors

# Preface and Acknowledgments

On the final evening of the global climate change negotiations in Copenhagen in 2009, Venezuela's lead negotiator Claudia Salerno Caldera pounded her fist on the table trying to get the attention of the Danish chair, Lars Løkke Rasmussen, before he left the podium. Having cut her hand, and with blood pouring from the wound, she exclaimed, "Even if we have to cut our hand and draw blood to make you allow us to speak, we will do so." Caldera was furious about what she and many other representatives of developing countries considered to be a highly unequal and ineffective framework for addressing climate change in the newly introduced Copenhagen Accord and a decision-making process that many considered to be a violation of United Nations procedure.

Despite the drama and conflict in Copenhagen, the negotiations have since moved forward to achieve general consensus on agreements developed since 2009. In Durban in late 2011, an emissions reduction framework with "a protocol, a legal instrument, or an agreed outcome with legal force" involving all countries was pushed back to the year 2020. What the international community has agreed to is an emissions reduction framework that, as currently configured, condemns the planet to 3.5–4.5 degrees Celsius of warming.[1] Scientists consider this temperature rise far above what will trigger catastrophic environmental events around the world. In the case of such temperature rise, some countries would completely disappear with rising sea levels, while others would face a diverse set of catastrophic consequences, from drought to flooding to heat waves and storm surges on top of sea level rise. It is now well documented that the poorest countries, which have the lightest footprint on the climate, are suffering worst and first from climate change and will continue to do so in a warmer world.

These are our core questions: How did we get to this point, and is there any way out? *Power in a Warming World* draws on nearly three decades of our experience as observers and participants in the UN global climate change negotiations. We have been in roles as government delegation member (Mizan), negotiation group research and writing support (Dave, Timmons, and Mizan), nongovernmental organization participants (Dave and Timmons), and researchers (all three). We have used these opportunities to take a close look at how global environmental inequality has been made, reproduced, and contested through this political process. In doing so, we offer a window into the complex global politics of power and consent that provides insights beyond climate change, involving unlikely divisions and alliances between players in both the global North and South. We demonstrate that environmental inequality has been preserved in these negotiations by both big systems processes and grounded social relationships of domination, accommodation, and consent.

Beyond material self-interest and the use of coercive force, there are three more currencies of power: identities, ideas, and institutions. National identities survive and are reshaped from experiences such as wartime alliances and colonialism, explaining the behavior of developed nations and developing nations, who continue formal and informal negotiations in historic blocs even when it may not be in their longer-term self-interest. But new identities are emerging that disrupt those enduring coalitions, and these get leveraged to gain and preserve new privileges. Ideas of justice and of viable policy solutions come into the negotiations from civil society, national delegations, and international institutions like the UN Secretariat or global trade organizations. The quirky institutional structure of UN climate negotiations, and the larger governance architecture that has emerged around the issue of climate change, shape what is possible and which interests, identities, and ideas gain and hold sway. In this way, we offer a challenge to much of the literature on international politics that defines power solely in material and coercive terms or, alternatively, rejects power altogether and instead focuses on institutions or ideas. And we provide a rare empirical account of how everyday relationships of inequality are reproduced and contested in the international realm, and in environmental politics in particular.

*Power in a Warming World* is geared to lay readers, climate experts, and upper-level undergraduate and graduate students in the areas of international relations, global politics, environmental sociology, geography, public policy, social movement studies, and environmental studies. The scholarly literature on climate change politics is gaining volume and substance. We build on four distinct groups in the literature that deal explicitly with contemporary global climate change politics. First, there are books that deal with normative or ethics-based arguments of justice in addressing climate change. A second group of scholars more explicitly deals with relationships of power and inequality in global politics and focus on obstacles to preventing progress. Roberts's previous MIT Press book (co-authored with Bradley Parks), *A Climate of Injustice* (2007), falls in that category, and we hope this book will be a useful sequel, updating its history from the 2009 Copenhagen negotiations and expanding it in new directions. A third, and related, series of books offers possibilities for overcoming gridlock. These include a focus on the need to leverage rational national self-interest rather than wishful thinking to overcome gridlock, how international law can be tailored to facilitate cooperation and effective action, and the need for an individually and collectively rational and fair climate treaty. The fourth group of scholars focuses on nonstate actors in global climate politics including civil society and market actors, multilateral development banks, donors, and cities across multiple levels of political organization.

We hope that *Power in a Warming World* addresses a key need by providing an analysis of power that is attentive to both macrostructural and microrelational processes that have shaped inequality and inaction in the contemporary UN climate negotiations and beyond. Following Barnett and Duvall, we define power as the "production, in and through social relations, of effects that shape the capacities of actors to determine their circumstances and fate."[2] This conception highlights that power is embedded in social relations and that its effects work to the advantage of some and the disadvantage of others. It also emphasizes that power determines the capacities of actors to control outcomes, but it does not prescribe to them a prefabricated desired outcome based on a narrowly defined conception of rationality.

As a whole, the research on contemporary global climate politics to date suffers from some important shortcomings. First, there has been a notable focus on short-term global political dynamics, while neglecting long-term global historical trends, especially steady and abrupt transformations in the structures of the global economy. Second, there has been a shortage of analysis of the social relations of accommodation and consent between actors and the role of ideas, institutions, and identities in the climate negotiations. Third, there has been an overwhelming focus on the politics of emissions reductions (mitigation), while neglecting issues of adaptation and climate finance, areas that have exploded in importance and where we have focused much of our attention over the past decade. Fourth, the literature continues to portray a simplified and no longer accurate view of North–South political alliances, which have grown far more complex over the past decade in the climate negotiations. As we'll discuss, there are now over a dozen negotiating groups, with ten or so in the South and several in the North. Finally, there have been only limited analyses of the relationship between transnational civil society, states, and market actors within the negotiations, an area we consider fundamental for progressive change. By addressing these shortcomings that we see in the literature, we hope to offer some new insights into why major changes have occurred in the negotiations, what accounts for their timing, and what possibilities exist for transformative change down the road, both within and outside the UN process.

We acknowledge the publications where earlier versions of this book's chapters appeared. A portion of chapter 2 was published in Ciplet, Roberts, and Khan (*Edgar Elgar Handbook on Climate Governance*, forthcoming 2015) and part appeared in Roberts (*Global Environmental Change*, 2011). Part of chapter 4 appeared in Ciplet (*Global Governance*, 2015). An earlier version of chapter 5 appeared in Ciplet, Roberts, and Khan (*Global Environmental Politics*, 2013). An earlier version of chapter 8 appeared in Ciplet (*Global Environmental Politics*, 2014).

We acknowledge the amount of focus on the United States in this book: we are two Americans and a Bangladeshi based in the United States for the year we worked most on this book. We are aware that this shades our understanding and influences our types of knowledge. We have also spent time researching this book in South America, Europe, and elsewhere.

Chapter 1 includes an outline of the book's chapters; what remains here is to acknowledge our remarkable support network in completing this exciting and sometimes exhausting project. First, we thank our spouses, Jennifer Ciplet, Holly Flood, and Parvin Khan, and our children: Eliza, Cora, and Marlon (Dave); Quinn and Phoebe (Timmons); and Farhana (Mizan). We express our sincere thanks to Brown University's Watson Institute and Center for Environmental Studies for support, including an office and support for Mizan during his 2012–2013 stay as visiting fellow, and our thanks to Patti Caton, Jeanne Lowenstein, and the Center for Environmental Studies (now the Institute for the Study of Environment and Society) for much support over the years. We thank the Graduate Program in Development for support for Mizan's stay at Brown in fall 2012. For the 2011–2012 work with the Climate and Development

Lab and Mizan's visits, we appreciate support from Anna Karina Wildman and Matt Guttman from Brown's Office of International Affairs, former director of Watson Carolyn Dean, Barbara Sardy, and Katherine Bergeron, former dean of the college.

We have learned from conducting research in support of the Least Developed Countries group, and we deeply appreciate former chair, Pa Jarju Ousman, and the International Institute for Environment and Development (IIED) climate change staffers, Achala Chandani and Saleemul Huq, for facilitating that. We have learned a lot from attendees at the several conferences we have organized at the Watson Institute at Brown, including representatives from a half-dozen key negotiating groups. We are also grateful to the Global Alliance of Waste Pickers, from whom we have learned a great deal. And we have learned from our constant work with and support of members of our Climate and Development Lab at Brown: Guy Edwards, Adam Kotin, Linlang He, Brianna Craft, Spencer Field, Keith Madden, Emily Kirkland, Kelly Rogers, Hanna Ross, Becca Keane, Graciela Kincaid, Daniel Sherrell, Cecilia Pineda, and other current members and alumni. Dave is also grateful for funding from the Switzer Foundation and the Horowitz Foundation for Social Policy, which generously supported him in this research.

We thank Clay Morgan and Beth Clevenger at MIT Press for their supportive and thoughtful editorial wisdom and guidance and their patience as the book got pushed back twice by three busy lives. Series editors Oran Young and Frank Biermann were incredibly supportive from the beginning and provided some useful tough comments at a key early stage when we could act on them. Anonymous reviewers of the prospectus and the first draft manuscript helped shape this final product. We thank Guy Edwards, Damien White and Brian Gareau for reading earlier drafts of the chapters. The errors that remain are, of course, our own.

Finally, a parting word on what we hope to achieve. Modestly, we hope to provide a picture not of an inevitable train wreck of human realpolitik with the geobiophysical climate system that supports us. Rather, we hope to create a useful framework to understand the roots of this political crisis as a tool to help identify pathways forward. The material interests of the global North and South on which we dwell at length here are crucial to understand and acknowledge in developing new frameworks for agreement. We hope this book can help to inform a new generation of global climate solutions. The time is short.

## Preface and Acknowledgments

1.  Climate Interactive's website forecasts 4.5 degrees Celsius of warming by 2100 if national pledges made by April 2013 are met (Climate Interactive 2013). A UN Environmental Programme report (2010) found that even if the Copenhagen pledges are met, the amount of greenhouse gases remaining in the atmosphere would "imply a temperature increase of between 2.5 to 5°C before the end of the century." Other calculations also show that voluntary pledges under the accord, even if implemented, will raise temperature more than 3 degrees Celsius (Rogelj et al. 2010). The International Energy Agency (2012) also predicts 4 degrees Celsius of warming with current pledges.

2.  Barnett and Duvall (2005, 42).

# 1

## Trading a Livable World

### Crisis in Copenhagen

It was 3:00 a.m. on December 20, 2009, during the final plenary session of the United Nations (UN) Copenhagen climate change conference. Representatives of 187 countries gathered as part of the official UN decision-making body on climate change, what is called the "Conference of Parties" (COP), to decide on how the international community would collectively address the problem of climate change. After more than a decade and a half of intense negotiations, many regarded this moment as the last chance to effectively tackle climate change to avoid catastrophic ecological tipping points.

In the hands of most of the delegates was the confidential twelve-paragraph Copenhagen Accord, which radically changed how the nations of the world would address the climate problem. Just hours before, the Accord had been secretly drafted by an unlikely alliance of five countries: the United States, China, India, Brazil, and South Africa. For many developing country delegates, this was the first time over two long weeks of negotiations that they had set eyes on this document.

Acknowledged finally by the chair, the lead delegate of the tiny low-lying island nation of Tuvalu, Ian Fry, turned on his microphone. He looked up at Lars Løkke Rasmussen, the Danish prime minister and chair of the proceedings, and exclaimed: "It looks like we are being offered 30 pieces of silver to betray our people and our future. Our future is not for sale. I regret to inform you that Tuvalu cannot accept this document."[1]

Fry was furious about what he and many other representatives of developing countries considered to be a highly unequal and ineffective framework for addressing climate change. Rather than strengthening the existing international legal process to combat climate change in the Kyoto Protocol, the Accord put forward a "voluntary" framework. Developing country delegates felt that the Kyoto Protocol was being stripped of its teeth; the bedrock of global climate change policy was being fundamentally shattered. Many also objected to a decision-making process they considered a violation of UN procedure.

As part of the Accord, developing countries such as Tuvalu were being offered promises of dollars: $30 billion over the coming three years and $100 billion a year by 2020. Despite these financial promises, delegates of several countries refused to offer their consent to the Accord, and it was not adopted as a legal agreement in Copenhagen.

One year later at the international climate change negotiations in Cancun, the tide had dramatically turned.[2] The main content of the Copenhagen Accord was integrated into the Cancun Agreements, adopted nearly unanimously.[3] In doing so, the international community set in motion a process, solidified in Durban and Doha the next two years, that would replace the legally binding Kyoto Protocol with a voluntary pledge-and-review system and delay core decisions on an alternative path forward until 2015.[4]

In this new approach, there is not an agreed-on aggregate figure for reducing greenhouse gas pollution or a system to ensure that the pledges made are deep enough to meet scientifically required targets. As currently configured, this framework will allow a temperature rise substantially above what scientists predict will trigger catastrophic environmental events around the world.[5] In the case of such temperature rise, several countries, such as Tuvalu, would completely disappear under water due to rising sea levels, and others would face similarly catastrophic consequences such as massive famine and disease outbreak. For example, global circulation models (GCMs) publicized just before the Copenhagen meeting suggested that Africa would warm 50 percent faster than global average temperatures.[6] It is now well documented that the poorest countries with the lightest footprint on the climate are suffering worst and first from climate change and will continue to do so in the future.[7]

Beyond the formal negotiation sessions of this UN regime, global efforts to address climate change have been correspondingly weak. Rises in temperature and sea level due to human-caused emissions of carbon dioxide from fossil fuel burning from 2000 to 2008 were higher than even the most pessimistic scenarios developed by the Intergovernmental Panel on Climate Change (IPCC).[8] Levels of carbon dioxide already in the atmosphere have locked in inevitable and dangerous levels of climate change.

Progress on providing financing to enable developing countries to adapt to climate change impacts and reduce their own emissions, including sharing green technologies with them at low cost, has also been very slow. The promises made by the Group of Twenty (G20) and G7 major nations to provide funds necessary for countries vulnerable to climate change to adapt to its impacts remain largely unfulfilled. Rich countries, in fact, spend staggeringly more on subsidies to fossil fuel industries, the main contributors to human-caused climate change, than on adaptation measures for those harmed the most.[9]

Thus, the global process to confront climate change is well characterized by the phrase *active inaction*. Despite decades of activity by nations shuttling around the world to give speeches and participate in negotiations, setting up entirely new multibillion-dollar carbon markets to "efficiently" reduce emissions, and developing a series of new institutions and funding agencies to address this problem, overall, things have continued to get worse. In this book we ask: *Why has the response to climate change in the contemporary era been so utterly inadequate and inequitable? And what needs to change if we are going to reverse course?*

We begin this chapter by presenting our main argument of the book. We then provide a summary of the problem of climate change as understood through the optic of global climate justice. We highlight what such a position demands of states: profound transformations of their economies and societies. The nature of those demands explains well why a number of wealthy nations are actively resisting taking action on what would seem so obvious a negative outcome: climate destabilization. We conclude this chapter with a brief discussion of the balance of the book and a sense of where we will end up on all this.

## Our Argument

The main point of this book is that the basic organization of powerful interests globally has undergone fundamental transformations since the Copenhagen round of negotiations in 2009, yet we still generally talk about climate inaction, as if things are the same as in 2007, 1997, or 1992 (the Bali, Kyoto, and Rio talks, respectively). Power relations are shifting in new ways, and will continue to shift in the future, particularly as the world warms. The book explores these shifting power dynamics and considers what impact they have had on our ability to take sustainable and equitable action, and how we can change course. We focus particularly on the struggles of marginalized and especially vulnerable states and civil society actors, and the processes that inhibit and facilitate their influence.

We draw substantially on the scholarship of Italian social theorist Antonio Gramsci to offer a strategic view of power relations, attentive to how emerging transnational political coalitions, including state, market, and civil society actors, are able to navigate a rapidly shifting world order to shape the global governance of climate change. As Brian Gareau found in his study of the ozone treaty, some voices are heard and others ignored as global power shifts.[10] This period offers distinctly new political opportunities and challenges for political coalitions.

We explore how competing international state and nonstate coalitions engage in a global political economic arena that is in the midst of major upheaval. This includes a fundamental reorganization of the interests of the global North and global South, a wounded and widely discredited neoliberal global political economic doctrine, an ecological system in crisis posing new limits on growth and exacerbating forms of social inequality, and new forms of transnational social mobilization and engagement.

Much of the literature on global environmental governance is segmented. Some writings focus on states and local governments, some on social movements, and a third group looks at the environmental initiatives of firms and industry organizations. We argue that any viable efforts to stabilize the climate and achieve justice for those most adversely affected by changes already underway must expand beyond isolated conceptions of states, institutions, markets, and social movements. We seek to understand how different groups—whether environmentalists, business lobbies, social movements, and others—mobilize in transnational coalitions and, in Gramscian terms, attempt to construct a new balance of forces or historic bloc to tip the scales in their favor. We focus on what strategies they adopt, how they collaborate, and how they adapt to changing times and new limits and opportunities.

On stopping climate change, we find ourselves with only a highly fragmented and uneven global governance system, which may more accurately be described as a nonsystem. Some cities, subnational states, and national governments are taking forward-looking action on climate change right now. Other places are almost entirely inert and appear paralyzed, building fossil fuel and economic infrastructure that will last for decades and lock our species into perilous pathways of greenhouse gas emissions.

We describe layers of struggles going on in the United Nations, but also in smaller groups of nations, inside nations, within and between firms and nongovernmental organizations. Despite the deficiencies of the UN process, we contend that it is still our best hope for landing an equitable deal to keep our climate system from spiraling out of control. This will happen only if the UN process is complemented with intensified social movement organizing in local and national contexts that fosters and forces a new calculus by states and firms to address the issue now.

In sum, we intend that readers of this book will gain knowledge about the events and actors at the UN climate negotiations, as well as the underlying forces that have shaped inaction on climate change. We hope to deliver a more nuanced understanding of the diverse mechanisms of power in global politics, and provide some insight into emerging opportunities for transformative social action to challenge *climate injustice*. In the boxed text below, we outline what we mean by this concept.

---

**Box 1.1**

What Is Climate Injustice?

*Climate injustice* can be broadly defined as heightened and disproportionate vulnerability to climate-related harm by disadvantaged social groups, who in general are far less responsible for the problem and are excluded from decision making about its resolution. Our conception of climate injustice takes three main forms, all of them closely tied to forms of social and environmental inequality: climate change–related causes, impacts, and responses.

First, the main causes of climate change often place socially marginalized groups at greater risk of harm. Most notably, fossil fuel industries have disproportionate impacts on poor and minority social groups at various sites along the chain of production and marketing of commodities. These include at the sites of extraction (like coal mines in Appalachia and the oil fields of the Niger Delta), transportation (in poor neighborhoods near rail yards and explosive pipelines), processing (near refineries and chemical plants), and disposal (like oil field waste and toxic materials disposal sites). Other causes of climate change, such as deforestation, also often have a disproportionate impact on socially marginal populations. Injustice here takes forms including loss of land and livelihood, endangerment to health, and disruption of cultures built around rain forests and their complex life forms.

Second, the negative impacts from a changing climate, such as vulnerability to famine, drought, hurricanes, disease, flooding, and displacement, are causally related to social forms of inequality.[a] As Adger and Kelly argue, "The vulnerability or security of any group [in relation to climate change impacts] is determined by the availability of resources and, crucially, by the entitlement of individuals and groups to call on these resources."[b] As a result, groups that suffer from various forms of social, environmental, and economic inequality face compounded vulnerabilities to climate change impacts. As Roger and Jeanne Kasperson put it long ago, "Recognizing and understanding this differential vulnerability is a key to understanding the meaning of climate change."[c]

Third, even the "responses" employed to reduce emissions of greenhouse gases also have collateral impacts for socially marginal groups. Examples include the environmental health impacts of mining silicon for manufacturing solar panels or of lithium for advanced batteries, the displacement of whole communities for the deployment of large hydroelectric reservoirs, and large-scale food shortages and price jumps due to switching cropland and diverting food products to produce biofuels.[d] Chapter 7 explores a surprising example along these lines, of how informal recyclers known as "waste pickers" have argued that new funding for disposal technologies that have been inaccurately deemed sources of clean energy, has deprived them of access to discarded materials, threatening their livelihoods. Climate justice movements have taken up the term *false solutions* to depict such responses, often with the critique that these "solutions" don't address underlying causes of climate change, including unbridled global capitalist growth.

Measures to help communities adapt to climate change can also have their own adverse unequal impacts. Adaptation decisions are inherently value laden, and reflect existing power dynamics, with the potential to mitigate risk for some, while exacerbating it for others.[e] And finally, a growing number of alarmed observers are pointing out the potential disproportionate impacts on marginal populations that are likely to result from climate-engineering projects that attempt to manage entire climate systems.[f]

**Sources**
a. Roberts and Parks 2007.
b. Ibid.
c. Kasperson and Kasperson 2001.
d. Zehner 2012.
e. Carr 2008.
f. Hamilton 2013; *Guardian* 2011a.

### From Climate Injustice to Climate Justice

The evolving history of climate negotiations has been told over and over again, sometimes quite clearly.[11] Rather than repeat it, this section lays out some of the basic dimensions of international climate justice by which that history can be assessed.

Most readers by now are familiar with the history of the relationship between the burning of massive amounts of fossil fuels, deforestation, and noncarbon emissions such as methane and the warming of the global climate. For those who aren't, we point to several excellent reviews.[12] What we do want to make clear is the enormity of this problem and the difficulty of finding clear solutions that are socially just, broadly acceptable, feasible, and fast. After decades of research, the consensus of at least 97 percent of climate scientists is that a relationship exists between fossil fuel use and the warming of the planet, that humans are causing the warming, and that action needs to be taken to reorient our economies to be not nearly as dependent on fuel sources that add to the warming effect.[13] A new series of studies is showing that improving measures of human well-being such as life expectancy and literacy are compatible with living at very low levels of carbon emissions, but high levels of income are not.[14]

The reasons for us to care about climate change are both immediate and self-preservationist, and moral and altruistic. They involve our self-interests and our sentiments about what kind of world we would like for ourselves, and for the world our children and grandchildren will inherit from us. The places where we live are experiencing more frequent and intense extreme events like hurricanes, heat waves, extreme rain and snow storms, and worsened droughts.[15] Food supplies are vulnerable, as are water and, potentially, national security.[16]

Most vulnerable of all are poor and marginalized people, and especially those in poor nations.[17] For example, those in the forty-eight Least Developed Countries (LDCs) are five times more likely to die from climate-related disasters than people in the rest of the world.[18] The great irony is that the populations most vulnerable to climate change impacts are those that have contributed the least to the problem. In the case of the forty-eight LDCs, the combined historical emissions from these countries make up less than 1 percent of the global total.[19] And while we are beginning to experience the unsettling of weather cycles around us, our children and their children yet to come into this world will likely experience impacts far worse than those we will see in our lifetimes. However, more and more, we are all facing this issue now. This surprising rapidity of the onset of perceptible climate impacts is changing the politics around the issue.

The implications of the mass of climate science for policy are critical. Following the 1992 Earth Summit in Rio de Janeiro, Brazil, states that had ratified the treaty (what are called "Parties") agreed that they would work together to "avoid dangerous climate change." This led to the simplifying notion that any rise in the global average temperature over 2 degrees Celsius (3.6 degrees Fahrenheit) would have dangerous impacts. Island states and African nations at the Copenhagen conference argued that even 1.5 degrees Celsius of warming would lead to dangerous impacts.

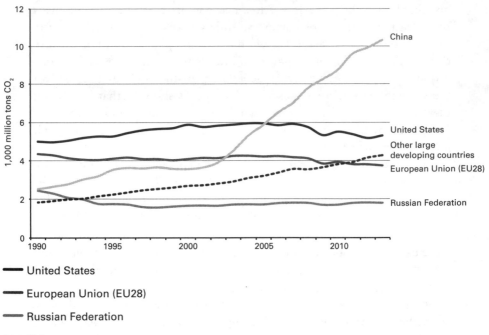

**Figure 1.1**
$CO_2$ emissions from fossil fuel use and cement production in the top 5 emitting countries and the EU, 1990–2013. Figure 2.2 in Jos G. J. Olivier, Greet Janssens-Maenhout, Marilena Muntean, and Jeroen A. H. W. Peters. "Trends in Global CO2 Emissions: 2014 Report." The Hague, Netherlands: PBL Netherlands Environmental Assessment Agency.

In 2007, the IPCC summarized the science and projected the steps that were needed. The biggest political bombshell is buried in a footnote of the 2007 IPCC report. To stay under 2 degrees Celsius of warming, the whole of human society must reduce its emissions by 80 to 95 percent by 2050. That means a complete restructuring of our economies away from fossil fuels in a short generation of just twenty to thirty years, far less than one lifetime. But the implications are even more striking and immediate than that: by 2020 there would need to be 25 to 40 percent reductions by the wealthy nations.[20] The point was reinforced by the 2010

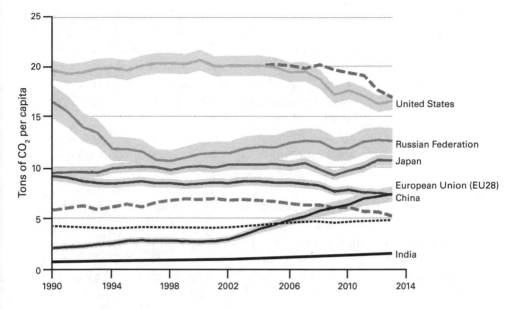

**Figure 1.2**
Per capita $CO_2$ emissions from fossil fuel use and cement production in the
top 5 emitting countries and the EU, 1990–2013. Figure 2.6 in Jos G. J. Olivier,
Greet Janssens-Maenhout, Marilena Muntean, and Jeroen A. H. W. Peters. "Trends
in Global CO2 Emissions: 2014 Report." The Hague, Netherlands: PBL Netherlands
Environmental Assessment Agency.

UN Environment Programme's report on the "emissions gap," which projected how global pathways of emissions growth had to quickly slow and then stop their growth, and then swiftly begin to reverse direction.[21] The report stated that by 2010, we were already behind if we were to stay on a safe pathway of emissions. New reports using a "carbon budget" approach show that we are burning through the atmospheric space available before we head over the 2 degree threshold.[22]

For those who fear that it will be wrenchingly difficult to end society's chronic dependence on cheap fossil fuels, these numbers are deeply threatening. They mean that states and their societies need to start making drastic changes now.[23]

*So what would it mean to address the climate challenge in a just fashion?* The concept of global climate justice points us first to a basic question: How can we as a species share the burden of not ruining this global public good of a stable global climate? The basic idea is that there is a limited amount of atmospheric space out there, which is being rapidly filled up with emissions. Simply put, if the rich nations move all the carbon from the oil fields, natural gas deposits, and coal hills to the stratosphere, then there will be no room for the poor nations to improve their living standards with very affordable and reliable fuels like coal, oil, and natural gas.[24] Reductions now mean less drastic steps are needed later, which will also be cheaper than waiting until climate impacts become extremely severe.[25] If no immediate action is taken, we will be cleaning up and rebuilding after terrible disasters at the same time as we are on a crash diet to shift quickly off fossil fuels.

For countries in the developing world, the writing is on the wall: due to shrinking atmospheric space in which to dump carbon, the time for them to make up ground in their national economic development through conventional practices of relying on fossil fuels is quickly coming to an end. This puts an intense squeeze on the development space available to the emerging countries, most acutely the biggest of these: China and India. As the think tank EcoEquity and the firm PriceWaterhouseCooper have put it, if we look at the curve of total emissions reductions needed and subtract what the rich countries have offered to do, we end up with a steep curve downward for emissions from the developing countries around 2020 or 2030 at the latest.[26] China, in particular, is in the crosshairs, since its population and economic growth rates are so enormous. China's skyrocketing emissions are clear in figure 1.1, showing that its total emissions from fossil fuel use in 2013 were nearly double that of the United States (which it had just surpassed around 2006). Most notable in figure 1.2 is that China's emissions per person are now nearly identical to the average of the European Union, around 7 tons of carbon dioxide per capita.

As we will discuss, the Copenhagen Conference in 2009 was an intense competition for development rights and atmospheric space between developed and developing countries. Developed countries emphasized the importance of climate issues, but climate change is largely a product of their own unconstrained greenhouse gas emissions during the past two hundred years of industrialization. So currently many developed countries downplay their historical responsibilities as the problem, demanding instead that all nations take measures to reduce greenhouse gas emissions. In other words, they emphasize the issue of climate change but downplay the issue of development. The rapidly developing countries like China and India, which we discuss at length in this book, place their development priorities above severe climate efforts. Some countries, like the United States, are often unwilling to discuss any acknowledgment of responsibility and compensation, while some others appear more willing to act on such principles.

To end this section, we operationalize the concept of global climate justice in a climate treaty with six criteria:[27]

1. A climate treaty would be just if it respected procedural justice, giving all nations equal voice and participation, and not giving wealthy or large nations monopoly power at key junctures in the talks.[28] It should also provide voice and participation to particularly vulnerable groups that are not states. Indigenous peoples should have equal voice, participation, and decision-making power as provided to states, given that they have sovereign territory rights in the international system.[29]
2. Science should guide our actions. The rate of emissions reductions should be based on our best understanding of atmospheric space, and we should respond when the bulk of scientific evidence suggests we may be straying into pushing the atmospheric system beyond tipping points that would cause major ecological and social disruptions. Several elements of the targets such as peaking levels, peak year, and rate of decline need to be set for that to happen.[30]
3. A just agreement would be based on an equitable sharing of the global burden in reducing emissions. Wealthier countries should go first and help poorer nations avoid massive growth of greenhouse gas emissions while still meeting their development goals.

4. Action should follow a gender-sensitive, participatory, and fully transparent approach, with priority to preventing and eliminating disproportionate impacts to vulnerable groups, communities, and ecosystems.

5. Whether a just international climate policy could include the trading of permits to emit greenhouse gases has split the environmental and environmental justice movements. A series of climate justice groups have taken a hard stand against trading of carbon permits as being an appropriation of the atmosphere for private benefit.[31] The remainder of this book could be spent debating this issue. However, it is enough for now to say that a just solution would not place heavy carbon taxes on the poor, dispossess communities of resources and decision-making power without their explicit prior consent, concentrate polluting industries in marginalized communities, or raise their energy costs disproportionately as compared to their income.[32]

6. The costs of adapting to climate change should be borne by those who proportionately caused the problem and should not come from the poor, who need precious funding for their other pressing needs like health, education, and basic infrastructure. This suggests that major financial flows will be needed from the global North to the South for climate adaptation and compensation for the most severe disasters (what's called "loss and damage" in the climate negotiations), such as whole countries being swallowed by rising sea levels.[33]

## Beyond Perfection

In developing our definition of global climate justice, we draw on Amartya Sen's clear-headed and useful view of justice. For Sen, justice means sustaining people's capability to have and safeguard what they value and have reason to attach importance to.[34] His approach to justice differs from "perfect world" theories of justice[35] by focusing on practical ways to enable society to reduce injustice and advance justice rather than focusing on the abstract components of a perfectly just society. As Sen writes, "A theory of justice must have something to say about the choices that are actually on offer, and not just keep us engrossed in an imagined and implausible world of unbeatable magnificence."[36] This is not to say that we think that efforts at achieving greater justice in climate change should be confined to the readily winnable or that we shouldn't seek transformative change; rather, these efforts must engage with existing political realities, not merely utopian ideals.

Second, Sen's theory of justice attempts to get beyond attending only to changing institutions and creating new rules (while neglecting the actual behavior of actors and assuming their compliance). This is a huge problem in climate efforts of many types; for example, years have been spent wrangling over constructing new legal frameworks, funds, trading systems, and penalties, which are subsequently ignored by influential parties to the very treaty to which they've agreed. Sen says we need effective institutions in practice, not only design.

Third, this perspective recognizes that there are often ambiguities between divergent approaches to organizing society that are all reasoned as just. For example, in the UN Framework Convention on Climate Change (UNFCCC) climate negotiations, there may be hundreds of methodologies for defining vulnerability in order to fairly allocate funding to help the most vulnerable peoples or nations adapt to climate impacts, all of which can be reasoned as "just" in one way or another.[37] Together, these perspectives on justice provide a useful theoretical entry point for developing a workable and realistic definition of global climate justice.

Fortunately, it turns out that we have a treaty with good language on several of these elements of climate justice: the UNFCCC. Negotiated in 1992 in Rio de Janeiro and eventually ratified by 194 nations, the UNFCCC agreed that Parties to the Convention should "take precautionary measures" and act in a way that would "avoid dangerous climate change" and according to equity and their "common but differentiated responsibility and respective capabilities." The treaty also promised "new and additional, adequate and predictable" funding for poorer nations to adapt to climate change. Each part of the convention language was carefully crafted but equally vague to avoid binding commitments, which were already being called for in 1991 in the run-up to Rio.[38] Although the words were impressive, they were just that: words. The difficult parts—the critical details—were pushed back until later. The Framework Convention remains in effect, and making it more concrete has been the focus of great attention, since it is the only international climate agreement in effect that the United States and several other key nations have ratified.

In the next section, we discuss why we continue to focus on nation-state level interactions at the UN, given the lack of progress to date.

### Why We Need the UN to Deliver

We are now over twenty years into the efforts of the United Nations to solve the problem of climate change, and clearly the problem remains.

With so little progress since Copenhagen, many national governments' foreign relations ministries, corporations, activists, and even some environmental NGOs seem to have lost interest in the process. All along, many other levels of governments have been involved in contributing to a solution to climate change: cities and municipalities have made serious efforts and pledges to address climate impacts and reduce their emissions, subnational states and regions have developed carbon trading schemes, supranational regions like the European Union have instituted comprehensive carbon trading systems, and a series of multilateral organizations have taken up the issue.[39]

Corporations and industry organizations have made pledges and developed climate programs, hundreds of universities across nations and across the world have pledged to "go carbon neutral" and have created climate alliances, and individuals have created some of their own small networks for mutual action in reducing emissions. Grassroots activists have played a major role in various countries in stopping fossil fuel projects before they are built. For example, in the United States, grassroots activist networks have had a role in stopping 150 proposed coal plants since 2007. In short, the effort to govern humanity's impact on the climate has become extremely multilevel, and national governments and their interactions in the UN can seem cumbersome, obstructionist, and obsolete. In this section, we ask, Is it time to give up on expecting national governments to work together at the UN given the inadequacy of this process to date?

At the beginning of the 2000s, a wave of activism swept across universities, churches, and cities in the United States, demanding their institutions make binding pledges to reduce their carbon emissions, sometimes to zero. Frustrated with the inaction of the George W. Bush administration, university students pushed their administrators to sign the Presidents' Climate Commitment, which varied slightly across schools but which all shared a pledge to go carbon neutral by 2050. This is an extraordinary thing to promise, since we essentially didn't have the technology for schools to do so affordably. Presidents were asked to plunge ahead down a road they could not see the end of. Six hundred US university presidents did so, an impressive number given the quantity of often-conflicting issues and pressures these top managers face.

The universities' voluntary pledges and creation of a voluntary association sought to establish new emergent norms. These new rules were laid out in reporting requirements, specific pledge targets, and best practices to get there. Annual rankings by the Sierra Club in its glossy monthly newsmagazine *Sierra* tout the "coolest schools" based on climate initiatives. Another group, the Sustainable Endowments Institute, created a green grade card of university practices on environment, especially investments. Finally the Princeton Review began including its assessment of environmental efforts by schools in its influential annual rankings of universities.[40]

Cities across America and across the world have made their own pledges of efforts, including deep reductions or even reaching carbon neutrality by future dates. ICLEI, an international network of cities seeking to boost their sustainability efforts, had its membership soar from a dozen US cities and towns in the late 1990s to over one thousand globally today.[41] After the failure of the UNFCCC in Copenhagen to seal an adequate deal, the cities created the Global Cities Covenant on Climate (dubbed the Mexico City Pact), which "aims to scale up cities' role and efforts in combating climate change globally."[42] As of March 2013, the initiative listed over thirty-five hundred voluntary greenhouse gas reduction commitments of local governments in the global North and South.[43]

The limits to these voluntary commitments and pledges lie in the weakness of the commitment. There is neither a base level of commitment that is required to be listed as a signatory, nor a clear mechanism should cities fail to file a plan in time to meet the group's deadline (eight months after signing up), nor any clear repercussions if a locality fails to meet its commitments down the road. The same can be said for the commitments of the university consortium's members.

In a research project organized through the UK Tyndall Centre for Climate Research, Diana Liverman and Emily Boyd attempted to calculate the meaning of all these overlapping pledges and claims of carbon reductions by institutions, cities, states, sectors, and nations.[44] Their focus was on California, where progressive universities, cities, counties, economic sectors, and the state all had made pledges to reduce emissions and made claims of how much they had done so. They found that it was impossible to verify that the pledges were being met, and due to their overlaps, some emissions reductions were being claimed several times. For example, if the University of California, Berkeley, decided to buy carbon offsets for travel by professors to a conference, that "reduction" might be claimed by the sponsoring department, the university, the city of Berkeley, a greater San Francisco regional pact, the Oakland county government, and the state. Some claims of the total number of commitments or volumes of carbon reduced could easily double or triple count real actions.

Sub- and supranational carbon trading schemes have also captured the attention of observers and led some to claim that a Kyoto-style global pact is no longer needed or even relevant. California has been a pioneer in the United States in creating and instituting its own carbon reduction plan, Assembly Bill 32 (AB32), the Global Warming Solutions Act, which was developed under Governor Arnold Schwarzenegger in 2006, and brought into effect after some delays in 2013.[45] Stepping up a level in scope, a couple of regions inside the United States have created regional trading systems, such as the Regional Greenhouse Gas Initiative, in which electricity utilities in eight northeastern states can reach target emissions reductions in part by trading carbon permits across boundaries.[46] Then stepping up to supranational regions, the European Union Emissions Trading System is in its second period after an ambitious but ill-fated start. In addition, under increasing social pressure, major institutions like the World Bank and European Bank for Reconstruction and Development are adopting their own climate-friendly policies to limit their lending to the construction of coal-fired power plants and other controversial projects.

In this context, many radical climate justice activists have grown weary of what they the call the UN's "Conference of Polluters" (a play on the term "Conference of Parties"). They have called on movements at various junctures to "Seattle the COP"—in other words, to shut down the UN negotiations, as activists did to the World Trade Organization negotiations in Seattle in 1999. Their argument is that the process is too corrupt and that fossil fuel companies are calling the shots, with professional NGOs providing legitimacy. They instead point to growing social movements and their successes in stopping fossil fuel projects around the world as the primary means for addressing climate change.

To return to our core question here, Are these layers of initiatives adequate to address the crisis of climate change? Perhaps framing the question in this way foreshadows our conclusion: we believe that no other level of solution is able to do what global interstate negotiations can, which is to assemble in a mutually agreed-on process the legal representatives of the people of the world to agree on binding commitments to address a problem facing all of humanity. To be clear, the UN process has yet to live up to its potential on the issue of climate change. It is a process that has been highly unequal in terms of whose voices matter most and far too friendly to fossil fuel interests. However, the UNFCCC is the only institutional body currently capable of realizing a representative, adequate, and equitable global climate agreement. While critical, the initiatives discussed above are not alone likely to yield the scale of emissions reductions needed to maintain a stable global climate system. We have limited time to act, and starting from scratch with a new global institution seems unlikely to avoid many of the UNFCCC's toughest issues.

The penultimate chapter of this book, where we develop a series of scenarios for climate change politics, explores outcomes of worlds with and without adequate UN action. In that chapter, we also envision possible futures where smaller groups of nations begin the process by making a deal in a smaller group of countries and then bring that deal back to the UNFCCC for negotiation. We want to be clear: more localized action on climate change, particularly those that are networked transnationally (what are known as translocal solutions), are valid and important, and they play an essential role in pushing states to adopt ambitious policies and practices of their own, which open new possibilities for international coordination. In fact, we do not see any hope of realizing an adequate international deal without far more aggressive and ambitious social movement organizing at the local and national levels, pushing states to act. However, we believe that the level of ambition that is needed to avert disaster can be achieved only if we simultaneously pursue an international process with robust forms of accountability and action at all lower levels.

## What Lies Ahead

In chapter 2, we draw on relevant theories, and particularly the scholarship of Antonio Gramsci, to offer a strategic framework that considers power relations broadly, including strategic, layered, and historical dimensions. We argue that the world is not the same as it was back in 2008 before Copenhagen. Rather, we explore recent shifts in power dynamics and consider how they might be having an impact on our ability as a species to take effective and equitable action on climate change. Our argument is that the world order has transformed along four main axes: its political economy, geopolitics, ecological conditions, and the capabilities of transnational civil society. Our approach focuses especially on struggles by marginalized and particularly vulnerable states and civil society actors seeking climate justice, and the processes that inhibit and facilitate their influence.

Chapters 3, 4, and 5 focus on the mobilizations of state actor coalitions in the negotiations. In chapter 3, we explore global shifts as they took shape in the pivotal negotiations in 2009 in Copenhagen. We argue that these negotiations were the beginning of a new equation of power instead of the conventional North–South divide. Since the late 2000s, a fragmentation of countries in the talks has emerged along new lines. Copenhagen provided a world stage where a new alignment of five of the largest countries (the United States and BASIC—Brazil, South Africa, India, and China) saw the opportunity to flex their muscles to keep open their international development space. And critically, these new global shifts involve the hegemonic decline of the world's formerly undisputed superpower—the United States, the surging Chinese model of development, and the coordination of various economic interests that are integral to causing climate change. As such, these negotiations were far more about territorial fights to establish new alignments of power in a rapidly transforming geopolitical order than about mitigating climate change.

Chapter 4 takes us to the aftermath of Copenhagen, to negotiations in Cancun, Durban, Doha, Warsaw, and Lima in the following five years. We ask: After the debacle in Copenhagen, how was low-income-state consent produced during the following years to an emissions reduction framework that is both highly inadequate and starkly inequitable? Drawing on our theoretical framework, we argue that consent was produced through three interlinked processes: material concessions, norm alignment, and structural conditioning. We demonstrate that the material concessions, in particular, have resulted in few substantive gains for low-income states, but they have been instrumental in securing the stability of the climate regime in the face of an escalating crisis of international climate change leadership. Overall, this analysis provides insight into the processes by which international environmental inequality has been reproduced in contemporary international climate politics.

Chapter 5 directs attention to climate change adaptation politics. We ask: With very little progress having been achieved in two decades of negotiations on reducing emissions and with some major climate change impacts now inevitable, what types of political conflicts are emerging around the issue of funding for adaptation? We identify and discuss three main related points of contention between countries on both sides of the North–South divide. We call these conflicts the Gap (in raising the funds), the Wedge (in who is prioritized to receive funds), and the Dodge (in using just governance institutions).

We direct attention in the second half of the book outside the UN process, to consider more broadly the governance challenges and geopolitical dynamics that are likely to emerge as drastic climate change impacts become inevitable in a rapidly warming world. We argue that the fragmented governance system that currently exists is hardly capable of effectively and equitably managing issues such as climate-induced migration, climate-related security issues, the disappearance of states under rising sea levels, fragmented intergovernmental structures for disaster management, geopolitical conflicts over the thawing Arctic, and the role of insurance companies and private sector actors in climate adaptation and disaster response. And there exist gaping governance, political, and social challenges related to large-scale technological attempts to engineer the climate.

In sum, our main argument in chapters 3 through 5 is that given the particular historical conditions of the contemporary world order, states or well-designed international institutions alone show very little promise of arriving at a sustainable, effective, and equitable climate treaty. The states experiencing the impacts of climate change worst and first are also the ones that have the least power in the negotiations, and they are particularly vulnerable to cooptation in the current historical context. Given the fears of downward trends in its economy, the United States shows little promise of emerging from its intransigent position in the negotiations, despite modest progress on greenhouse gas regulatory policy at home. As an emerging global leader with an economy that has largely thrived during the Great Recession, China offers perhaps the most promise of being in a position to shift the negotiations in a more promising direction, but this seems unlikely in the short term, when ambitious action is desperately needed. The United States and China are also two of the countries with much to lose from an effective climate change treaty, given that they sit upon 18 and 9 percent of global carbon reserves respectively—carbon dioxide that would result if each country's proven fossil fuels were burned.[47] This brings us to analysis of the role of business coalitions and civil society in international climate change politics.

In chapter 6, we engage with scholarship that argues that a new class of business interests, which defines climate change mitigation in their financial interest, is our best hope as the catalyst for the change necessary to address climate change. As climate change science has become increasingly difficult to dispute, fossil fuel associations have become less visible in the international negotiations. However, while there has been fragmentation and diversification in the approach of different business actors in international climate politics, there is little evidence that the obstructionist forces of fossil fuel lobbies have waned in power. These industries still compete with renewable energy on a highly unequal playing field, receiving massive subsidies from the very governments that negotiate international climate treaties. The shift of some in the industry to a carbon market approach, once admonished by some of the more obstructionist fossil fuel companies, has not proven a real threat to fossil fuel interests, and the biggest actors have continued to rake in record profits with no sign of slowing down.

We explore the processes that have ensured the continued dominance of fossil fuel industries in international climate politics and argue that the change that is needed will not be simply designed by business coalitions with their eye on new markets. The existing investments made by fossil fuel interests in exploration and refining products are simply too high, their profits too astronomical, their instrumental and discursive power too great, and the diversity of their product too limited to easily let go of their advantaged position and grip on power.

In chapter 7, we shift attention to the role of civil society broadly in climate change politics. We argue that the literature on civil society in international climate change politics has not fully accounted for the causes of the failure of civil society to influence mitigation action in the contemporary period. We highlight three main deficits. First, despite the diversification of actors involved in the negotiations, resources and links to power still rest overwhelmingly in the hands of professionalized NGOs that take a more reformist and market-based approach. Second, civil society has failed to take a coordinated and viable strategy for building strength in domestic contexts to realize influence at key hinge moments in the international negotiations. And third, civil society has primarily devoted its attention at the international level to the UN climate processes, while often neglecting less accessible but highly relevant international governance frameworks, including international trade regimes, financial institutions, and scientific bodies.

Chapter 8 shifts to the efforts of particularly marginal and vulnerable movements to gain rights in the climate regime (the climate-related institutions and the treaty). We explore the engagements of three distinct transnational advocacy networks in the UN climate change regime: networks working for gender equality in climate governance, Indigenous peoples, and waste pickers (informal sector recyclers). These networks have all sought to gain certain rights in the climate regime to redress forms of marginalization and inequality related to climate change and climate change responses, often referred to as climate injustice. We develop an analytical framework for *regime rights*, and identify and assess four types of related struggles: those for recognition, representation, increased capabilities, and rights that extend beyond the negotiation halls (what we call extended rights). The analysis suggests that while some moderate gains have been achieved by these networks, more substantive change will likely necessitate networks to pose a far greater threat to regime and international political stability, thus forcing responses beyond palliative measures at the margins.

In sum, chapters 6 through 8 argue that while there has been much diversification and fragmentation of both business coalitions and civil society in international climate politics, the interests of fossil fuel actors continue to reign supreme. The most vulnerable civil society actors, such as Indigenous peoples and waste pickers, have found limited ability to influence the regime in their favor. Professionalized environmental NGOs, while often having strong ties to both powerful state and business actors, have been unable or unwilling to build coalitions capable of pushing for real change. As a result of the largely fragmented and unequal condition of civil society, those on the inside, while flush in resources, have limited leverage to put actual pressure on states that would cause the transformational change needed to address such a difficult issue as climate change.

This brings us to the final chapters of our book. Here we look ahead at the potential scenarios for taking action in the coming years and, importantly, to discuss what might provide the catalyst we need to prevent catastrophic warming. Chapter 9 outlines six potential scenarios in a warming world. We develop a framework for analyzing change across two axes, level of democracy and level of sustainability. The scenarios range from quite catastrophic to very positive outcomes. We point out key decision points, power relations, and political and material conditions shaping the direction we as a global society will head.

In the concluding chapter, we summarize our main arguments, discuss our theoretical contribution, and consider how we might achieve global climate justice in the turbulent waters ahead. Given the need for independent actors whose focus is on the global public good of avoiding catastrophic climatic change, we argue that a transformed approach of civil society is our best hope for realizing an equitable, sustainable, and effective international climate treaty and advancing global climate justice. We introduce the concept of *linking movements* to specify the changes that we think will be necessary in this area. To begin, however, we lay out our strategic theory of power in the next chapter.

# 2
## Power Shift

**Another Lens**

The scholarly literature on climate change politics is gaining volume and substance; we'd like to encourage its movement in a new direction. Four perspectives in the literature on developments in global climate change politics are prominent: structuralist, institutionalist, market pragmatist, and ecosocialist perspectives.[1] Although all of these perspectives offer valuable insights, none is sufficient for explaining international inaction on climate change or offering explanations of potential turning points in the contemporary period.

We argue that a fifth lens, building on the work of Italian social theorist Antonio Gramsci and his followers to create a strategic perspective on power, offers a more promising direction. However, this concept has not been sufficiently developed as a framework to investigate transnational power relations and social change. In seeking to do so, our framework has three main components: a strategic view of how hybrid coalitions (state, market, and civil society actors) mobilize to shift the balance of forces on a given issue, a layered view that considers the fragmented governance institutions on which such coalitions engage, and the historical dimensions of world order within which such struggles are embedded.

The framework that we develop here helps to bridge key contributions of existing perspectives to understanding climate action and inaction. Importantly, this approach is also useful for bridging the divide between theories that emphasize the agency of actors and those that emphasize big political and economic structures, a weakness often identified in analyses of global politics.[2]

In the following section, we explore existing theories of power relations in climate change politics, including the strategic approach. Next, we develop our strategic theoretical framework in three parts: strategic, layered, and historical dimensions. We then delve deeper into the historical dimensions of power relevant to climate change as they have changed in the contemporary period. This includes discussions of the changes in areas of global political economy, geopolitical relations, ecological conditions, and transnational social movement organization. In the conclusion, we summarize our arguments and look ahead to the next chapters.

### Power and Climate

Perhaps the most common lens through which scholars have analyzed international climate change politics is a *structuralist* perspective. This approach views climate inaction as primarily determined by enduring unequal power relations between states, including political, ideological, military, and economic asymmetries. Realist iterations focus on how powerful states in an anarchic international system use their relative might to push for policies that support their domestic economic interests and military advantages.[3] This scholarship emphasizes the unwillingness of powerful states to make meaningful short-term sacrifices to advance long-term protection of the global commons. Nuanced iterations have investigated the role of particular domestic politics and political structures in shaping participation and nonparticipation in the UN Framework Convention on Climate Change (UNFCCC).[4] Other structuralist works, including Roberts and Parks, have drawn on world systems theory to consider the importance of the unequal insertion of countries into the global economy and how this has structured worldviews that inhibit cooperation on climate change policy.[5]

Second, the *institutionalist* approach focuses on the importance of international institutions for resolving global environmental conflicts. Neoliberal institutionalist and regime theorists, while maintaining realism's focus on power differentials between states, point to the cooperative and mutually beneficial dimensions of multilateralism in terms of increasing absolute rather than relative power.[6] From this perspective, powerful states are likely to participate in international institutions and agreements when the perceived benefits outweigh the costs. Earth systems governance, a growing body of scholarship in this camp, pays particular attention to the overlapping and fragmented nature of global governance institutions in governing climate change.[7] They call for new and reformed global institutional structures more capable of coordinating state interests, providing incentives, and holding states accountable for their actions.

A third approach is what we call the *market pragmatist* approach. Scholars in this area view the solution to climate change in leveraging and reforming powerful private sector actors through market incentives and penalties. From this perspective, the role of state actors is minimized, with the primary focus placed on business interests. In one camp, there is a celebratory view of market actors being driven by a new sustainability imperative as inexorably motivating companies to act more ecologically responsibly.[8] A more tempered view in the scholarship is that given the immense economic and political power held by fossil fuel industries and other actors that benefit from the existing model, a realistic solution must involve the emergence of a powerful new business class that views their economic interest in mitigating climate change.[9] This second view is sympathetic with more radical calls for change from below, but recognizes that we have limited time to act to address the climate crisis. It views system-level change as unlikely in this short window of opportunity, and therefore calls for leveraging the markets that are already in place, like it or not.

A fourth approach, which we refer to as the *ecosocialist* perspective, explicitly rejects the idea that solutions to climate change will be found within the same capitalist economic model that created the problem in the first place. This approach urges us to look beyond incrementalist and reformist approaches. Its proponents argue that only by aggressively challenging social inequality and confronting the imperative to endlessly grow our economy above all other social goals will we realize a viable solution to the climate crisis.[10] Scholars and activists in this camp emphasize the importance of social movements and community-led change efforts, particularly those that confront powerful interests such as the fossil fuel industry; offer more democratic or consensus-based forms of decision making; empower the poor and most vulnerable; and offer fundamentally alternative models of organizing social life. Of the four approaches discussed here, this last is most closely representative of a climate justice perspective.

All of these perspectives offer important windows into the climate crisis, the obstacles that we face in realizing a political solution, and the possibilities for shifting course. However, to some degree, all of them present an insufficient theory of large-scale social change that we believe is necessary to address climate change. Structuralists offer an overdetermined view of enduring power relations in climate change politics, leaving scant room for change, with limited attention to how political processes and social systems shift in profound ways over time. Institutionalists reduce political change to mostly a design problem, while often neglecting how existing institutions serve as a terrain for social struggle and are freighted with baggage that will be difficult to overcome. Market pragmatists often present historical social and economic structures as inevitable and down-play or completely neglect the ability of social movements to shift how we as a society define our preferences, systems, and values. Ecosocialists, with their emphasis on bottom-up change, often neglect the importance of building far-reaching institutions and the need for social movements to construct alliances that extend beyond the marginalized or like-minded, to bureaucrats, entrepreneurs, and fractions of the ruling classes.

These are likely crude and oversimplified depictions of these differ-ent theoretical perspectives; there is indeed much variation within these camps and much more to their arguments. The point that we emphasize is that the walls between these perspectives are rather high. They exist largely in silos, and we think our understanding of climate change inac-tion would benefit from a theory of power and social change that at-tempts to bridge some of their unique contributions. In order to do so, we point to a fifth perspective: a *strategic* approach to transnational power relations and social change.

A strategic approach, in the words of neo-Gramscian scholars David Levy and Daniel Egan, is attentive "to the capacity of agents to com-prehend social structures and effect change, while simultaneously being constructed and constrained by them."[11] As such, it attempts to overcome a divide often found between theories that emphasize actor agency and those that emphasize political and economic structures. Importantly, this approach rejects what we see as a narrow and false reading of Gramsci's work, including by some neo-Gramscian scholars themselves, which views political outcomes as a crude reflection of capitalist interests.[12] Much, but not all,[13] of neo-Gramscian work in this area to date has focused on the power of business interests in intergovernmental processes, to the neglect of analysis of the role of civil society. However, Gramsci and some of his contemporaries view durable power relations and processes of social change as far more complex than business interests merely getting their way due to their structural position, and place the role of civil society at the center of theory.[14]

This approach is also transnational in nature,[15] and it points to pub-lic, private, and hybrid networks strategically carrying out crucial func-tions and influencing state behavior in the global governance of climate change.[16] Importantly, this offers a concept to move beyond isolated con-ceptions of states, institutions, markets, and social movements discussed above. To this end, in the following sections, we draw on Gramscian and other relevant social theory to develop our strategic theoretical framework in three parts: strategic, layered, and historical dimensions (as shown in table 2.1 below).

**Transnational Coalitions**

First, our approach takes a strategic view of power. This approach is concerned with how competing transnational coalitions struggle to preserve and challenge the principles and practices through which global society is organized. This builds from Italian sociologist Antonio Gramsci's central concept of hegemony. Gramsci argued that a social class exerts supremacy in two ways: as domination, or rule by coercion, and as intellectual or moral leadership, or rule by hegemony.[17] Hegemony is the process by which a ruling coalition of actors—which he calls a "historic bloc"—is able to embed itself in society, generate an ideological framework capable of achieving consent, and negotiate the conditions of its continued rule. From this perspective, stable rule is contingent not on pure coercion (although coercion is always present in some capacity), but on leadership of a ruling group and the consent of an interclass alliance to the conditions of this rule. Hegemony is never complete, and therefore the ruling group must continually struggle to earn the consent of the subordinate class.

In contemporary society, a transnational historic bloc typically comprises actors including a managerial elite from multinational corporations, professionals from NGOs and academia, and governmental agencies.[18] And sure enough, we see these coalitions defining the range of reasonable solutions in global climate change politics. These actors engage in strategic efforts—often in coalitions with well-defined identities—to leverage material, discursive, and organizational resources in what Gramsci refers to as the "war of position" to challenge, establish, or maintain hegemonic rule. Thus, a Gramscian lens encourages us to direct attention to the tactical struggle by politically and economically marginalized actors to resist domination through control of culture, ideas, and identities in the realm of civil society.[19]

The question is how do those in power keep their power and privileged access to resources even when doing so damages the life chances of other groups? As Gramsci says, "The state is the entire complex of practical and theoretical activities with which the ruling class not only justifies and maintains its dominance, but manages to win the active consent of those over whom it rules."[20] Thus, a historic bloc does not merely bully its way through politics, taking what it wants; it also works to build consensus around particular ways of understanding the world and the problems that we face. In our capitalist society, rising inequality persistently undermines the legitimacy of the system, but historically dominant blocs offer certain material concessions or small privileges to actors who might otherwise resist these systematic forms of inequality. At the same time, the stability of political rule of a historic bloc is always under threat by forces seeking to challenge dominant ways of thinking and doing, forming what Gramsci calls a counterhegemonic movement.

We find this perspective highly useful for analyzing climate politics. We see power as a process that involves intense political struggles by competing coalitions over what constitutes legitimate leadership, as well as assertions of dominance through wielding ideological, economic, and military might.

A handful of contemporary scholars of global environmental politics have built on this analysis to emphasize, in particular, the processes by which powerful state and nonstate coalitions "manufacture consent" and accommodate certain interests of materially weak actors in environmental governance. For example, David Levy and Peter Newell argue that environmental treaties and regulatory regimes are shaped by "microprocesses of bottom-up bargaining" while also being constrained by the political world they are born into—the "existing macro-structures of production relations and ideological formations."[21]

This is a nuanced position: while being attentive to material forms of inequality and the overall domination of capitalist ways of thinking and acting, this perspective challenges the idea that states or international organizations such as the UNFCCC are themselves mere transmission belts for transnational capital. Rather, "sensitivity to a strategic dimension of power suggests that intelligent agency [action] can sometimes outmaneuver resource-rich adversaries."[22] At the same time, nonmaterial currencies of influence such as ideas and norms are themselves closely bound by relationships of material inequality. This is not to say that ideas don't have some level of autonomy, but struggles over what is "rightful" extend from and at times challenge, rather than transcend, historical power relations.

Thus, we take Levy and Egan's interpretation that any system of hegemonic rule simultaneously employs a blend of coercive and legitimate forms of influence,[23] and we focus our analysis of power on these two interrelated and often complementary forms of social control. In terms of coercive power, we define this broadly to encompass structural, institutional, and instrumental forms of influence. This is what Gramsci refers to as political domination. What is key is that coercive mechanisms of influence rely on material force rather than legitimate forms of action or reason. Coercive forces can include military intervention or posturing; threats to cut foreign aid; exclusionary political clubs; laws that promote economic inequality; institutionalized forms of privilege such as weighted voting rights; practices of ideological dominance such as racism, sexism, and homophobia; asymmetrical access to information and knowledge; historically conditioned barriers to competing on an equal playing field in the global economy; illegitimate or manipulative forms of debt; intentionally misleading propaganda; efforts to obstruct transparency; and narrow agenda management, among many others. All of these play crucial roles in determining the (arguably disastrous) outcomes of the global climate negotiations, and much of what follows in this book describes their impact.

However, what we think is as interesting and far less discussed, including for the case of climate politics, is the creation and use of noncoercive legitimate power.[24] Our analysis considers the processes by which various actors, both state and nonstate, strategically act and interact to contest, construct, solidify, and leverage shared ideas of what is socially acceptable.[25] For example, how did it become acceptable to commodify the atmosphere? How did it become a radical and marginal position to attempt to keep global levels of greenhouse gases at what are known to be truly safe levels? Why was it considered okay to allow the largest polluters to continue to get the largest share of rights to the atmosphere, only making incremental adjustments? Why were more just approaches such as giving equal shares of the atmosphere to every human never given even minimally adequate time in the discussions?

The key is to understand how and why some courses of action have been embraced by state representatives, civil society groups, and industry as politically feasible, while other options have been viewed, often preemptively, as off the table. This is not to say that what is possible in the negotiations has been totally predetermined and fixed in concrete, but the range of what has been seriously discussed or considered as a likely course of action in the UN negotiating halls has always been extremely narrow.

As in other intergovernmental processes,[26] we have seen civil society, often in coordination with state and market actors, play a critical role in shaping global climate governance as a reflection of dominant historic systems of power and conditions of world order (which we discuss below). This is consistent with what global governance scholars in the tradition of sociologist Michel Foucault refer to as a process of "governmentality."[27] As articulated by Sending and Neumann, this literature points to the fact that "civil society does not stand in opposition to the political power of the state, but is a most central feature of its exercise."[28]

This is not to say that transnational civil society is unified or one-dimensional. Rather, it is a political space in which competing ideas and identities are leveraged in a struggle over what counts as just or legitimate knowledge and policy. However, transnational civil society is deeply unequal in terms of resources and access to political power. Those actors who are flush in resources and access are often least willing to challenge those ideas deemed hegemonic, and thus off-limits to negotiation in a given political process. In the current historical context, this includes a commitment to market-based solutions, the imperative to sustain economic growth as the primary goal, and a narrow view of the particular types of scientific knowledge and expertise that are deemed relevant to the policy-making process (for example, quantifying and measuring carbon reductions, but not the social impacts of a particular policy).

Importantly, low-income states and civil society groups did not accept the inequitable deals in the contemporary climate regime solely because they were coerced into doing so; rather, they have also celebrated aspects of these agreements as rightful or legitimate, and they have played an important role in "co-producing" the terms of legitimate governance.[29] Legitimacy is evidenced by compliance that is not coercive but rather when an actor or institution's decisions, actions, identities, and interests are socially sanctioned or there is a generalized perception that its normative precepts are rightful in a political community.[30] Legitimacy thus represents "not just the capacity to act, but the right or entitlement to act."[31] This conception reflects Gramsci's discussion of hegemony as a contested and dynamic process of moral or intellectual leadership. To be clear, what is deemed legitimate in a political community is always embedded in existing material power relations. For example, some actors are always excluded from governance processes, and usually these are the materially weak or marginal, while others are heard first and with the greatest attention to their words.

But the politics of legitimacy also sometimes offers new openings for actors to challenge existing relationships and practices of exclusion and inequality on moral or reasoned grounds and to demand fairer, more rightful, more equitable, and more inclusive approaches to organize society. This is particularly the case when the political regime is in a state of transition after previous agreed-upon social norms have been broken. We are interested in these cases as well because to say that the weak always lose in climate politics is to miss important parts of the story. We will describe several cases in climate politics when poor and vulnerable nations, and more radical civil society networks, have scored surprising victories, but these are often on issues where it may be easier for dominant nations or industry groups to give ground. Sometimes the materially weak (as manifest in state alliances such as the Least Developed Countries group or the Alliance of Small Island States, or as part of civil society as in the case of waste pickers and Indigenous peoples' networks) have extracted nontrivial concessions from the wealthy and powerful nations. We have observed firsthand the governments of the wealthy nations acting very eager to prove that they have "followed through on their promises." For example this has been the case with key promises made at Copenhagen to deliver a pile of "fast-start finance" to developing nations to deal with the most urgent climate impacts and to begin to make their economies and land management more environmentally sustainable. Simple realism, structuralist, or realpolitik approaches fail to explain these outcomes.

In short, a strategic approach encourages us to look at the transnational balance of forces[32] that form around any particular set of issues, the coercive and legitimate strategies that particular coalitions use to exert influence to shift or preserve this balance, and importantly, the institutional and historical context in which coalitions engage. This brings us to our next dimension of a strategic view of power and social change: a layered approach.

**Fragmented Climate Governance**

Second, we take what we call a layered approach to power. In particular, we consider the importance of the global governance architecture in shaping power relations. Following Bierman et al. in the intellectual tradition of earth systems governance, the term *global governance architecture* is used to describe the "overarching system of public and private institutions that are valid or active in a given issue area of world politics."[33] This framework usefully draws our attention to understanding how conflicting and fragmented global governance structures both constrain and enable certain strategic transnational coalition struggles to address a global problem.

So, for example, if one is concerned about addressing global health disparities, disputes over pharmaceutical patent laws in the World Trade

**Table 2.1**
Strategic Power Analytical Framework

| Concept | Focus | Theories and concepts drawn from in the framework |
|---|---|---|
| Strategic | Power is defined by transnational hybrid coalition struggles over what constitutes legitimate leadership, as well as assertions of dominance through ideological, economic and military force. | Hegemony, historic bloc, counterhegemony, war of position, and negotiated consent (Gramsci [2012] 1971) <br> Strategic power (Levy and Newell 2002, 2005) <br> Legitimacy (Hurd 1999; Reus-Smit 2007) <br> Transnationalism (Andonova, Betsill, and Bulkeley 2009) <br> Governmentality (Foucault [2003] 1976; Sending and Neumann 2006; Okereke, Bulkeley and Schroeder 2009; Gareau 2013; Goldman 2007; Lipshutz 2005) |
| Layered | Conflicting, fragmented, and historically derived global governance structures enable distinct forms of social mobilization and institutional approaches for addressing global problems while constraining others. | Earth systems governance, global governance architecture, fragmented governance (Bierman 2007; Bierman, Betsill, Gupta et al. 2010; Bierman, Pattberg, Van Asselt 2009) <br> Multilevel governance (Betsill and Bulkeley 2006) <br> Transnational governance (Andonova, Betsill and Bulkeley 2009) <br> Liberal environmentalism (Bernstein 2001; Gareau 2013) |
| Historical | Power struggles are embedded in historically specific structures of world order, the class formations that undergirds this order, and the fissures or antagonisms that threaten its stability. | Critical theory and historical dialectic (as compared to problem-solving theory; Cox 1986, 1992) <br> Transnational historical materialism (Levy and Egan 1998) <br> Systemic cycles of accumulation, hegemonic transition (Arrighi and Silver 2001) <br> Transnational capitalist class (Sklair 2001) <br> Ecological unequal exchange (Roberts 2001; Roberts and Parks 2007, 2009; Rice 2007; Jorgenson and Clark 2009; Shandra, Leckband, McKinney et al. 2009) |

Organization (an organization with no explicit mandate to address global health) may be of equal or greater importance to this issue than developments in the World Health Organization (an organization with an explicit mandate to address global health). The same can be said for the problem of climate change. Numerous institutions and other forms of international organization shape how action on climate change is governed, far beyond the UNFCCC, the usual sole focus of such studies.

Governance arrangements on any issue vary wildly from elite clubs and democratic state decision-making institutions to public-private partnerships, civil society certification programs, and for-profit businesses. Even seemingly cohesive institutions like the UNFCCC that have a formal voting structure also have fragmented internal dynamics that privilege certain types of knowledge and participation in differing enclaves of the regime. Decision making is sharply different on the Adaptation Fund Board, for example, than it is in the core negotiations on emissions reduction targets.

To make it even more complicated, extending outside a given institution such as the UNFCCC, each subissue in an issue area has its own set of institutional arrangements and actors, often with no clear rules about which governance body is to hold ultimate authority. This is true whether the issue is climate finance, forest management, carbon markets, emissions accounting, or technology transfer, among others. Thus, the global governance architecture concept helps us to overcome tendencies in the climate change literature to elevate the importance of the developments in the UNFCCC, which in reality may have limited real-world consequences compared to other institutional bodies. Moreover, this lens encourages us to analyze the negotiations within the UNFCCC with attention to the ways that this regime is itself fragmented and a loose configuration of parts, each with different informal rules of governance rather than a cohesive, well-coordinated whole.

The earth systems governance approach also has its weaknesses. In particular, our observation is that its proponents tend to focus on institutional architecture, while neglecting how actor coalitions engage in struggles to use such architecture to their advantage. There is an implicit assumption that problems such as climate change can be solved through fairer or more effective governance arrangements alone. This leads us to a core contribution of our framework. We are concerned with not only how institutional arrangements constrain or enable certain possibilities for action on climate change, but also how different actor coalitions engage with this complex architecture to reproduce, secure, or challenge relationships of privilege and inequality in the international system. This builds from the neo-Gramscian strategic theory of power, which focuses on how various state and nonstate actors negotiate the terms of a historic bloc in global environmental governance.[34] Thus, our analysis focuses on identifying (1) the major competing coalitions and how they define their interests and identities and (2) how these coalitions strategically navigate and engage with the complex tapestry of global governance that relates to the issue at hand.

### States, Markets, Ecosystems, and Civil Society

Third, our concept of strategic power relations is attentive to how these governance structures and coalition engagements are conditioned by broader historic structures of the shifting world order. As Robert Cox argues, analysis of the process of multilateralism "begins with an assessment of the dominant tendencies in existing world order, and proceeds to an identification of the antagonisms generated within that world order which could develop into turning points for structural transformation."[35] From this viewpoint, multilateralism, or global governance for that matter, can be understood as a "site of struggle between conservative and transformative forces."[36] Moreover, such struggles are historically specific and cannot be understood in abstract or ahistorical terms.[37]

We argue that analysis of strategic power relations and social change should consider the historic structures of world order, the historic bloc that undergirds this order, and the fissures or antagonisms that threatens its stability. We use Cox's formulation of being attentive to three main areas: the interstate system, the global political economy, and the biosphere or global ecosystem.[38] He argues that "these three components are both autonomous in having their own inherent dynamics, and, at the same time, interdependent with each other."[39] We also add a fourth area of analysis: transnational civil society. Whether it is professionalized nongovernmental organizations or emerging transnational social movements, we see this fourth area as a key structure of world order that is sometimes deeply influential and has undergone profound changes in the contemporary period.

These four areas of analysis direct our attention to the ways that a dominant global political order is vulnerable to attack and reform, which may adapt to accommodate bottom-up challenges, as well as to identify the historically specific issues that remain firmly off the negotiating table. This is what Gramsci refers to as the "decisive nucleus of economic activity."[40] Much of the status of power elites and hegemonic nations depends on where they sit on global cycles of boom and contraction and international systems of extraction of resources, labor power, and wealth that are inherent in the workings of global capitalism. Each conjuncture requires different strategies and resources to remain in power, and each offers different windows of opportunities to those wishing to change the system.[41] Numerous works have pointed to a process of ecological unequal exchange, whereby pollution and other environmental "bads" are concentrated in the poorest countries, as part of structurally unequal trade relations in the world system.[42] Such historically bound conditions of inequality also play a critical role in shaping power relations in the negotiation of global environmental governance.[43] In the next section we begin to discuss these factors in their concrete historical context after 2008, that is, after the global economic crisis and during and after the fateful Copenhagen conference.

Overall, to recap, we propose a strategic, layered, and historical framework for analysis of power relations in global environmental politics. This directs our attention to three corresponding areas of analysis: (1) the legitimate and coercive forms by which coalitions strategically interact to exert or challenge social control, (2) how such power struggles interact with fragmented and variegated global governance architectures, and (3) the ways in which historic structures of world order condition, and are conditioned by, these coalition struggles. What is key from this perspective is that "capital's international hegemony is not uncontested in the international sphere; rather, it secures legitimacy and consent through a process of compromise and accommodation that reflects specific historical conditions."[44] This reflects Gramsci's core insight that hegemonic resiliency rests not in its rigid and unresponsive structures of domination but, rather, in the adaptability of lead actors to make certain accommodations to those they profess to lead and, by doing so, manage to stay in power.

## Historical Shifts Relevant to Climate Change Politics

In what follows, we discuss the particular historical shifts underway relevant to climate change politics. The world order is shifting rapidly in ways that create major new obstacles to human action to address climate change, and some important new opportunities. We discuss the main features of the contemporary world order and the fissures or challenges to this order. Following and building on Robert Cox's framework, we describe four fissures: those related to its political economy, the crisis in the interstate system, the apparent early collapse of critical ecological support systems, and the evolution and fragmentation of transnational civil society focused on climate change. Our discussion begins with the changing nature of global political economy, beginning with the Great Recession of 2008.

**Shifting World Order**

To understand the importance of the Great Recession of 2008 as it relates to the state of the global political economy and, in turn, to climate change, we need to rewind briefly.

Climate change is a global problem, and any adequate solution must be global. This is tricky, since it has long been assumed that the strongest environmental movements and states with the most monetary resources to address the issue are in the world's wealthy nations of Northern Europe, North America, Japan, and Australia (the North). These nations are also far more buffered from climate change impacts by their technology, level of development, and geography since most are out of the path of hurricanes and in regions less at risk of drought. However, in the past few years, a handful of developing countries have surpassed the North and are now emitting more greenhouse gases.[45] The imbalance is becoming more extreme; newly industrialized countries are increasingly a critical part of the problem and therefore crucial parts of any adequate solution.

Gaining the enthusiastic participation of newly industrializing states in the climate negotiations has been an ongoing source of difficulty in the climate negotiations. In *A Climate of Injustice*, Roberts and Parks described how the global nature of climate change provides the world's developing nations greater leverage in negotiations than they have in economic or trade talks because global participation is more necessary to address global environmental public goods issues. Therefore, developing nations are more likely to resist coercive tactics in environmental talks than they are in other issue areas such as trade or intellectual property protection, where they are in weaker bargaining positions.[46] They argued there that the roots of distrust, which often reveal themselves in the negotiations, go far beyond the issue of climate change. That is, having been frustrated in other negotiations such as those over the World Trade Organization or, before that, over issues like the "right to development," or over intellectual property or domestic content rules, some nations will be more likely to withhold their participation in climate talks than would otherwise be the case. As Adil Najam put it, "It is tempting to dismiss the South's persistent distrust of the North as the paranoia of historical baggage. However the South's anger is directed ... by what it sees as subjugation today, and its inability to influence what might happen in the future."[47] The history is fraught, and resentment rears its head in the climate talks.

Where do these resentments come from? Here's a brief reminder: most Latin American nations gained political independence in the nineteenth century and Asian and African countries did so in the twentieth century. However, nominal independence has often been just that—incomplete, forcing a fragile new nation to attempt to govern itself while its economy is dominated by outsiders and built on unstable footings. Worst of all is a tiny and poorly prepared host government attempting to control the actions of huge corporations from another country that are extracting its limited natural resources or building vast monoculture plantations, and with the muscular support of its home government. In many cases, that support has included military intervention or the threat of military action. And regardless of where firms are from, having been brought into a world economy already dominated by bigger players and nations has made it exceedingly difficult for developing countries to diversify their economies, industrialize, and move up the value chain to more profitable products, to create taxation systems that are stable, and to avoid corruption. These nations (referred to as the "periphery" in World Systems Theory) export their mineral and agricultural wealth "at the price of bananas" while importing all kinds of manufactures and services at top dollar.[48]

A group of African and Latin American scholars observed that worst of all for national development was a heavy reliance on natural resource extraction, especially by foreign firms.[49] These left nations subject to extreme economic booms and busts as prices soared and collapsed as substitutes or more sources were found. Resource boom income has frequently led to citizen demands for social program spending, which often leads to civil unrest when cutbacks occur during busts. Resource dependence often drove repressive labor conditions and savage inequality between the owners and managers on the one hand and, on the other, the laborers and those left behind. With so much of a government's revenue coming from just one taxpayer (the extractive firm), this underdevelopment failed to support the accountability systems needed for stable democracies.[50]

In efforts to reverse the trap of dependency on an unfair global economic system, a series of countries took on more active approaches to substituting their imports with manufactures made at home by protecting their local industries behind high tariff barriers. Many built governmental programs to bring broad-based education and health programs to the poor and train workers to compete globally. Many of these development drives were financed by variable rate loans borrowed from banks flush with cash deposited by the oil-producing nations after the 1973 and 1978 oil price spikes. Often the loans went to dictators, some of whom advanced their nations' economies, but many of whom also siphoned off substantial funds into personal accounts.

This worked for a while, until interest rates spiraled quickly, and Mexico, Brazil, and Argentina set off a wave of debt crises and hyperinflation in developing countries. To break through key shortages in capital for local investment, many developing nations had taken advantage of low-interest variable rate loans in the 1970s; they were devastated in the 1980s when the rates ballooned, they faced a debt crisis, and international lenders like the IMF told them that if they wanted future access to credit, they would have to slash government spending. These were called structural adjustment programs, and they often destabilized national politics and set back national development plans by years. Government programs were slashed and state enterprises auctioned off as part of fiscal austerity to free up funds to pay off new loans taken out from the International Monetary Fund (IMF) to pay off the toxic commercial loans.

To sell this bitter medicine, an ideology called neoliberalism was promulgated, which held that free markets and state intervention prioritizing investor interests were the pathway to economic success and development. John Williamson described this as the Washington Consensus, because the major agencies such as the World Bank and IMF (headquartered there), combined with the US international agencies and Washington-based think tanks, convincingly sold the idea that state intervention in the economy was necessarily a bad idea and that nations should go back to their "comparative advantages" of producing and selling only the things they were good at. The government cuts were devastating to local economies, and the 1980s was described in Latin America as the lost decade for economic and social development.

Another change was happening as well: rather than making money from producing things, finance had increasingly become the source of people's wealth. The finance-led growth regime and the globalization of that capital corresponded with a sharp shift rightward in economic ideology, a rise in the power of finance business in relation to manufacturing industries, a rise in inequality within nations,[51] and a shift of influence in global supply chains from manufacturers to buyers, designers, financiers, and distributors.[52] A diverse bloc of actors—including transnational corporations, globalizing bureaucrats and politicians, globalizing professionals, and consumerist elites—has come to identify their interests in relation to the ideological and organizational basis of financialization.[53] Their capital is extremely footloose, able to leave locations that do not offer them advantageous terms. Less discussed is that while a commitment to financial expansion is central to this regime (and not manufacturing), in material terms, like other regimes since the birth of capitalism and the industrial revolution, it has been fueled by a nearly complete reliance on fossil fuels.[54]

Scholars of global environmental governance have argued that the shift to neoliberalism, including a focus on enhancing investor rights, privatization, and expansion of the "free" market since the 1990s, has infiltrated the governance of global environmental problems such as the ozone layer, climate change, forest protection and use and biosafety.[55] In practice, they argue that this has meant movement away from top-down regulations rooted in the precautionary principle (first, do no harm) and the polluter pays principle (those responsible for pollution should bear the cost of remediation), to "liberal environmentalism" including market-based mechanisms rooted in principles of efficiency, flexibility and capital expansion.[56] Nevertheless, there have been notable developments and antagonisms in the foundation of this order. In the context of stagnated growth and state shrinking after the Great Recession in 2008, various movements have threatened the long-held gospel of open markets and its governance institutions as part of the Washington Consensus. Such challenges include demands from developing countries, particularly BRIC countries (Brazil, Russia, India, and China) and the "pink wave" in Latin America (Venezuela, Ecuador, Bolivia, and others) to reform the governance structures of international financial institutions and the global distribution of power.[57] There have been widespread calls among world leaders for a more regulated global financial system[58] and resistance to externally imposed austerity measures and resulting political turnover in some European economies.[59] Public debates have arisen about the extent to which these crises constitute an existential challenge to the legitimacy of neoliberal globalization.[60]

In addition, despite decades of discourse that neoliberalism would increase prosperity for all countries, basic development needs have continued to be unaddressed in the poorest countries. This reality has been heightened by the global recession, which since 2009 has caused substantial and persistent stagnation for the forty-eight Least Developed Countries (LDCs).[61] And many of these countries have become more, not less, structurally disadvantaged in the global economy: between 2001 and 2010, the LDCs as a group became increasingly reliant on a limited number of exports.[62] This trend implies that they are faced with a threat of becoming increasingly commodity dependent and vulnerable to external shocks, just like in previous periods of instability.

However, more poor people in fact are living in middle-income countries rather than in the LDCs. Larger and middle-income nations in South America find themselves also becoming more dependent again on the export of volatile natural resources, increasingly now to resource-hungry China.[63] The Washington Consensus is no longer seen as holding the answers, and some nations are shifting to more state intervention in an attempt to turn themselves around.

Despite these and other challenges, overall there is little evidence to claim the death of the finance-led growth regime nor neoliberalism.[64] In various states, we have seen trends toward austerity during this period rather than a reassertion of the welfare state. What is clear is that internationally there are notable trends toward a more heterodox and multipolar form of political economic organization.[65] In this context, "Emerging societies are increasingly fulfilling core functions on the world stage—acting as development role models, providing stable markets, loans, aid and security, with China as a leading force."[66] Moreover, the basic forms of organization of the global capitalist system are being transformed. This includes a fall in the concentration of economic activity in the international system and, within this, the rise of concentrations outside the previous core.[67] In particular, "the systemic change has to do with the unfolding of a deterritorialized global capitalism made up of flows, fluxes, networked connections and transnational production networks, but marked by inequality, instability and new patterns of stratification."[68] In simple terms, one can no longer look only to Wall Street or High Street to attempt to understand this decentered and complex new global economy.

These developments have major implications for international climate change politics. On the one hand, the clear shortcomings of neoliberal doctrine revealed through the global financial crisis offer new openings for challenges to the intensification of various forms of inequality. We are also seeing weak oversight by state and international regulatory bodies over economic affairs and unbridled corporate power over national political processes. In climate change politics, we have seen states such as Bolivia, Indigenous peoples' networks, and civil society organizations raise bold critiques of the neoliberal development model and its deleterious impacts on global ecologies.

On the other hand, we see a doubling-down on neoliberal policies in various states, which have adopted austerity measures, rolled back the regulatory state, and become less willing to cooperate in international regulatory agreements. Shifts in governments in this direction have had notable impacts in rolling back or minimizing the ambition of domestic and international climate change policies in the United States, Canada, Australia, Japan, and states across the European Union. In addition, we view these shifts in political economy as driving a new geopolitics within and between the global North and South, which we discuss in the next section.

### US Decline and China's Rise

A framework we find useful to understand the shift in the dynamics of the global economy and of climate negotiations is to consider the massive upheaval in the global political economic system going on over the period that included Rio (1992) and Copenhagen (2009). The late sociologist Giovanni Arrighi and Beverly Silver wrote a series of pieces that describe historical rises of a world power to hegemonic domination over other nations and their subsequent decline.[69] In each case of rise and decline, financial capital plays a key role by creating flexibility of accumulation for the hegemonic state's elites and diversifying their income when different types of activities in certain locations become more and then less profitable.

In the US hegemonic cycle, the profitability of manufacturing in the core nations dropped sharply in the late 1970s, 1980s, and 1990s as job-heavy production shifted to cheap labor zones such as Mexico and China. In this way, the fiscal crisis was put off, as it was in previous hegemonic cycles of the Dutch and British empires, as financial power sustained each hegemon beyond its time. Each at the end of its cycle of hegemony experienced a final boom when it pursued its "national interest without regard for system-level problems that require system-level solutions."[70] Although Arrighi and Silver were not writing about climate change, this is a useful comparative view.

In this international leadership vacuum, Arrighi and Silver argue that hegemonic crises have been characterized by three distinct but closely related processes: "the intensification of interstate and inter-enterprise competition; the escalation of social conflicts; and the interstitial emergence of new configurations of power."[71] They note that the final stages are "complete hegemonic breakdown and 'systemic chaos' ... a situation of severe and seemingly irremediable systemic disorganization." The problems are both internal and external, which involve areas outside the control of the hegemonic state: "As competition and conflicts escalate beyond the regulatory capacity of existing structures, new structures emerge interstitially and destabilize further the dominant configuration of power. Disorder tends to become self-reinforcing, threatening to provoke or actually provoking the complete breakdown in the system's organization."[72] This is supported by Robert Keohane's work, which describes how "the degree of international cooperation will be directly proportional to the degree to which one actor dominates international politics."[73]

To bring this back to interstate climate politics, in his landmark book, *The Long Twentieth Century*, Arrighi describes how, in the face of military and financial crisis in 1973, the United States retreated from the world stage and "U.S. strategies of power came to be characterized by a basic neglect of world governmental functions." He continues that "it was as if the ruling groups within the United States had decided that, since the world could no longer be governed by them, it should be left to govern itself."[74] Arrighi argues that in this vacuum, oil producers organized an effective way to gain huge rents from petroleum. In 1973 and 1978, the Organization of the Petroleum Exporting Countries (OPEC) utilized embargos to attempt to modulate production and to keep prices up.[75]

Two things happened with that money that are important to the story of the rise and fall of the Kyoto Protocol. First, Arab oil producers gave foreign assistance of at least $100 billion accumulated since that period.[76] We do not know whether one of the goals of Arab aid has been to secure support for their position in other negotiations, such as to keep key recipients from dissenting from OPEC views in G77 negotiations during climate change talks. However, many of the poorest nations in the world are highly dependent on OPEC oil and financial assistance, and they fear angering those donors on an issue such as this, where OPEC views are very strong. If Arab donors did use aid that way, they would not be alone: anecdotal information suggests Japan has secretly used aid for votes on the International Whaling Commission, and the WikiLeaks release of US diplomatic documents in 2010 showed that payments from the promised Copenhagen funding were provisional on recipients signing the Copenhagen Accord. Money is leverage in UN negotiations—one may not get everything one wants, but large aid donors are often successful in influencing positions. In the case of OPEC (and the United States), their position has been in resisting binding limits on their carbon dioxide emissions, and this has persisted for more than two decades through tough negotiation.

Second, the oil boom money from OPEC governments was often loaned (through Western banks) to other developing countries with adjustable rates, and these rates skyrocketed when the Reagan administration in the United States adopted a tight fiscal policy to regain control of inflation at home.[77] This created a debt crisis that set back many developing countries for a decade. This failure of development to measure up to expectations certainly strengthened the G77's cohesiveness in the climate negotiations, even as the interests of its members on this issue and others diverged.[78] Meanwhile, China's economy (and energy use/carbon emissions) has grown exponentially since 2001, threatening US global economic hegemony, which was continuous since World War II.[79] Each year the United States grows more indebted to China (as do many EU nations and those of much of the rest of the world). India also has the ability to undermine US labor competitiveness in a large number of job categories long thought to be secure.

Arrighi and Silver argued that the rich countries cannot compete with the ascendant nations in East Asia because of profoundly different developmental paths (especially wage rates), and they cannot be restructured "without causing social strains so unbearable that they would result in chaos rather than 'competitiveness.'"[80] Arrighi and Silver end with the ominous warning that "if the system eventually breaks down, it will be primarily because of U.S. resistance to adjustment and accommodation. And conversely, U.S. adjustment and accommodation to the rising economic power of the East Asian region is an essential condition for a non-catastrophic transition to a new world order."[81]

In his 2009 "Post-Hegemonic Climate Politics?" piece, Matthew Paterson argues that the United States has been surpassed and Europe has been leading in the area of global climate policy. However in Copenhagen, we saw the rise of BASIC, especially China, as the real challenger to US hegemonic power. As Arrighi and Silver write, the hegemon is typically the only power with the ability to lead the world in protecting global public goods. This suggests that the United States is leaving the next economic hegemon (seemingly China) with the climate mess to clean up. However, despite this, there was some progress on this issue in late 2014 when the US and China agreed to a joint pledge to cut greenhouse gas emissions and partner on clean energy development. While this partnership on its own is not sufficient to maintain stable global temperatures, it demonstrates that these two powerful states are currently embracing more of a leadership role on the issue of climate change than in the past.

We do not subscribe to the idea that history always repeats itself in cyclical waves. Rather, in line with neo-Gramscian thought, we are concerned with how state preferences are continually redefined as a result of social conflict and shifting ideological constructs. But Arrighi and Silver offer a framework to understand the importance of structural constraints on states in periods of hegemonic decline. As Arrighi and Silver put it about economic issues, "An equally essential condition is the emergence of a new global leadership from the main centres of the East Asian economic expansion. This leadership must be willing and able to rise up to the task of providing system-level solutions to the system-level problems left behind by U.S. hegemony."[82]

Whether China will be the next global hegemonic power is of course uncertain. And though it has the ability to mobilize extraordinary resources and has invested heavily in renewable energy sources, the extent of its leadership's devotion to addressing climate change remains uncertain because it has economic growth as its top priority. The next global hegemon, whether China or another nation, in the context of a warming world will be faced with an even more intense ecological crisis to manage.

However, while recognizing geopolitical shifts in the current context, some suggest that the shift toward multipolar governance, or multilateralism, may be overblown.[83] As Wade argues, "The United States remains the dominant state, and the G7 [Group of 7] states together continue to exercise primacy, but now more fearfully and defensively."[84] Our position is agnostic on this: things are shifting, but the timing and ultimate outcome are quite uncertain. China's very particular model of state-directed capitalism with huge labor reserves and constrained political freedoms may prove unable to sustain both economic and political domination internationally. Or it may not.

An additional geopolitical trend relevant to this discussion is the fragmentation of coalitions within the global South.[85] While in some contexts, including the UN climate negotiations, the G77 plus China coalition still at times formally negotiates as a single body, there are growing tensions within this bloc as the class position of states in political, economic, and ideological terms grows increasingly variegated. This is also compounded by a trend of regionalism and relative economic prosperity in areas such as Latin America, which poses a challenge to the identity of "South" in geopolitics.

In sum, we view a very different world of geopolitics from that prior to 2008. The US position as global hegemon is increasingly precarious; China's rise is now a forgone conclusion, but it faces stability concerns; and the global South has fragmented along various lines, with new interstate class dynamics threatening longstanding ideals of developing world solidarity. We argue that these new dynamics have been pivotal in international climate change politics and will continue to be so for the foreseeable future. It is impossible to understand the outcome of these negotiations without attention to this reshuffling and highly volatile global political and economic landscape.

### Ecological Collapse and Development Space

A third development in the contemporary world order is frequently overlooked in such discussions: the impact of degradation, exhaustion, and collapse of ecological systems on political processes. Some of these changes are still out in the future: scientists are saying louder and louder that environmental crises are coming, but it increasingly seems the case that we are already experiencing the first waves of climate change–related extreme weather events.

New access to unconventional fossil fuel reserves is generating new geopolitical and social conflicts, which have important implications for international climate politics. The supply of fossil fuels is finite, but the reserves now seen as proven are vast, even if the cost of accessing these reserves grows higher the farther down or offshore we have to go to get them, or if rocks or sand need to be squeezed to extract energy. New access to unconventional energy sources such as shale gas and tar sands have extended access to global fossil fuel resources significantly. And new technologies are continually developed to access deeper, though often more energy- and resource-intensive, energy supplies such as methane hydrate at the bottom of the ocean. Thus, claims about impending "peak oil" or "peak fossil fuels"[86] may be largely overblown, as long as we are willing to pay higher prices (both financial and ecological) for our energy.

Such unconventional fuel development is restructuring geopolitics in important ways. An obvious example has been Canada's reversal of leadership in both domestic and international climate change policy in the name of developing its own vast reserves of highly greenhouse gas–intensive oil buried in the Alberta tar sands. The United States provides another example. At the behest of Barack Obama, the US Environmental Protection Agency has taken strong measures to limit domestic coal use.

The downtick in coal use is being largely replaced by the production of relatively cheap and abundant shale gas made accessible by technologies such as hydraulic fracturing and horizontal drilling. The United States has increasingly structured its domestic climate policy around these developments and has made the supposed transition from coal to natural gas a key argument of its commitment to leadership in international climate change negotiations. There are also numerous other examples of new energy players emerging with the discovery of extensive unconventional fossil fuel reserves.[87] Climate change itself, through melting of deep ice sheets, is also leading to geopolitical conflicts for newly accessible oil reserves in places such as Antarctica and the Arctic sea floor. Finally, such access to new fossil resources worldwide is stimulating new social conflicts as people living on fossil fuel deposits are resisting the appropriation of their land for clear-cutting, strip mining, drilling, and pipeline development.

Yet our dependence on energy from fossil fuels, and growing use of carbon-intensive nonconventional sources, has contributed to a scarcity of atmospheric space. Bill McKibben and the 350.org movement recently crisscrossed America beseeching us to "Do the Math." The bottom line of that math is that there are now five times as much fossil fuel reserves on the books of oil and coal companies (ready to be tapped) than can be safely burned and still avoid dangerous climate change.[88] The International Energy Agency puts that figure at one-third of the world's energy that can be burned for us to stay within 2 degrees Celsius, but the point has gained broad acceptance: a vast majority of the oil must stay in the soil, and nearly all of the coal must stay in the hole.

This simple math in theory should have important implications for investment in fossil fuel resources. This seems especially true when one considers the likelihood of a proliferation of climate policies in the future, putting a price on carbon and thus further increasing the cost of fossil fuel development. Indeed, in the case of coal, there have been notable developments with both private and public investors being less willing to support the development of new coal-fired power plants. Major international development banks such as the World Bank, the European Bank for Reconstruction and Development, and the European Investment Bank have all adopted policies to severely limit their coal lending. The governments of the United States and United Kingdom have adopted similar policies for their development agencies and operations abroad. Universities such as Stanford, at the behest of student activists, have recently taken up policies to limit their investments in coal. Finally, there are indications that private banks such as Wells Fargo and JP Morgan Chase are cutting ties with the coal industry.[89]

These are important developments and demonstrate that it is not inevitable that all known fossil fuel resources will be developed without pushback. However, there is a risk that investments in coal will merely be displaced by heightened investment in natural gas. Natural gas has its own significant greenhouse gas impacts, sometimes comparable to or worse than that of coal due to the escape of methane, a highly potent greenhouse gas, into the atmosphere. Overall, despite growing evidence of the unsustainability of reliance on fossil fuel development, there is little evidence that what Tim DiMuzio calls the "petromarket civilization" is being meaningfully transformed. As he argues, "Investors are nowhere near betting on a future outside of fossil fuel energy—a future that would require *immediate* and *intensive* capital investment for there to be any chance of a successful energy transition without protracted political, social and economic dislocations."[90] Even when we have seen growth in national non–fossil fuel energy use, this has not translated into substantial reductions in fossil fuel use.[91]

Despite increasingly robust evidence of potentially catastrophic anthropogenic climate change,[92] we have failed to reduce or even stabilize global emissions levels during two decades of international negotiations[93] or provide the financing adequate for the countries most vulnerable to a changing climate to adapt to its impacts.[94] This creates new potential tensions among allied countries in the global South, where countries compete over scarce financial resources to adapt to climate impacts. As we will explore in chapter 5, this has become a competition over which countries are perceived as the most vulnerable countries to climate change and therefore most entitled to financial support and compensation.

During this period of what can be called "active inaction" on climate change, we have seen the establishment of new multibillion-dollar carbon markets (some currently at dire risk of collapse), where carbon allowances are traded on financial markets. This approach of climate capitalism has sought to create the possibility of economic winners from decarbonization, mainly financiers.[95] Fossil fuel interests remain mostly unchanged or even have become more profitable, with coal in certain contexts being a notable, and very important, exception. As we discussed in the cases of Canada and the United States, the changing global energy landscape is restructuring geopolitics in important ways, influencing the positions that states are taking in international climate change negotiations. How the tension between rapidly constricting atmospheric space and growing access to unconventional fossil fuel resources will play out is entirely unclear. But as we will argue in chapter 6, what is clear is that a sufficient solution to the climate problem will not emerge from new business interests alone. Transnational civil society will have to play a key role.

### Civil Society Gone Transnational

Analysis of shifting opportunities and challenges for civil society in the contemporary order is an essential fourth dimension to add to Cox's analysis of multilateralism. Many scholars have attempted to theorize about the particularly transnational nature of civil society.[96] For example, Kaldor argues that civil society is no longer confined to the borders of the territorial state. Rather, global or transnational civil society is a new form of politics in the world that is both an outcome and an agent of global interconnectedness.[97] However, as many scholars have noted, the normative and celebratory view of civil society as rights-bearing citizens that serve to counterbalance the market and state is rarely substituted for what exists in reality in terms of power imbalances, inequality, corruption, co-optation, and exclusionary practices.[98] In this section, we explore the opportunities and challenges that such a shift in territorial boundaries poses to civil society in relation to the problem of addressing climate change.

The contemporary context offers three key opportunities for transnational civil society to emerge as a political force. First, globalization offers opportunities created by the use of new technologies and tools for movement organizing that transcend space and time in ways that used to be unimaginable.[99] Climate activists hold Skype meetings with partners around the world, launch global petition drives in hours from dispersed home offices or coffee shops, send e-mail blasts reaching tens of thousands of recipients, and post to Facebook pages, Tumblr accounts, or Twitter feeds with thousands of "followers." These are not without major limitations or problems, but they are new and sometimes result in meaningful governmental responses and provide new opportunities for mobilizing constituencies and advancing ideas.

Second, contemporary globalization offers opportunities to forge new social movement identities that cross various boundaries, including those based on demands for collective rights.[100] In terms of the environment, we have seen the beginnings of this in domestic contexts in the United States and European Union in the form of "blue-green" alliances that bring together environmental advocates and labor. Transnationally, for example, new movements have begun to coalesce that seek to address environmental and labor concerns associated with electronics such as the iPhone. These movements are targeting harmful activities across the commodity chain, from extraction of materials, to assembly, transportation, distribution, and disposal. In such cases, movement identities are transcending those of a particular national, local, class, cultural, or other interests to embrace ideas of solidarity that are grounded in universal and broadly conceived ideas of shared rights.[101] These are what Manuel Castells calls "project identity" movements, distinct from movement building that is reactive in nature.[102]

Third, geographer David Harvey points to political openings to challenge obvious contradictions and failures of the neoliberal model. We have seen this with Occupy movements, which have spread from US cities internationally, calling for a more equitable economic system. On the issue of climate change, transnational groups such as the volunteer-only organization Rising Tide and the massive peasant social movement Via Campesina have sought to link climate change to the excesses of global capitalism.

However, the contemporary globalized context also offers transnational civil society historically unique challenges as it seeks to regulate the worst excesses of global capitalism, such as avoiding the worst impacts of runaway climate change. The first identified major pollutants such as sewage and urban air pollution were local in scale and allowed for cause and impact to be readily connected. This made addressing them far easier compared to climate change where impacts (flooding in Tuvalu or Bangladesh) can be half a world away from their cause (my driving to work or your taking a flight to London). Some of the most difficult challenges for climate social movements include the fact that they are fighting a more diffuse and less geographically constrained system of power, the new technological and communication tools by which power is organized, and the increasing irrelevance of civil society identities organized primarily around the state. Having a national movement using existing tools and targeting old adversaries is no longer enough, but not having fear-based national movements has also proven deadly to efforts to address climate change.

In certain cases, some civil society organizations have actually been the promoters of key tenets of the neoliberal project rather than its main force of resistance. For example, many of the most powerful actors that occupy and dominate the transnational space of civil society on climate change subscribe to and promote a neoliberal rationale of the self-regulating market. Major organizations playing a role in the climate talks and advocating market-based solutions are Resources for the Future, Nature Conservancy, Environmental Defense Fund, and the Harvard Belfer Center.

Though their reasons varied at different times for their support for market solutions to climate change, most were making pragmatic decisions to attempt to be included in decisions they saw as inevitable. These groups therefore have come to define "rights" in terms of access to the market or individual property rights rather than that of inclusion in the economy, collective decision making, access to basic human services, or traditional forms of interacting with ecological systems.[103] Thus in the climate change movement, a coalition of powerful NGOs was responsible for paving the way for trading in carbon credits (the commodification of greenhouse gas emission reductions) to become the industrialized world's primary policy response to global climate change.[104] Jonas Meckling (2011) describes how a segment of businesses and of environmental NGOs went from loggerheads and stalemate over climate change to a compromise posture that was welcomed by policymakers since it did much of their work for them.

Overall, the literature on transnational civil society points to the need to consider the ways in which new forms of social movements are complex and layered entities where conditions of power and inequality are reproduced as well as challenged (see the discussion in chapter 7). The transnational nature of social movements and advocacy networks may actually heighten power imbalances, as actors in these networks, due to geographical difference, have disparate access to resources and decision-making capabilities and are exposed to consequences of network actions in different ways. Thus, in civil society networks targeting climate change in the UN process, the "big green" groups such as Environmental Defense Fund, Sierra Club, Greenpeace, and World Wildlife Fund far overshadow the smaller groups and those headquartered in developing countries.

Alternatively, the transnational space may provide opportunities for grassroots initiatives from the global South to bubble up and shape the particular ways in which capitalism is constrained in the global North.[105] The cases of Indigenous peoples and waste pickers in chapter 6 offer examples of such movements.

To sum up, the global political-economic context presents new and reconfigured opportunities and challenges to social movements seeking to influence the governance of the climate system. The relationships of power and inequality within transnational civil society, and its particular relationships with both the state and market, are fundamental to the norms adopted, preferences formed, strategies advocated, and access granted to particular actors. We should not simply assume that these changes flatten the playing field within civil society as some might claim[106] or make them consistently a force of transformational change. Nor should we assume that these changes lessen the corrupting influence of money and power on the transnational public sphere: groups compromise and mute their critiques and avoid disruption to keep their governmental, private foundation or corporate funding, their contracts with governments, their access to government officials, or even their jobs. This is true in both the global North and South. Rather, empirical and theoretical research should focus on uncovering the actual and existing conditions and relationships within transnational civil society and how it relates to the contemporary configurations of global power. These are issues that we will return to in our discussion of transnational civil society in the climate change negotiations in chapters 7 and 8.

### Agency in the System

In this chapter, we have developed a neo-Gramscian strategic view of power relations and social change. This approach enables us to bridge unique contributions of existing theories of global political change and overcome the divide prevalent in this scholarship between those who focus on individuals and groups as agents and those focused on big constraining structures. In doing so we have offered a framework to analyze the terrain on which any efforts at transformative change in international climate change politics must engage. This has directed our attention to how strategic coalition struggles for power and influence take place within a complex and layered global governance system and historically specific conditions of world order.

Following and building on Robert Cox, we have discussed major shifts in the contemporary period in four areas: the global political economy, geopolitics, ecological conditions, and transnational civil society. In the post-Copenhagen period, the global political context has shifted in important ways since the signing of the Kyoto Protocol a dozen years before. In each area, important tensions are largely structuring the limits and possibilities for action on climate change moving forward. First, this context includes a wobbly and wounded neoliberal doctrine due to the 2008 Great Recession. This has led to discursive openings for state and nonstate actors to call for new forms of regulatory and social action, but also responses by several key states to pursue austerity measures and limit state intervention toward environmental goals.

Second, we have discussed notable geopolitical tensions in the contemporary period. This includes the hegemonic decline of the United States, the rise of China and an increasingly multipolar geopolitical landscape, and the fragmentation of the global South's identity along various lines, with new interstate class dynamics threatening longstanding ideals of developing world solidarity. Third, we have discussed the impact of ecological constraints and degradation, and development in energy technologies, on international political processes. On the one hand, there is a changing energy landscape with the development of unconventional fossil fuel technologies, which extend our ability to access reserves previously considered out of reach. On the other hand, there is a scarcity of atmospheric space, which requires keeping most of these proven fossil fuel deposits in the ground in order to avert catastrophic climate change. In this context, the overall trend in fossil fuel energy extraction has remained constant despite rapid growth in renewable energy and carbon markets, with the notable exception of coal investment, which shows important signs of change.

Finally, we added a fourth area of analysis to Cox's conception of world order, transnational civil society, which has undergone profound changes in recent years. The contemporary globalized context offers transnational civil society historically unique challenges as it seeks to regulate the worst excesses of global capitalism, such as avoiding the impacts of runaway climate change, while fighting a more diffuse and less geographically constrained system of power.

We will explore how each of these issues has played out in international climate change politics in the chapters that follow. At least initially in the lead-up to Copenhagen, this multipolar geopolitical context included two powerful leaders, President Obama and Prime Minister Hu Jintao, each of whom indicated some willingness to take leadership on the issue of climate change. What was not clear was what direction they were willing to lead the world on the issue and how this set of actors would combine in the context of Copenhagen in December 2009. In chapter 3, we explore how these particular historical dynamics played out in the fateful Copenhagen negotiations.

# 3

## Beyond the North–South Divide?

### Welcome to Hopenhagen

Copenhagen in December 2009 was supposed to be the glorious conclusion of two years of preparation, following the delicate language agreed to in the Bali Action Plan in 2007 to settle a successor treaty for Kyoto. This had to happen by 2009 because Kyoto would run out at the end of 2012, and the process would need that long to make the transition. The two tracks of negotiation—one for Kyoto Protocol signatories (nearly all the world but the United States) and the other for all members of the Convention including the United States—were to settle texts that would be agreed on and merged during the two-week summit.

The UN added more negotiating sessions in 2009 in Bonn, Bangkok, Barcelona, and Singapore to address lingering disagreements in the text and prepare the ground for what was expected to be a major international event. A record number of heads of state promised to attend, leading to expectations that the major issues had been resolved so they could sign something glorious and pose for the cameras as solvers of this global crisis. The recent election of Barack Obama in the United States, with his roots in Indonesia and Kenya, created new hopes of his being a catalyst to break down old and enduring divisions between countries. Civil society groups and scientists all put what pressure they could on their states to deliver a robust mitigation framework in what was dubbed "Hopenhagen" in a branding campaign.

Things did not turn out as hoped. One serious problem in Copenhagen was the inept leadership of the Conference of Parties (COP) presidency throughout the negotiations.[1] President Lars Rasmussen was naive about UN negotiating procedures and offended many delegates unnecessarily and at crucial moments.[2] From the beginning, tensions appeared to arise along the old battle lines of North versus South. In the first week, the

Africa Group threatened to walk out and boycott the talks. In response to difficult drafting during the first week, the Danish presidency prepared a text to break through some of the roadblocks, but the text was leaked and many countries balked at the breach of UN protocol in its drafting. Negotiations were set to end on Friday of the second week, but through the Tuesday before, the texts coming up through the two negotiating tracks were riddled with brackets, signifying areas of disagreement. Within those brackets were starkly different visions of how the big issues of climate change should be addressed.

Copenhagen showed that the world had become a vastly more complex place since 1997 when the Kyoto Protocol was adopted. A series of new groups had emerged, based on different combinations of interests and solidarities (see tables 3.1 and 3.2 for comparison of major negotiating groups in Kyoto and Copenhagen). The expanded EU was now twenty-seven countries, and the increased diversity among them weakened their ability to argue forcefully and nimbly. About eight new groups could be described as coalescing around issues of vulnerability and compensation, often in competing directions: the Least Developed Countries, the mountain landlocked developing countries, the Alliance of Small Island States (AOSIS), the Central American Integration System and the Arab States, the Coalition of Rainforest Nations, and the Alliance of Bolivarian States (ALBA). Critically, a new coalition of large, emerging economies made up of Brazil, South Africa, India, and China (BASIC) formed as a major political force of its own. The G77 was splintering quickly.

The fragmentation of positions and the gaps in perspectives in the Copenhagen draft texts on mitigation were vast. The options for emissions reductions targets were all over the map, based on country submissions made just after Copenhagen and on the earlier texts.[3] Even the temperature increase targets couldn't be agreed: global average temperature increases of [1][1.5] and [2.0] degrees Celsius were all still in the text.

Confirming the fragmentation of the G77, President Mohamed Nasheed of the Maldives argued at Copenhagen, "Developed countries created the climate crisis. Developing countries must not turn into a calamity. Therefore, I invite the leaders of big developing countries to recognize their responsibilities. I urge them to come forward at Copenhagen with quantifiable and verifiable actions to reduce emissions 30 percent below business as usual by 2020."[4] These major reductions were explicitly described as unacceptable by all the BASIC countries, to which Nasheed was referring.

This chapter argues that climate politics is no longer primarily about North–South politics. There have always been deep divisions over the issue between blocs in the North, but what's new are the number of emerging coalitions based on new solidarities and divisions in the South and some new geopolitical alignments that cross the North–South divide. We find these new alliances to be some of the most interesting developments in the negotiations as we look for some political window of opportunity when a new deal, global compact, or grand bargain to solve this pernicious problem might be struck. The issue remains extremely salient: some very powerful nations in the North, as well as representatives of some particularly vulnerable countries in the South, hold that the Bali firewall (a term we explain below) between the North and South must be torn down for any global climate deal to be made. An equally steadfast group of developing nations is digging in again to keep this wall between North and South standing.

This chapter provides the history of the major coalition divisions in the negotiations and explains how and why the old North–South alignments shifted in Copenhagen. We simplify the journey into two themes. First, we explore the old world order, when the North–South rift grew and was reinforced. We describe the forces that created it: enduring gaps of wealth and of responsibility for the problem of climate change. We also discuss a third dimension that has not been well theorized, which threatened the unity of the G77 leading into Copenhagen: the emergence of an identity rooted in disproportionate vulnerability to climate change impacts, which has been embraced by the poorest countries and low-lying island states. This identity of disproportionate vulnerability challenges the main link of solidarity in the G77: the idea that all countries in the G77 share a common predicament in the global system, with the North to blame. The next two sections focus on the new developments in Copenhagen. We discuss the emergence of the alliance between BASIC and the United States that seized control and rewrote the rules there. We then explore how and why China, India, Brazil, and South Africa joined forces in Copenhagen; the tensions that exist among rising powers; and what this means for G77 solidarity moving forward. In the chapter conclusion, we revisit how this analysis connects to our theoretical framework and set the stage for the next chapter.

## The Old World Order

As Roberts and Parks explored at length in *A Climate of Injustice* (2007), the roots of the G77's unity lay across many issues far beyond the climate talks in their lack of access to meaningful participation in the global order, the deep inequity in their well-being compared to the wealthy nations, and their agenda for Third World solidarity.[5] These are the underlying forces that held the coalition together until Copenhagen in spite of their diverging material interests. In the climate talks, they shared interests in pressuring the historically wealthy or developed countries (what are called "Annex 1" countries in the negotiations) to act according to their historical responsibility for having created the problem and their capabilities to address it. Developed countries also advocated to maintain their own sovereignty from outside intervention (especially from limits on their ability to pursue national economic development),and for the provision by wealthy countries of adequate funds and the most modern technologies for dealing with climate change.[6]

From the very first UN Conference on the Human Environment in Stockholm in 1972, and again at the huge 1992 Rio Earth Summit, developing nations represented by the Group of 77 and China (G77) had a suspicion that the industrial countries were using environmental concerns to stop them from reaching their development goals. At the beginning of the climate negotiations in the early 1990s, the G77 was a reactive coalition because of its suspicion of the environmental negotiations as an agenda of the industrial countries. Put bluntly, poorer nations thought that green concerns were a ruse to keep them poor, a conscious or unconscious effort by the wealthy nations to keep the poor nations from usurping their place atop the global hierarchy. If individual environmentalists were well meaning, their concerns could still be used by those who would set up impossible restrictions on developing countries against their using their own natural resources to climb out of poverty. Rich countries had relied on cheap and abundant fossil fuels during their own rush to industrialization, and there was currently no foreseeable development alternative.

Addressing climate change means reducing consumption of cheap fossil fuels and switching to more expensive renewables like wind and solar; it also can mean not clearing rain forests to create farmland to expand the national economy. For this reason, the G77's initial approach to this new agenda was wait and see, learn and react, or reject.[7] If they were to address climate change and other environmental concerns, they needed to be compensated for lost economic gains and helped with new green technologies. When it came time to draft the UN Framework Convention on Climate Change (UNFCCC) before the 1992 Rio Earth Summit, the G77 and China succeeded in its goal of avoiding responsibility for making emission reductions.

However, from the beginning of the climate negotiation process in 1989 under the newly established Intergovernmental Negotiating Committee, the Alliance of Small Island States (AOSIS) was very active in attempting to insert binding commitments for greenhouse gas emissions reduction. Their compulsion was obvious. The first report of the Intergovernmental Panel on Climate Change (IPCC), published in 1990, indicated an ominous development: sea-level rise due to climate change would condemn many low-lying areas and island states to a watery death. In this effort, AOSIS found a willing partner in the EU, which, being influenced by public opinion and strong social movements, also showed great interest in controlling greenhouse gases from the beginning. Yet small island developing countries continued to stand behind G77 statements and positions in the negotiations, which were generally for slowing the progress of aggressive climate treaties. They did so because their voice was so easily ignored when they spoke alone: if they could get some of their positions into G77 statements, they felt they had some chance of influencing a treaty in which they would otherwise have little chance of having impact.

The Group of 77 and China is actually a bloc of developing nations now numbering over 134 countries. As Antto Vihma put it, the G77 is "a product of the North/South divide and the political economy of the late 20th Century. It is broadly based on a 'self-definition of exclusion' from world affairs."[8] That is, the vast global South, consisting of all of Latin America, Africa, and nearly all of Asia, felt that they had been left behind over decades of efforts at economic development and globalization.[9] Brought into the world economy through colonial conquest and continuing to be dependent on the production and export of minerals and agricultural products whose prices fluctuated wildly or tended to go downward, they saw themselves as trapped in structurally disadvantaged positions.

This vast group of countries at the bottom of the global division of labor and the wealth pyramid it created was described starkly as the "periphery" by Latin American economists from the "dependency school."[10] The periphery was seen in a (losing) role in relation to the "centre" or "metropole," where wealthy countries were able to draw resources and cheap labor from around the world to manufacture high-value products they could export back to the periphery. The terms *core* and *periphery* were adopted and elaborated on by North American sociologists in the world systems theory tradition.[11] They then added a region that sat between the top and bottom countries: the semiperiphery. The key characteristic of the semiperiphery was that it acted as a middleman between the core and the peripheral nations. These semiperiphery nations led the exploitation of the other countries in their regions to bring those resources to the world market, managing labor and investments there. In doing so, they developed decidedly bimodal or mixed economies, with extremely modern sectors and vast internal regions continuing to live in premodern conditions. In rural areas subsistence living and family ties made labor cheaper to reproduce than in urban and industrial zones.[12]

The G77, which incorporates the periphery and semiperiphery nations, has never been a homogeneous group. A key tension in the group has been between AOSIS and the Organization of the Petroleum Exporting Countries (OPEC). At the first meeting of the COP in Berlin in 1995, when a majority of G77 countries supported binding reductions of emissions, OPEC advocated against them, even for the industrial countries. The G77 took stands against any taxes on carbon, insisting instead that they should be compensated for lost business since measures to respond to climate change would severely affect their economies by slashing their ability to sell oil. The idea of compensation of oil producers for lost revenue is enshrined in Article 4.8 of the Convention, which included special consideration for economic vulnerability to climate change response measures. Year after year, Saudi Arabia was criticized by environmental activists in the Climate Action Network as a "fossil of the day"—a mock award given to the most obstructionist actor on each day of the negotiations. Saudi Arabia reportedly collaborated secretly with its great customer and partner, the United States, in derailing important proposals in the negotiations.[13]

In 1997, at the third meeting of the COP in Kyoto, the host nation, Japan, was eager to have its name on a serious and lasting protocol. Again the EU and AOSIS were active in pushing for reducing emissions. The EU proposed a 15 percent reduction by 2010 from a 1990 baseline level by the wealthy countries (listed in the treaty's Annex 1). There needed to be an effort to reduce global levels of carbon and other greenhouse gases, and a consensus formula was needed on a fair way to share the burden. China and India strongly put forward proposals for there to be a per capita allocation of emissions, since both nations were well below the global average needed to stop dangerous warming.

A couple of other groups canvassed formally or informally in the Kyoto round. One was the Eastern European nations, which wanted to be able to count the steep drop in emissions after their economies collapsed from the fall of the Soviet Union. That group seemed to gain what it wanted, except that in later years, Russia would have to fight for rights to sell those emissions reductions to other nations in the Annex 1 group that had to meet reduction targets. Finally, there was a small group of countries that sought to use the talks as a way to raise their diplomatic visibility, who were able to speak to actors on both sides of the North–Side divide as brokers in the middle. However, essentially at Kyoto, the main groups were the EU-17, JUSSCANNZ (a coalition of non-EU developed countries including Japan, the United States, Switzerland, Canada, Australia, Norway and New Zealand), and the G77. The opposing positions of OPEC and AOSIS mostly canceled each other out, though OPEC wielded considerably more economic power (table 3.1).

**Table 3.1**
Negotiating and Ad Hoc Groups at Kyoto, 1997

| Kyoto negotiating and ad hoc groups ($n = 8$) | Goals/positions at Kyoto |
| --- | --- |
| EU17 (European Union) | High ambition: EU offers one overall target. −15% of 1990 levels by 2010; with individual nations taking different goals, and establishes EU-ETS carbon trading system |
| JUSSCANNZ (Japan, United States, Switzerland, Canada, Australia, Norway, New Zealand) | Low ambition: −3% to +10% of 1990 levels by 2012; want flexibility in meeting their targets such as through tradable and bankable permits, joint implementation, removals by sinks |
| EIT (Economies in Transition—Central and Eastern Europe, Russia) | Seeks baseline year just before economic collapse, providing potentially saleable credits ("hot air") |
| "Brokers in the middle" (Philippines, Argentina, South Korea) | Intermediate positions |
| G-77 and China | Per capita emissions standard, historical responsibility, no binding commitments on themselves, technology transfer, and adaptation assistance from rich nations |
| Rapidly developing nations (Brazil, China, India) | No limits to economic growth |
| OPEC | No limits on emissions and compensation for any non-exploitation of petroleum reserves |
| AOSIS | 20% reductions by 2005 |

Pressure from the US-led veto coalition gave rise to a patchwork of promises with the target of achieving an average of 5.2 percent of Annex 1 group emissions by 2012, rather than the 15 percent proposed by Europe. The general sentiment was that the Kyoto Protocol was inadequate to address the problem of climate change, but it was seen as a first step and all that was politically feasible at the time.

Why did the global South accept this inadequate Kyoto decision developed by the North that excluded them entirely? Some reports suggest that various countries from the South, led by China, tried to tone down the AOSIS pressure for high ambition, lest these southern countries would also have to assume binding commitments as part of the agreement. The other major developing countries that were concerned about giving up sovereignty and their right to rapid economic growth agreed with them and presented a united position. In addition, Atiq Rahmin of the Bangladesh Institute for Advanced Studies argued that expectations for the flow of money to the South through the Clean Development Mechanism, a new funding institution introduced in Kyoto, secured the support of scores of the world's poorest nations.[14] Never comprehensive or adequate, the Protocol agreed to in Kyoto's convention center was defended as a stepping-stone to a better agreement in the future, a first step in a longer process.

## A New Identity of Vulnerability

Beyond responsibility for climate change and having the capability to do something about it, a third dimension that has emerged in the negotiations has become increasingly prominent since the turn of the millennium: vulnerability. The figures are stark: over ninety countries and their people have contributed an almost negligible amount to the problem of climate change, but they are already being hit first and hardest by the impacts, and they face these disasters with the least capacity to adapt to the changes.[15] What's more, their high levels of social vulnerability weaken their bargaining position in climate diplomacy because they have limited leverage to push more powerful actors to adopt their positions (an issue to which we return in chapter 4). As Desmond Tutu put it in 2008, a system of "adaptation apartheid" is already developing in the form of increasing investments in protections against climate-related disasters in industrial countries, while efforts in the most vulnerable countries have always been grossly underfunded.[16]

Formed back in 1989 at the beginning of the Convention negotiation process, AOSIS was active and aggressive from the start. This group was particularly active in demanding ambitious, science-driven, legally binding emissions reductions targets and compensation funding for climate impacts. The group's forty-four members are spread across the South Pacific, Indian Ocean, and the Caribbean, Africa, the Mediterranean, and the South China Sea. AOSIS's unity comes from the fact that more than nearly any other countries, their physical survival as states is at stake due to gradual sea-level rise from climate change. Within G77 meetings, the group openly demanded that the major emitters from the G77 also assume emission reduction commitments. The group was the first to propose a draft text during the Kyoto Protocol negotiations calling for cuts in carbon dioxide emissions of 20 percent from 1990 levels by 2005.[17] The group demanded the establishment of an international insurance pool for climate victims; the UNFCCC finally adopted the related agenda item of loss and damage in Cancun in 2010.

Similarly, the Least Developed Countries (LDC) group formed as a negotiating bloc in 2001. Its members are forty-eight of the world's poorest nations that were disenchanted by the G77 positions, which they believed did not adequately reflect their interests, particularly on issues of adaptation. For several years, the group struggled to gain the capacity to negotiate effectively, since so many of its members had tiny delegations of only one to three officials, often staffed by meteorologists and environment ministry officials. Furthermore, many of the LDCs are quasi-states,[18] relying heavily on foreign aid and governance relationships, often with their colonial powers. As emerging powers such as China and India have been busy building their new role outside the G77 and OPEC has worked to block or decelerate the whole climate negotiating process, LDCs and AOSIS struggled for influence to mobilize the G77 bloc and lost much time in not having their needs addressed.[19]

Although the eighty nations in AOSIS and the LDCs are vulnerable to climate change, the postures and positions of the individual countries often differ substantially across the groups. For example, the particular states take varying stances on whether to challenge the positions of OPEC and BASIC, depending on who is chairing the groups, and the particular areas of conflict. The island states Tuvalu and the Maldives have gained attention for being far more ambitious and aggressive in their stances than have Saint Lucia or Samoa. Like the economically more powerful states in BASIC and OPEC, many LDC countries also pursue both bilateral and minilateral diplomacy with single countries or smaller groups to promote their individual and group interests.[20]

The moral force of a nation's extreme vulnerability to climate change is now often pitted against the need for development in emerging economies.[21] For example, at one of the key informal meetings in the 2011 Durban negotiations, in response to the Indian environment minister's statement arguing for their right to development for meeting basic needs, the delegate from Grenada, representing AOSIS, reportedly retorted, "Why should we sink when you develop?" Some analysts argue that AOSIS and the LDCs lack power to fight the most powerful interests, so they have turned against their partner developing countries, blaming India for inaction, where per capita emission is lower than most of the AOSIS members. The G77 bloc, best understood now as effective only as a defensive assemblage of developing nations concerned that the rich countries will force them to slow or stop their national economic development to protect the environment, was severely fractured heading into Copenhagen.

## The Bali Firewall

The tropical heat of the island of Bali, Indonesia, was enough to lead the United Nations to send a memo to participants of the 2007 COP there to leave behind their three-piece suits and instead don elegant, flowery, short-sleeved Indonesian-style guayabaras. The conference was held in the exclusive and guarded beach resort area of Nusa Dua at the southern end of the island, away from the bustle and shanties, roadside shops, and temples that line roads on much of the rest of the island.

In the halls, the multilateral system seemed near a breaking point. The time to negotiate a successor treaty to the Kyoto Protocol was running out, and a plan was needed to get that done by 2009. Advocates in the Climate Action Network, the largest and oldest civil society network in the negotiations, wore buttons styled after the London Underground signs, reminding delegates to "Mind the Gap." This was a reference to the need to ensure continuity as the first commitment period of the Kyoto Protocol was set to expire. After seven years of George W. Bush in the White House, the United States was still dragging down the negotiations as the richest large emitter to be a non-participant, but his presidency was winding down, and both leading candidates to succeed him expressed being in favor of international action on climate change.

The most dramatic moment of the conference was in the public final plenary session, when Kevin Conrad, the feisty delegate of the tiny Papua New Guinea delegation, asked for the floor and boldly told a recalcitrant and isolated United States to "lead, follow, or get out of the way." The house burst into applause, and by the end of the session, lead US negotiator Paula Dobransky had new instructions from Washington to go along with the rest of the world. But what was the United States afraid to accept in Bali? Late the night before, some brilliant text was put forward that rested on neologisms like MRV (Measurable, Reportable, Verifiable) and the location of each bit of punctuation. In addition to pledges to address mitigation, adaptation, finance, and other issues, it laid out an almost indescribably delicate balance of expectations on the North and South nations.

Most central, the Bali text cemented different expectations for the developed and developing countries; this is the "Bali firewall."[22] Subparagraph 1bi said that all developed country Parties needed to fairly divide up the task of reducing emissions and transparently show how they had done so. In the next subparagraph 1bii, developing countries agreed to take what steps they chose, based on their needs for development and only *if* they got adequate funding and technology. As with the North, these countries had to be transparent about what they had done, agreeing to a process to be sure commitments are "measurable, reportable and verifiable" (MRV). So what was agreed was a strict line between groups in the global North and South—a firewall that would be defended for years by many developing countries. Nowhere does the Bali action plan describe whether or how countries might move from one group to another, either up or down. Nor is there clarity on how a scientifically adequate solution might be met or clear rules for compensation for countries losing revenue from reducing their emissions sharply.

On the way from Bali in 2007 to Copenhagen in 2009 were meetings in Bangkok (2×), Bonn (3×), Accra, Poznan, and Barcelona. The issues that arose and the twists in the negotiations were endless. However, for this chapter's task, we need to skip ahead to describe the alliance that emerged in Copenhagen between the United States and the new BASIC coalition to write the Copenhagen Accord, with massive implications for international climate policy and developing country unity in the negotiations. This is the subject that we turn to next.

**Table 3.2**
Negotiating and Ad Hoc Groups at Copenhagen and Beyond, 2009–2013

| 2009–2010 negotiating and ad hoc groups ($n = 15$) | Goals/positions in 2010, after Copenhagen |
| --- | --- |
| EU-27 (European Union) | More ambitious emissions reductions |
| Umbrella group (usually Australia, Canada, Iceland, Japan, New Zealand, Norway, Russian Federation, Ukraine, United States) | Build on the Copenhagen Accord, generally not committed to continuing the Kyoto Protocol |
| Environmental Integrity Group (Mexico, Lichtenstein, Monaco, Republic of Korea, and Switzerland | Much more ambitious emissions reductions, based on the science |
| G-77 and China | Continuity with Kyoto and REDD+ (deforestation protocol); adaptation fund and technology transfer; 1.5% of Annex 1 GDP in climate funding |
| BASIC (Brazil, South Africa, India and China) | Voluntary emissions promises through "nationally appropriate mitigation actions" |
| OPEC | Compensation for response measures (not burning fossil fuels), no binding limits on emissions |
| AOSIS | 1.5°C average global warming maximum; sharp emissions reductions, "fast start" adaptation funding |
| Least Developed Countries | Direct and easy access to 1.5% of Annex 1 GDP for adaptation funding through the LDC Fund and Adaptation Fund |
| CACAM (Central Asia, Caucasus, Albania and Moldova) | Unknown |
| Coalition of Rainforest Nations | Attention to and approval of REDD+ (compensation for reducing deforestation) |
| Arab states | Compensation for "response measures" (see OPEC) |
| ALBA (Cuba, Venezuela, Bolivia, Nicaragua, Honduras, Dominica and Saint Vincent and the Grenadines) | Dismantle carbon trading mechanisms, compensation for climate debt |
| African Group | 1.5% of GDP for adaptation funding |
| Group of Mountain Landlocked Developing Countries | Adaptation support |
| Central American Integration System | Predictable, sustained, additional adaptation funds |
| AILAC-Independent Association of Latin American and Caribbean Countries | More ambitious emissions reductions |

## Hijacking Copenhagen

The key moment at Copenhagen was when President Barack Obama burst into a room where the leaders of all four BASIC were meeting in private, and together the group of five nations set aside the existing negotiating texts entirely and drafted their own deal. The draft mentioned the goal of keeping global mean temperatures under 2 degrees Celsius rise, but they avoided any binding emissions reduction targets to achieve that and any mention of the time when perilously rising emissions would peak and begin to fall.

Most crucial, the Copenhagen Accord they drafted entirely shifted the approach taken by the global community in the face of climate change. The Kyoto Protocol approach was top down, with binding national commitments based on levels of emissions and capabilities of countries (usually understood to be roughly their level of income per capita). The Copenhagen approach that the United States and BASIC put forward was entirely bottom up, with nations pledging and reviewing their own choice on what emissions reductions they would undertake. Despite the text in the Copenhagen Accord that commits to recognize "the scientific view that the increase in global temperature should be below 2 degrees Celsius," calculations are that the pledges condemn the world to 3.5 to 4.5 degrees Celsius of warming.[23]

China and the United States, the big emitters, consciously avoided a time frame for a midterm emissions reduction target, though the EU and the low-emitting G77 countries were strong supporters of such a commitment by 2020. Indeed, the dynamics between the Chinese premier Wen Jiabao and Barack Obama at Copenhagen in 2009 represented an interaction between a rising and a declining hegemon, on an issue they both would have preferred to avoid: binding emissions reduction targets on greenhouse gases.[24]

In addition to the new pledge-and-review approach, the other part of the Accord was a nonbinding pledge of "fast-start funding" of $30 billion during the three years starting immediately after the conference (2010–2012), and wealthy nations put forward a long-term goal of "mobilizing jointly" $100 billion a year by 2020.[25] These pledges were crucial in making the Copenhagen Accord the basis for a new climate governance architecture, and we discuss them at some length in the next two chapters.

The bold move (some would say hijacking) at Copenhagen showed the ascendant power of the BASIC group and its ability to work directly with the United States and entirely cut their G77 colleagues and the EU out of the decision making. The way the Copenhagen Accord was cobbled together was unprecedented, for heads of state and governments rarely get directly engaged in, let alone lead, international climate change negotiations. The Accord was quickly brought to a hand-picked group of twenty-eight countries to rubber-stamp, with almost no time to review it thoughtfully and no opportunity to revise it. In this group of twenty-eight were nearly all the wealthy OECD countries and just one representative from each of the developing world regions: Africa, Latin America, AOSIS, and Asia.

Why didn't those four developing country representatives block the process at that point? For the Africa group, the representative was Meles Zenawi from Ethiopia, who Wikileaks cables link to having had direct negotiations with the United States on the terms of aid in the agreement.[26] For Latin America, Colombian president Alvaro Uribe was there, a strong ally and major drug war military aid recipient of the United States. All were heads of state, unaccustomed to the negotiations and disagreements, and unwilling to sour their nation's relations with so powerful a set of donors and investors as the United States and China over such a distant issue as climate change, which many saw as a second-tier issue diplomatically.

Then in the age of instant electronic media, the document was reported publicly, before the other 160 countries had been able to see and review it. The story was reported with the claims that the broader group of nations had already approved it. Obviously this didn't sit well with the nations that were not consulted and now made painfully aware that they were not included in the circles of people and countries that mattered.

The final all-night plenaries at Copenhagen were fiery, with a few feisty speeches by the countries willing to risk upsetting the global order and the major aid and investment players, the United States and China. For ALBA, Pablo Solón of Bolivia adamantly resisted the Accord as scientifically inadequate and unjust in process. G77 chair Lumumba Di-Aping argued that the Accord was tantamount to a death sentence for the people of Africa. He said, "2 degrees Celsius [global average temperature rise] is certain death for Africa, [and] is certain devastation of island states. ... The more you defer action, the more you condemn millions of people to immeasurable suffering."[27]

Ultimately, to end the tumultuous meeting in the face of a lack of consensus, the Copenhagen Accord was simply taken note of by the COP (a nonbinding action) and countries were to subsequently sign on one by one. This forced a year of shuttle diplomacy by Mexico. In the next chapter, we take up the complex process by which that (near) consensus was manufactured; first, we explore the emergent BASIC coalition in detail.

## Splintering the South

The world had waited for years, at each negotiation cycle, hoping the United States would someday step into a position of leadership. Obama's bold move in Copenhagen looked like leadership; however, the most decisive players in Copenhagen were the four BASIC countries: Brazil, South Africa, India, and China. What brought BASIC together, and how stable a coalition is this? What is the nature of its solidarity besides being the leaders of the rising semiperiphery? Understanding this bloc is vital.

BASIC was formed in October 2009, just before the Copenhagen conference. However, for years before that, the big four emerging economy countries had been working outside the climate negotiations in different combinations. India, Brazil, and South Africa in 2003 created the IBSA (India, Brazil, South Africa Trilateral) Dialogue Forum with a broad agenda to reform the UN Security Council, where all three seek to become permanent members.[28] Similarly, Brazil, Russia, India, and China (the BRICs) worked as a group with the goal of counterbalancing US dominance in the world economy.[29] Ironically, the idea for the BASIC group was created by a EU project whose primary focus was on "linking national and international climate policy by enhancing and strengthening institutional capacity on climate change for Brazil, China, India and South Africa."[30] They continue to meet quarterly and before major UN climate negotiations to coordinate their positions and devise collective strategies. At the negotiations, they also sometimes hold joint press conferences, where media attention tends to reflect their outsized influence.

These countries nevertheless are highly diverse in their interests, which are sometimes even conflicting. Their economic base, their energy infrastructure, and their emission levels all vary greatly, as do the nature of their states and their approaches to making and meeting greenhouse gas emissions reduction goals.[31] In the cases of China, India, and South Africa, coal and other fossil fuels dominate their energy mix and electricity production, but in overall volume, China consumes more coal.[32] Brazil's main concern is to stop deforestation, which has been the largest source of its carbon dioxide emissions since the late 1980s, and the nation depends far more on hydroelectricity and biofuels than the other nations do. This makes Brazil far more efficient than its BASIC partners in terms of carbon emissions per unit of GNP. Despite massive investments in renewable energy and a commitment to stop its emissions from growing by the year 2030 (as part of a climate accord agreed to in partnership with the United States in 2014),[33] China appears to be pursuing a relatively high-carbon pathway of development in the near term, while Brazil's commitment to its lower-carbon one appears to be weakening.[34]

The economic and geopolitical influence of these four countries is bringing them to the center of global politics; for example, they were invited in 2005 to participate in the G8+5 Dialogue (which included Mexico) and to participate in 2007 in the Major Economies Meeting/Forum on Climate and Energy formed by the United States. Sensing that they have growing clout, BASIC countries have started to look at climate diplomacy as a crucial plank in their foreign policy where they are both under pressure and able to exert significant leverage on the world stage.

Reviewing reports from their quarterly BASIC meetings, two clear patterns emerge. First, they repeatedly call for a second commitment period under the Kyoto Protocol, meaning binding commitments are required only of the rich countries and not of themselves. Second, they make a public effort to build bridges with and show support for the rest of the G77. Each publicly touts the value of building South–South economic and political relations as a way to diversify away from dependence on the wealthy nations. However, while publicly defending the continuation of the Kyoto Protocol into a second commitment period for the OECD countries, BASIC members are taking a pragmatic view on alternatives, such as by joining bilateral and minilateral groupings that might bring them access to new technology or markets.[35] Moreover, some observers argue that BASIC countries are ultimately an "obstructive grouping" that can agree only on avoiding mandatory international emission reduction commitments, and that they publicly support the G77 while simultaneously undermining it with their actions.[36]

Understanding the fracturing of the G77 requires that we understand especially the power of China and the complex position of India. We described China's rise briefly in the previous chapter, but three points are particularly relevant to its role as leader of the G77. First, China is a bit like someone of very modest means who has quickly grown wealthy. It seems to have one foot in its old peer group, the G77, while simultaneously looking for ways to play in the realm of the big players—in this case, the Organization for Economic Cooperation and Development, that is, the EU and other wealthy states.

Second, China in the 1990s and early 2000s was very different from China today. Economically, its 7 to 10 percent annual growth and state-led capitalist transition has rocketed the nation to the highest levels of economic power. By some measures, China has just surpassed the United States and is now the world's largest economy, it is the workshop of the world in manufacturing, and it already is the holder of the world's greatest currency reserves and of other nations' debt.[37] China is a critical example of the rising semiperiphery, building its own development pathway that combines strong state planning and influence in the economy with (arguably repressed) capitalist labor relations and vast cheap rural labor reserves built and sustained by collective production and consumption during the Communist era.

Third, China has increasingly seen climate negotiations as an important area of foreign policy to show that it is capable of addressing global problems and as an avenue for asserting leadership among developing countries.[38] For this reason, China from the beginning worked for a united "G77 and China" strategy,[39] perceiving its own role as speaker for the group.[40] China is heavily investing in renewable and nonrenewable energy resources and infrastructure development in Asia, Africa, and Latin America, and unlike the West, it is reported to have declined to make its investments conditional based on government reform.[41] It seems likely that China's involvement as an investor is responsible for some of the recipient countries' supportive responses to Chinese positions and leadership in climate change negotiations.[42]

Importantly, in 2014, China agreed to a joint announcement with the United States to mitigate climate change. Specifically, China pledged to peak its rising emissions trajectory by the year 2030 and to increase its non–fossil fuel share of energy to 20 percent.[43] The United States, for its part, pledged to reduce its emissions in 2025 by 26–28 percent below what the levels were in 2005. While these pledges by the world's two biggest polluters are far from adequate, particularly in the near-term, the partnership represents an important political breakthrough, perhaps setting a precedent for more ambitious actions and agreements down the road. It also showed China's self-identification alongside a superpower, not making joint announcements with its BASIC or other G77 partners.

Similarly, India's position has also undergone great changes in climate diplomacy. For many years, its approach in the negotiations embodied anti-imperialism, and it has often demanded greater justice, equity, and democratic decision making at the global level. The nation was fairly impervious to pressure from developed countries to compromise:

Indian diplomats year after year repeated the same mantra that their nation was too poor to take action on climate change and that the rich countries should act first, based on their high per capita emissions. Their justification is unassailable: over 300 million people in India have no access to electricity at all.[44] Citing its huge populations in poverty, India was unwilling to take on binding emissions limits at any time in the future.

Nevertheless, in recent years, India has come under increased pressure of becoming diplomatically isolated due to the size of its economy and emissions (now the world's third largest emitter),[45] despite its very low emissions per capita (ranked 147th among all countries in the world).[46] Indian policy toward climate diplomacy appeared to shift in 2009 away from the entrenched idealistic approaches seeking an equity-based approach to a more pragmatic one. At Copenhagen, to nearly everyone's surprise, India pledged to reduce its emission intensity between 20 and 25 percent below its 2005 level by 2020. The pledge was built in part on its domestic energy program, adopted the year before.[47] There appear to be three reasons for this rather seismic shift.

First, other countries in the South were beginning to move. Since BASIC is so big and growing so quickly, AOSIS and the LDCs took increasingly bold and public positions, calling on the BASIC countries to take on promises to reduce their emissions. Breaking from the G77 wall of silence, first Peru and then some Central American countries, then South Africa, and then Brazil made pledges to reduce emissions.[48] Before the Poznań meeting in 2008, Peru and then Mexico, Costa Rica, and Brazil announced emissions reductions. Mexico's national plan, revealed in Poznań, was especially notable, promising to reduce national emissions by 50 percent by 2050.[49] At Copenhagen in 2009, President Luiz Ignácio Lula of Brazil announced ambitious greenhouse gas emissions reductions, measures he had to go home and sell to a suspicious senate. But the point is that no longer was it okay for India to say that only Annex 1 countries should be expected to do serious mitigation.

That fall, a controversial letter from India's environment minister, Jairam Ramesh, to the prime minister, leaked to the press, questioned India's association with the G77 and suggested a move away from its stalwart approach to preserving the Kyoto Protocol's and Bali's strict firewall between Annex 1 and the others.[50] Ramesh was calling for India to take a more ambitious role in mitigating climate change. This created a stir among the Indian environmental community, including senior negotiators, and the uproar escalated in the coming months.[51]

Second, there were pragmatic reasons. India began seeing benefits from constructive engagement, such as substantial diffusion of Clean Development Mechanism projects in, for example, cement plants and sugar mills.[52] It saw improvements as well in the nation's energy efficiency and its benefits for reliability of electricity service to homes and businesses. India also saw the joint promotion of renewable energy projects with the West as a new area for constructive engagement. A core practical concern is the country's extreme vulnerability to climate change impacts, such as its dependence on glacier-fed water supplies from the Himalayas, its vast populations on semiarid lands with scarce irrigation, and its dense population in the coastal belt vulnerable to sea-level rise and intensifying monsoons. In monetary terms, the issue is clearly salient: the Indian government claims that 2 percent of its GDP is already being spent on adaptation to climate change impacts, which could result in a loss of 9 to 13 percent of GDP in real terms by 2100.[53]

However, ministries of foreign affairs have their own dynamics and goals quite detached from such domestic matters, so a third key factor may have been most decisive: India's aspiration to play a larger role in global governance. India's desires in this direction can be seen in its push for a greater role in international financial institutions like the World Bank and the International Monetary Fund, and most dramatically in its ambition to gain a seat on the UN Security Council. India may also have read that with power come responsibilities, and postures befriending Annex 1 countries were required to get there, even if it entailed the risk of alienating a number of G77 members. Indeed, a common characteristic of the BASIC group is that each nation is a regional power at risk of alienating many neighbors as it attempts to reach the world stage. However, their actions in the area of climate politics suggest that each is diminishingly concerned about alienating their regional neighbors and the rest of the G77.

China and India appear to be following the street wisdom of holding your rival close in order to avoid being knifed in the back. Just weeks before Copenhagen in October 2009, China and India signed a memorandum of understanding (MoU) to coordinate their approach to climate change negotiations and some domestic policies. They formed a joint working group to meet once a year and cooperate on renewable energy and research into the effects of climate change on Himalayan glaciers. Delhi also sought reassurance from Beijing that China will not sign a bilateral deal with the United States that runs contrary to G77 goals. At the MoU signing ceremony, Xie Zhenhua, China's vice chairman of national development and reform commission and top climate negotiator, tried to allay such concerns: "We regard India as a sincere, devoted friend and the MoU ... on climate change will take our cooperation on the issue to a new high."[54]

Before moving on, it is important to describe a bit about the other two partners of BASIC, to understand their divergent interests. During the bargaining over Kyoto, Brazil was nearly alone in pushing for rich countries' responsibilities for emissions reductions to be based on their total historical emissions of the gases since the start of the industrial revolution, and it also advanced important ideas for funding channels for climate action in the South.[55] It stood alone against several forest-related issues, including a widely supported proposal by the Coalition of Rainforest Nations to be compensated in advance for protecting standing forests.[56] So from the beginning, Brazil has often acted on its own, not even representing its region, let alone the G77.[57]

South Africa, the smallest partner of the BASIC group in all parameters, after being quiet in earlier negotiations, began to see itself as something of a bridge builder in the run-up to Copenhagen and episodically since then. In an effort to generate some momentum, at COP15 in Copenhagen, President Jacob Zuma offered a voluntary greenhouse gas emissions reduction pledge. This was before India and China's pledges and not something they appreciated because they were becoming isolated. To some, South Africa's own alignment with BASIC and its role at COP15 were viewed as a betrayal of the African cause.

So we see in BASIC an alliance of the very large developing countries (the emerging emissions powers) forged in the days just before the Copenhagen summit as they began to see more value in working together than in being tethered to their former peers in the splintering G77. They faced growing demands from both wealthy nations and the small islands and LDCs to reduce their emissions. By working together, these growing giants could resist these pressures and find allies who would also want to resist a treaty that would limit their economic growth.[58] As Copenhagen shrinks in the rearview mirror, the tensions between BASIC countries have led to declining energy around the coalition: environment ministers continue to meet quarterly, but higher-level officials rarely do.[59]

### An Emerging Climate Order

Copenhagen was significant not only for its failure to bring the world an ambitious climate deal. In Gramscian terms, this conference was a microcosm of major shifts in the balance of forces in the broader world order. With a global recession at the forefront of state leaders' concerns around the world, Copenhagen was approached by the world's biggest emitters as an opportunity to assert their dominance on the world's biggest stage. It was clear that atmospheric space to dump carbon emissions was rapidly shrinking. The consequence of business as usual meant certain catastrophe down the road. But rather than step back and work to divide the remaining atmospheric space fairly, the United States and BASIC dug in their heels to ensure they would not be burdened by onerous restrictions on their economic growth.

Thus, in their intense competition, a falling and rising global hegemon (the United States and China, respectively) found common ground by institutionalizing their joint inaction. Based on historical experience with such economic transitions, Arrighi and Silver observed that the typical characteristics of this phase of hegemonic decline are sharply increased competition, social conflict, and systematic chaos, where the existing political structures tend to fail to address the problems they face.[60] The rest of the world, and particularly the most vulnerable countries, saw the writing on the wall: a voluntary pledge-and-review framework would mean a world where the biggest polluters would continue to pollute at an alarming rate; there would be scarcely any development space left for the poorest countries to follow the same path, and simultaneously, they would increasingly experience the worst impacts of climate change, with limited resources to cope.

Thus, while there had always been tensions in the G77, it was now clear that climate politics no longer hinged solely on a North–South divide. But the poorest and most geographically exposed countries did not simply sit back and accept their fate. Rather, they became increasingly vocal in linking their survival to demands for a common, ambitious, and binding mitigation framework where all actors, North and South, make appropriate commitments. In Copenhagen, a group of developing countries refused to go along with the new deal. The G77 would no longer be relatively unified behind the Bali firewall, China and India had emerged as major net emitters (with India quite low in per capita terms), as well as rising economic powers, and the ties of solidarity between them and other developing countries had weakened.

We view these broader economic and historical developments as essential to what was happening inside the conference center in Copenhagen. A strategic perspective of power relations here has focused on new state coalition alignments and fragmentations and how these coalitions have navigated, leveraged, and challenged emerging shifts in the world order. Here we have discussed ecological, geopolitical, and political economic dimensions that have been central to this story. We have yet to engage with other transnational actors in this story, which also played an important role at this stage: these are private sector and civil society actors, which we discuss in chapters 6 to 8. And importantly, the UN process is only one, albeit central, relevant institutional venue where decisions related to climate change are made. In chapter 9, we offer insight into how this process fits within the broader complex and fragmented global governance web.

In the next chapter, we continue this story as it moves from Copenhagen to Cancun, Durban, Doha, and Warsaw in the following years. There we take up a puzzling question: With such a bad deal for most of the world's countries, why did 193 countries, and particularly the LDCs and AOSIS members, come to a near unanimous decision to accept the Copenhagen framework in Cancun and beyond? And what does this tell us about strategic power relations as they have evolved in international climate negotiations? We turn now to these questions.

# 4

# Manufacturing Consent

## Politics of the Possible

6:23 a.m., December 11, 2010. Cancun, Mexico, and multilateralism has survived another year. Nearly all faith in the UN system to address climate change had been lost with the debacle in Copenhagen the year before (see chapter 3). Yet no other organization was in a position to lead people to a cooperative solution, even as global temperatures rose and climate disasters mounted.

Two Latin American women, Christiana Figueres of Costa Rica, and especially Patricia Espinosa, foreign secretary of host nation Mexico, led the delicate task of rebuilding trust in the UN process after Copenhagen. In the giant Ceiba Hall of Cancun's Moon Palace, in the final sessions on Friday and through the night until Saturday at dawn, the tone was transformed from the acrimony felt in Copenhagen the year prior. Nation after nation took the floor in Cancun to congratulate Espinosa and her team in running an inclusive and transparent process through the entire year, rebuilding trust, and creating what many argued was a balanced document.

One after another, tough customers like Brazil, the United States, China, and India took the floor to say that although the Cancun Agreements were not perfect, they were balanced and a step in the right direction. After hinting there were goddesses in the room, India's delegate said, "What you have accomplished today has given us the confidence to move forward." Smaller and poorer countries in the Africa group, the Least Developed Countries (LDC) group, and small island states all endorsed the compromise texts, as did all Latin American countries except Bolivia, which argued that the agreement did not meet even the most basic considerations of adequacy or justice.

In this way, a year after the divisive conclusion to the negotiations in Copenhagen, the main content of the Copenhagen Accord was integrated into the Cancun Agreements, which were adopted all but unanimously. In doing so, the international community set in motion a process, agreed on in Durban the following year, that would mostly replace the legally binding Kyoto Protocol with the voluntary pledge-and-review approach that was plainly insufficient and unjust.[1]

In the new approach, there is not an agreed-on aggregate figure for reducing greenhouse gas pollution. As it is configured, this framework will allow a temperature rise above what scientists predict will trigger catastrophic environmental events around the world.[2] In the case of such temperature rise, several low-income states may suffer from severe droughts, famines, and floods, and others may completely disappear below water.[3] It is well documented that the poorest countries with the lightest footprint on the climate are suffering the impacts of climate change most severely.[4] In the LDCs, for example, people are five times more likely to die from climate-related disasters than the global average, yet their emissions are less than 1 percent of the global total.[5]

In this chapter, we ask: why have low-income states agreed to an emissions reduction framework that is both scientifically inadequate and inequitable and which transformed the course of international action on climate change?[6] Limited scholarly attention has been directed to understanding how consensus was reached on this new mitigation framework with potentially devastating impacts for low-income states. We argue that a sophisticated conception is lacking of how weak, vulnerable, and low-income states have come to offer their consent, particularly when agreements do not represent their core interests. Scholarship in this area has tended to emphasize structural factors while neglecting agency of actors in conditioning cooperation outcomes, or vice versa.

Building on the neo-Gramscian theory of power introduced in chapters 1 and 2, we argue that consent was produced through three main processes: material concessions, norm alignment, and structural conditioning. This analysis views international cooperation as a process of strategic power relations with both coercive and legitimate forms of influence. Moreover, cooperation has not been merely top down, as structuralists would have us believe: rather, it has been co-constituted by strong and weak states, in coordination with nonstate actors.

In the next section, we argue that with its attention to North–South conflict, the literature on international climate change politics has neglected the politics of consent in this context. We outline how our analysis of low-income-state consent to contemporary climate treaties supports a neo-Gramscian view of power articulated in chapter 2. Here we again engage with a bit of theory and lay out our theoretical framework for negotiated consent in table 4.1. This builds on what we have discussed previously. We then explore the major relevant developments in the climate negotiations in Copenhagen, Cancun, Durban, Doha, Warsaw, and Lima. In the final sections, drawing on this discussion and our theoretical framework, we address our main question of how states particularly vulnerable to climate change impacts came to offer their consent in contemporary international agreements. In table 4.2, we categorize and assess the gains by low-income states and the extent to which gains are consistent with coalition demands. In this way, we offer an approximation of coalition gains in each area, ranging from weak to high. We conclude with a discussion of what the politics of consent means for securing an equitable climate treaty in the near future, revisit our theoretical argument of the book, and look ahead to the next chapters.

### Rethinking Cooperation

The focus of the literature on international climate change politics, including our previous work, is largely on factors that have prevented North–South cooperation or have led to comparatively distinct domestic policies. Far less attention has been directed to understanding the processes that have facilitated the seemingly unlikely outcome of 193 countries coming to consensus on legal frameworks that have major implications for all involved, and potentially devastating impacts for low-income and geographically exposed states, in particular.

Few works focus specifically on the engagement of low-income or small states in the contemporary climate change regime.[7] Importantly, there has been no attempt to investigate the particular mechanisms of low-income, vulnerable, or small state consent despite the fact that low-income states far outnumber wealthy and newly industrialized states in formal votes in the climate regime. This shortcoming in the climate change literature reflects a weakness in broader international relations and international political economy scholarship, which has largely neglected the role of low-income states, and the processes through which consent is produced, in various approaches to cooperation theory. Cooperation theory, which has realist, constructivist, and institutionalist iterations, is concerned with the forces that constrain or enable collective action, cooperation, or multilateralism among states.[8] International cooperation is defined as "when actors adjust their behavior to the actual or anticipated preferences of others, through a process of policy coordination."[9]

While offering helpful insights, particularly about the strategies through which low-income states mitigate their material disadvantages, scholarship in this area does not offer a cohesive theory to explain how low-income-state consent was produced in the contemporary climate regime.[10] Building from the neo-Gramscian framework discussed in chapter 2, we argue that the politics of consent of low-income states has three primary dimensions.

First, redistributive *material and institutional concessions* from wealthy to low-income states have been instrumental in the negotiation of their consent to contemporary climate change treaties. As we will discuss, wealthy states have made promises of substantial sums of money to countries particularly vulnerable to climate change impacts, developed a new funding institution to distribute such funds more democratically, and begun the process of negotiating a loss and damage mechanism to respond to climate change impacts that cannot be readily adapted to.

This is representative of Gramsci's assertion that hegemony rests on certain, albeit limited, material concessions to the economic-corporate interests of the subordinate class. Such concessions, which have been largely ignored in the various strains of cooperation theory, are essential to maintain social cohesion in a class-divided society, giving the appearance that the forms of governance promote the general interest and are thus legitimate. As Gramsci argues, "The fact of hegemony presupposes that account be taken of the interests and tendencies of the groups over which hegemony is to be exercised, and that a certain compromise equilibrium should be formed—in other words that the leading group should make sacrifices of an economic-corporate kind."[11]

The second, and related, mechanism of consent is that of an ideological and organizational nature. While material and institutional concessions have been essential to gain low-income state consent, a process of what we call *norm alignment*—involving diverse coalitions of state actors, in coordination with transnational civil society—has served to generate these concessions and legitimize their acceptance. Norm alignment can be defined as a conflict-ridden and always incomplete process by which states with competing class interests come to some agreement on what are legitimate terms of consent.

This is representative of Gramsci's assertion that while economic concessions are a requirement for the leading class to exercise hegemony, equally important is maintaining an alliance through "cultural, moral and ideological leadership over allied and subordinate groups."[12] As such, "hegemony presupposes that account be taken of the interests and tendencies of the groups over which hegemony is to be exercised."[13]

Moreover, from a strategic Gramscian perspective, ideology is not merely a reflection of economic class interest or the economic structure but rather a terrain of struggle over the terms of legitimate social and political organization.[14]

From this view, policy coordination in multilateralism is not simply an exercise of top-down leadership, bullying, material concessions, or institutional leverage. Rather, it is made possible by what Gramsci refers to as a "war of position"—a tactical struggle by subordinate actors to resist domination through control of culture, ideas, and identities in the realm of civil society.[15] This includes attention to the ways in which certain normative constructs, defined as "collective expectations about proper behavior for a given identity,"[16] emerge as demands and how such constructs are leveraged, advanced, resisted, and legitimized by competing coalitions. Moreover, this approach is concerned with how elite interests adopt, co-opt, and align such normative constructs with existing hegemonic structures, thereby diffusing radical challenges to power.[17]

Third, while the processes of material concessions and norm alignment direct attention to the agency of low-income states to shape the terms of consent, a neo-Gramscian perspective also points to the ways in which such agency is constrained by historically derived structures of inequality as part of what he calls the "decisive nucleus of economic activity."[18] Various forms of what we call *structural conditioning* have limited the efforts of low-income states to influence climate policy, extract concessions, and withhold their consent. These include historically specific limits to organizing capability, disadvantageous positioning in the global economy, worldviews, ecological conditions and production relations, dependence on specific capital interests, and disproportionate vulnerability to forms of coercion such as bribes, divisive strategies, and threats. Such structural conditioning also limits the ability of low-income states to transform dominant ideological structures and define the terms of the policy agenda in international governance.

Overall, a neo-Gramscian perspective holds that international regimes may produce certain gains for low-income states not afforded to them outside such processes. But the framework of negotiated consent presented here builds on this insight to consider the ways that a dominant class may adapt, within certain structured limits, to accommodate strategic bottom-up challenges through a continual and conflicting negotiation of interests. This is what Gramsci calls "unstable equilibria" of compromise.[19] With these processes in mind, we turn to an analysis of how consent was achieved to the new mitigation framework in Cancun and beyond.

**Table 4.1**
Mechanisms of Negotiated Consent

| Mechanism | Function | Relevant Gramscian concepts |
|---|---|---|
| Material and institutional concessions | Redistributive sacrifices of an economic-corporate kind that promote social cohesion in a class-divided society, giving the appearance that the forms of governance promote the general interest | "Sacrifices of an economic-corporate kind" (Gramsci [2012] 1971, 161) |
| Norm alignment | Conflict-ridden and always incomplete processes by which states with competing class interests, in coordination with nonstate actors, come to agree on what are socially sanctioned and legitimate terms of consent | War of position (Gramsci [2012] 1971) Unstable equilibria (Gramsci [2012] 1971) |
| Structural conditioning | Historically conditioned constraints on efforts to influence policy, extract concessions, and withhold consent | "Decisive nucleus of economic activity" (Gramsci [2012] 1971, 161) Dimensions of world order including the interstate system, the global political economy, and the biosphere or global ecosystem (Cox 1992) |

## Conflict in Copenhagen

As we noted in chapter 3, there were cracks in the seams of the G77 coalition in the years leading into Copenhagen, but one had to look a bit close to find them. The G77 still mainly represented the interests of the South in the negotiations, despite tensions between groups such as the Alliance of Small Island States (AOSIS) and the Organization of the Petroleum Exporting Countries (OPEC).

As for the LDCs, for years they had been primarily focused on realizing gains for adaptation and finance. Given limited capacity in the negotiations, they favored focusing their attention on issues that are specific to their own group, particularly issues related to funding, capacity building, technology, and adaptation planning. When it came to mitigation, they presented strong demands in statements but generally fell into line with the broader G77 coalition despite obvious conflicting interests with China's rapidly escalating emissions and OPEC's concerns to avoid strong commitments on emissions reductions and demands for compensation for any slowing of oil sales.

Many of the demands of low-income states increasingly reflected the concepts of carbon debt, climate debt, and ecological debt, which were introduced into international climate change politics in the late 1990s by NGOs such as Christian Aid and Acción Ecológica.[20] These concepts argued that the global North owes the global South a climate debt, which is far greater than the current Third World financial debt due to its disproportionate use of atmospheric space without payment.[21] In subsequent years, the LDCs, AOSIS, G77 and China, and a coalition of more than thirty Western NGOs, policy institutes, and think tanks began to aggressively push for remuneration of the ecological and climate debts,[22] in addition to calling for wealthy states to take the lead on cutting emissions.

While gaining limited traction on these issues for the first part of the decade, these groups were effective in Bali in 2007 in elevating finance for adaptation to climate change as a core issue in the negotiations and for identifying LDCs, African countries, and small island developing states (SIDS) as those "most vulnerable" to climate change, and thus most in need of financial assistance.[23] Notably, the Bali Action Plan called for "improved access to adequate, predictable and sustainable financial resources and financial and technical support, and the provision of new and additional resources" for both mitigation and adaptation.[24] It also established a new negotiating track to develop a binding mitigation framework to supersede the Kyoto Protocol first commitment period, set to expire in 2012.

Although not posing a direct challenge to the G77 positions at this time, building from these gains in Bali, low-income states including LDCs, AOSIS, and Bolivia came into the pivotal negotiations in Copenhagen in 2009 with their own set of ambitious demands. As outlined in table 4.2, these included a legally binding treaty that would keep average global temperature rise below 1.5 degrees Celsius, $400 billion of "fast-start finance" from wealthy countries to enable those hardest hit by climate change to adapt to its impacts, and an equitable share of the atmosphere to ensure adequate "development rights."[25]

Table 4.2
Have Low-Income State Demands Been Met?

| Issue area | Low-income-state demands (2009 submissions to UNFCCC)[a] | Language in Copenhagen Accord and Cancun Agreements | Degree that commitments have been met |
| --- | --- | --- | --- |
| Equitable sharing of atmospheric space and rights to development | "Recognizing also the just, fair and equitable right of developing country Parties in particular Africa to achieve development making use of the atmospheric space." (Africa)<br><br>"Emissions reductions focused on the equitable allocation of the global atmospheric space … and the need of developing countries to achieve their first and overriding priorities of economic and social development and poverty eradication." (Bolivia) | "We should cooperate in achieving the peaking of global and national emissions as soon as possible, recognizing that the time frame for peaking will be longer in developing countries and bearing in mind that social and economic development and poverty eradication are the first and overriding priorities of developing countries."<br><br>"Take action to meet this objective consistent with science and on the basis of equity." | Weak: A 2011 meta-analysis report finds that developing country mitigation commitments coming out of Cancun far exceeded that of developed countries.[b] |
| Ambitious mitigation action | "Limit global average temperatures to well below 1.5 degrees Celsius above pre-industrial levels." (AOSIS)<br><br>"Well below 350 parts per million of carbon dioxide equivalent" (AOSIS)<br><br>"All Annex I Parties to the Convention shall … undertake ambitious national economy-wide binding targets for quantified emission reduction commitments of at least 45% of the 1990 levels by 2020 and adopt policies and actions accordingly to achieve these targets." (Africa) | "Stabilize greenhouse gas concentration in the atmosphere at a level that would prevent dangerous anthropogenic interference with the climate system, we shall, recognizing the scientific view that the increase in global temperature should be below 2 degrees Celsius, on the basis of equity and in the context of sustainable development, enhance our long-term cooperative action to combat climate change." | Weak: Current emissions pledges are estimated to lead to 3.5-4.5 degrees Celsius global average temperature change. A new mitigation approach will be decided on in 2015 based on a voluntary, rather than top-down approach, (what's called "intended nationally determined contributions").[c] LDCs and AOSIS, along with the EU, pushed through a second commitment period for the Kyoto Protocol for select actors representing 15 percent of global emissions. |

| | | | |
|---|---|---|---|
| Adequate climate finance | 5 percent of GNP of developed country parties (Africa)<br><br>$400 billion in public finances for fast-track financing (Africa)<br><br>$150 billion of Special Drawing Rights by the IMF for this purpose. (Africa)<br><br>"Grant-based, long-term and over and above existing official development assistance commitments." (Africa)<br><br>"Additional to and different from the ODA [official development assistance]." (AOSIS and Africa)<br><br>"Parties agree to provide financial and technological support … in a transparent, expedited, sustainable and predictable manner, with direct access, under the overall guidance of the Conference of the Parties." (AOSIS) | "The collective commitment by developing countries is to provide new and additional resources … approaching USD 30 billion for the period 2010-2012. …"<br><br>"Developed countries commit to a goal of mobilizing jointly USD 100 billion dollars a year by 2020 to address the needs of developing countries."<br><br>"Adequate" "new and additional," "predictable," "scaled up," and "transparent." | Moderate: US$33 billion committed to developing countries during the fast-start period is far greater than previous allocations.[d] However, estimates are that less than a third of fast start climate finance was new and additional to existing aid commitments.[e]<br><br>Financial promises made do not have a clear relationship with developing countries' needs.<br><br>No clarity on what proportion of funding will be publicly raised as compared to privately raised or grants as compared to loans. |
| Balanced adaptation support | "At least 50% will be for adaptation activities in developing country Parties." (Africa)<br><br>"New, substantial and sustained public funding [for adaptation] from developed countries, with an annual scale not less than 2.5% of the GNP of developed countries." (Africa)<br><br>"Provision of financial resources by developed countries equivalent to at least 3% of their GNP." (Bolivia) | "Balanced between mitigation and adaptation." | Moderate: Between 20 and 22 percent of climate finance has been reported as allocated to adaptation; this is hardly "balanced."<br><br>The share of adaptation finance increased modestly after the first year of the fast-start period.[f]<br><br>The Cancun Adaptation Framework, established in 2010, was instrumental in deciding two years later to establish institutional arrangements to address "loss and damage associated with climate change." |

Table 4.2 (continued)

| Issue area | Low-income-state demands (2009 submissions to UNFCCC)[a] | Language in Copenhagen Accord and Cancun Agreements | Degree that commitments have been met |
|---|---|---|---|
| Equitable multilateral fund | "Developing countries, especially the particularly vulnerable developing countries, shall be provided with the necessary financial, technological and capacity-building support by developed country Parties through the Multilateral Fund on Climate Change (MFCC)." (AOSIS) "The Executive Board shall have an equitable and balanced representation." (AOSIS) | "New multilateral funding for adaptation" will be delivered through "a governance structure providing for equal representation of developed and developing countries" and a "significant portion of such funding should flow through the Copenhagen Green Climate Fund." | Moderate: Green Climate Fund created with a board with equal representation of developing and developed countries. $10 billion pledged in 2014 to the Green Climate Fund over a four year period.[g] The World Bank serving as the interim trustee, despite developing country objections. Only 2 percent of all climate funds channeled through UNFCCC and Kyoto Protocol Funds.[h] |
| Fair allocation of funds | "[An adaptation framework] shall enhance and support adaptation to climate change in all developing country Parties and in particular African countries, Least Developed Countries and Small Island Developing States." (Africa) | "Funding for adaptation will be prioritized for the most vulnerable developing countries, such as the least developed countries, small island developing States and Africa." | Weak: Adaptation funds are not systematically allocated according to designations of vulnerability.[i] |

a. Proposal by the Alliance of Small Island States, December 2009; Proposal by the African Group, December 12 2009; Bolivia proposal, December 7 2009.

b. Kartha and Erickson 2011.

c. United Nations Framework Convention of Climate Change 2014.

d. Ciplet, Fields, Madden et al. 2012. Also see Buchner, Stadelmann, Wilkinson, Mazza et al. 2014. Bruchner et al. estimate that in 2013 there was $5–11 billion in official development assistance committed from developed to developing countries for climate change.

e. Oxfam America 2012.

f. Ciplet, Fields, Madden, et al. 2012.

g. UN and Climate Change 2014b.

h. Ciplet, Roberts, Khan et al. 2013.

i. Ciplet, Roberts, and Khan 2013.

Indeed, the leadership and capacity of the LDCs grew steadily stronger over the years.[26] This, combined with the adept legal skills and ambitious demands of AOSIS, meant that the presence of the low-income states in the negotiations had become highly visible since the pivotal conference in Bali in 2007. On the eve of the negotiations in Copenhagen, the G77, despite major shifts in broader geopolitical relations, seemed as strong and capable as ever in challenging the interests of the North.

On December 9, 2009, just three days into the negotiations in Copenhagen, a leaked Danish text appeared on the website of the *Guardian* newspaper. The lead delegate of Sudan and chair of the Group of 77 plus China (G77) negotiating bloc, Lumumba Stanislaus Di-Aping, said on learning about this document, "The text threatens the success of the COP on two counts. From a procedural perspective, the UNFCCC [UN Framework Convention on Climate Change] is the only legitimate platform for negotiations and is the only place where all nations of the world are negotiating in an open and transparent manner. A more serious problem is substance of the text."[27]

In particular, Di-Aping referred to the fact that the text was understood to advocate a new treaty, replacing the Kyoto Protocol, under the UNFCCC, with a new set of obligations to developing countries (with the exception of the poor and most vulnerable countries).[28] Thus, the leaked text was largely viewed as representative of developed countries' interests, and not inclusive of the demands of developing countries. Several developing country delegates expressed feeling betrayed by the Danish prime minister, whom they believed abused his role as chair in the negotiations by being partial to the interests and views of developed countries.[29]

News of the leaked text encouraged protests among African civil society groups and others in the main hall of the Bella Center where the negotiations were being held. They chanted, "Two degrees is suicide and genocide [for Africa]," referencing the commitment in the document to maintaining temperatures to below 2 degrees Celsius, when the African, AOSIS, LDC, and G77 negotiating blocs had advocated a safer 1.5 degree Celsius temperature limit.[30]

As Sudan's Di-Aping explained on BBC radio, "It has become clear that the Danish presidency—in the most undemocratic fashion—is advancing the interests of the developed countries at the expense of the balance of obligations between developed and developing countries."[31] On December 17, two days prior to the official end of negotiations, Rasmussen clarified that the dual tracks of the negotiations would produce the final texts of the negotiations.[32] Despite this, on the final evening of the negotiations, with an unprecedented 115 heads of state and government present, an unlikely alliance of five countries—the United States, China, India, Brazil, and South Africa—secretively met to develop an alternative "Copenhagen Accord" text.[33] Through this process, a new mitigation framework that departed from the legally binding nature of the Kyoto Protocol was introduced as part of a take-it-or-leave-it package tied to unprecedented levels of finance. Importantly, the Copenhagen Accord promised $30 billion for the 2010–2012 fast-start period and $100 billion a year by 2020 for developing countries.

Moreover, funding for adaptation was to be "prioritized for the most vulnerable developing countries," including African countries, LDCs, and SIDS. The accord also included a provision for a new Copenhagen Green Climate Fund. This echoed low-income countries' demands for recipient country ownership over climate funds through a centralized fund under the Convention. Several phrases that couch these promises are largely representative of low-income-country demands for finance over the years.[34]

Several developing country delegates refused to offer their immediate consent to the Copenhagen Accord, citing problems with both the process and content. For example, Di-Aping said about the funding in the Accord: "It is not enough to buy coffins for everyone who will die because of climate change in Africa. I would rather burn myself than accept these peanuts."[35] Many delegates were largely caught off guard that China, India, and the other BASIC countries (Brazil and South Africa), their longtime coalition partners in the G77, had joined the United States in what they saw as a climate change coup d'état, undermining the democratic basis of the United Nations with a backroom deal.

Representatives of civil society organizations in the Climate Action Network had a similar response to the new text introduced in Copenhagen. For example, Kumi Naidoo, executive director of Greenpeace International called the Accord a betrayal of the vulnerable. He explained:

This is, what's on the table from this cluster of countries, is a betrayal of the poor, the betrayal of vulnerable countries, the betrayal of small island states, and it's a betrayal of the future of all children and our grandchildren on this planet. If this is the best that we can get then essentially what the most powerful countries, and particularly the United States is saying is that we are issuing a death warrant for people in small island states. … And we must ask the question right now of why is there such a lack of urgency? If those impacts were happening in Manhattan, if those impacts were happening in Berlin, if those impacts were happening in Paris, would the urgency be what it is? Why? We have to ask that question bluntly to these leaders. Is it because the people in question who are first on the front line of the struggle are poor, is it because they are not militarily strong, is it because they don't have any resources like oil underneath them, or is it because of the color of their skin? That is the question that needs to be posed firmly to these leaders.[36]

The rebellion of outspoken leaders of developing and low-income states in Copenhagen came at a time when the lead delegate of the United Kingdom, Ed Miliband, publicly stated that developing countries should register their support for the Copenhagen Accord; "otherwise we won't operationalize the [climate change] funds."[37] Since this statement was connected with the promise of $100 billion a year by 2020, it sounded to many delegates in the South like coercion.

Other threats were reported happening behind the scenes. Venezuela's lead delegate, Claudia Salerno also expressed in frustration to the plenary in Copenhagen that she had been threatened that there would be no green climate fund unless Venezuela offered its support to the Accord.[38] It would be revealed in leaked cables that there had been several covert efforts, both during and after the negotiations, of developed countries to co-opt low-income states into registering their support for the Accord. In particular, WikiLeaks cables link the United States with using aid to attempt to persuade Prime Minister Meles Zenawi of Ethiopia, Ambassador Abdul Ghafoor Mohamed of the Maldives, Bolivia, Ecuador, and Saudi Arabia to support the Copenhagen Accord.[39]

Zenawi's role was especially important since he was the only representative of Africa in the group of twenty-eight countries that were shown the Copenhagen Accord before it was made public. Zenawi caused quite a stir during the second week in Copenhagen when he, under allegations that he was receiving humanitarian and military aid for his cooperation,[40] announced in a joint press conference full agreement with France and other developed countries.[41] He said, "On almost all of the issues, I was preaching to the converted ... and therefore, in a very brief period, we have come on almost every issue to a complete understanding of each other's position, and in support of each other's position."[42]

But in fact the announced plan contradicted demands on emissions reductions and finance held by the African, LDC, and AOSIS negotiating blocs. Many felt that Zenawi, the key representative for the African Union in the meeting of twenty-eight countries, drove an important wedge into the more ambitious demands of low-income states in this announcement.[43]

Colombia's president, Alvaro Uribe, was the only representative of Latin America and the Caribbean, outside of Brazil in its role as BASIC member.[44] Uribe, from the Colombia First Party, was sharply aware of his nation's heavy reliance on over $5 billion in drug war funding from its patron, the United States, and that nation's intense interest in the Accord being accepted "as is."[45]

Later, in spring 2010, the United States would announce that $5.5 million in climate aid was officially cut to Bolivia and Ecuador because of their continued opposition to the Copenhagen Accord.[46] Kate Horner of Friends of the Earth said, "The US is acting like a bully, strong-arming the most vulnerable countries to get them to sign onto an ineffective and unfair deal that will not move the world closer to a just climate agreement."[47] One African delegate was quoted as saying "The pressure to back the West has been intense. … It was done at a very high level and nothing was written down. It was made very clear by the EU, UK, France and the US that if they did not back them then they would suffer."[48]

Wealthy states also made efforts to build strategic alliances with low-income states, thus dividing coalition opposition. According to a US diplomatic cable released by WikiLeaks, EU's climate commissioner, Connie Hedegaard, reportedly told the US deputy special envoy for climate change, Jonathan Pershing, that "the AOSIS 'could be our best allies' given their need for financing."[49] The insinuation was that AOSIS countries might be bought off from the rest of the G77 group and thereby endorse US positions in the negotiations. It was revealed in 2014 that the United States had engaged in spying to monitor communications between key countries during the Copenhagen negotiations.[50]

However, despite the efforts of powerful states to accommodate, bribe, threaten, and divide the interests of low-income states, due to the vocal opposition of various delegates and an atmosphere of confusion, the Accord was not adopted as a "decision" of the COP at Copenhagen; rather, it was merely "taken note of." This meant that the Accord at the time of adoption was not a legally binding agreement. The Copenhagen Accord had come as largely a surprise to many low-income-state delegates, and several had yet to adjust their expectations to this new political reality.

Having failed to get the Accord adopted as a full consensus-based decision, the United States then pushed for it to be adopted as a plurilateral agreement. In this way, members could agree to take part in the Accord on a voluntary basis, in contrast to a multilateral agreement, to which all members agree to be a party. Thus, the United States encouraged all countries to commit their support to the agreement in the coming year.[51] This practice, several delegates argued, was a violation of institutional codes of practice in the UNFCCC.[52] Through this strategy, the United States sought to provide momentum to demonstrate heading into Cancun that there was broad support for the Accord. Despite the agreement not having legal status, with the strong showing of support, they believed that it would be increasingly unlikely that states opposed would be able to shift the negotiations in a different direction.

## Concessions in Cancun

At least 141 countries, representing 87 percent of global emissions, indicated a willingness to associate with the Accord months after Copenhagen.[53] With this strong showing of support, though without legal status, it seemed increasingly unlikely that opposing states would be able to shift the negotiations in a different direction at the next round of negotiations in Cancun. On the lead-up to the talks in Cancun in late 2010, only 8 countries, representing just 2.09 percent of global emissions, officially expressed that they were not willing to ratify the Accord.[54]

This was buttressed by extensive efforts at diplomacy by the Mexican hosts working to establish common expectations among delegates prior to the conference. Far more adept than their predecessors and successors, the Mexican presidency under Felipe Calderon and his Ministry of Foreign Affairs was focused on success. Patricia Espinosa was an excellent leader of negotiations. However, the occurrences in December had been carefully prepared for by a year of globe-trotting diplomacy by Mexico's legendary diplomat, Luis Alfonso de Alba, who spent over 240 days on the road that year.[55] De Alba had built up considerable trust around the world serving as head of the UN Commission on Human Rights in Geneva. For the Mexican government preparing to host the 2010 climate negotiations, he successfully traveled the world and listened to delegations from all the regions, hearing what they most wanted but also successfully lowering expectations. As de Alba put it to us, he was "addressing individual concerns and looking for compromises" and seeking to "build confidence in the process."[56] He held informal meetings of the original negotiating tracks in Mexico City, Berlin, New York at the UN, India, "and a bunch of other meetings with the African Group and AOSIS."

The mood in Cancun was much more conciliatory than in Copenhagen. Although the low-income states were not neglecting the issue of emissions reductions altogether, they focused much of their attention on securing climate finance on the terms established in the Accord, which was viewed as an easier task to accomplish.[57] This included a focus on initiating a green climate fund, establishing an adaptation framework, and holding wealthy countries accountable for each particular clause related to finance in the Accord. It was the first time that such commitments had been put in writing and with numbers attached, and low-income states, along with civil society actors, were eager to hold the EU, United States, Canada, and others to their promises.[58]

In December 2010, the main content of the Copenhagen Accord was then integrated into the Cancun Agreements, which received almost unanimous support from rich and poor countries alike. This included the voluntary pledge-and-review framework introduced in Copenhagen. All but Bolivia, which adamantly opposed the agreements, conceded this as a necessary step. The institutional procedures themselves regarding consensus were reinterpreted when Bolivia objected to the Cancun Agreements. The COP president, Patricia Espinosa, declared that "the rule of consensus doesn't mean unanimity."[59] The move was controversial for some parties but seen as necessary by others. As a result, the talks came to an end with a new set of agreements that did not include Bolivia's demands.

Despite the immense inadequacy of the mitigation framework adopted, the media and mainstream environmental NGOs largely celebrated Cancun as a great success in diplomacy and compromise.[60] The outrage in Copenhagen had been largely forgotten or glossed over, with many participants feeling cautiously hopeful that the UNFCCC had survived to fight another day.

Having agreed in Cancun to a text with high levels of ambiguity—most notably that it left the door open for a "legally binding outcome in the future"—many of the conflict-ridden decisions were pushed to the following year in Durban.

### Delay in Durban

For most of the two weeks of the negotiations in the sprawling industrial South African city of Durban, the main discussions focused on the potential for a new mitigation agreement. Rumors spread throughout the convention center that China would shift course and offer its willingness to be subject to binding emissions commitments, thus breaking down the Bali firewall between the global North and South. These hopes proved unfounded. Civil society actors from the Climate Action Network, for their part, wore "I [heart] KP" (for "Kyoto Protocol") shirts in droves. More radical civil society actors flooded the main corridors of the convention center, at one point blocking foot traffic and noisily calling for climate justice. The LDCs and AOSIS also called for Parties to commit to a second Kyoto commitment period. The LDCs continued to demand improved levels of climate finance, particularly for adaptation, which they asserted were far below what was promised in Copenhagen and Cancun. Despite evidence to the contrary,[61] several wealthy countries maintained that they were meeting their finance promises. Several argued that their own efforts should be particularly commended in the context of a struggling economy.[62]

Leading up to the final days of the negotiations, the outcome in Durban was entirely unclear. The only certainty was that the United States had not wavered in its support of a voluntary, nonbinding framework that applied to all nations. The negotiations were supposed to come to a close on Friday night, but the delegates were nowhere near agreement on a path forward, and a new coalition had formed during the meeting between the EU, the LDCs, and the AOSIS. Despite US objections, this "Durban Alliance" fought to keep language for a second commitment period for the Kyoto Protocol in what would become the Durban Platform for Enhanced Action.[63]

The victory of this coalition in the end was arguably more symbolic than substantive. Only the EU and three other actors, with a combined 15 percent of global emissions, would commit to legally binding emissions cuts in the interim period under the aegis of the Kyoto Protocol.[64] The strength of this alliance was that it mobilized the countries seen as most vulnerable to climate change, the LDCs and SIDS, with the major actor that had committed to the most ambitious action to address climate change: the EU.

Instead of calling for pledges from countries, a compromise position in the end between India and the United States called for "another legal instrument or an agreed outcome with legal force under the United Nations Framework Convention on Climate Change applicable to all Parties." What this means in practice is highly contested. India argues that the principle of equity and the need for countries to act according to their "common but differentiated responsibility and respective capability" in the convention still applies, and therefore the main burden for taking mitigation action still lies with the North. The United States holds that the Durban Platform signifies movement toward mutual common action. By this, it means tearing down the Bali firewall between the global North and South. Other developing countries, including the LDCs and AOSIS, have been mixed on this point. While some have taken aggressive stands calling for China to make its own commitments, others have stood behind the G77 position of South versus North.

Crucially, agreement on the details of such a measure was put off until 2015, and the start of implementation was pushed in Durban until 2020. This represented an eight-year period of institutionalized inaction for many of the world's biggest polluters. The Durban Platform for Enhanced Action notes the goal of keeping global average temperature rise "to 1.5 or 2 degrees C" above preindustrial levels, and notes that current commitments are insufficient for reaching that goal.[65] However, we believe that despite this, the large delay in action makes it unlikely that such a target will be feasible. It simply leaves the tough decisions to act to address climate change for future leaders, who will have brutally steep reductions to achieve in very little time.

Despite the inadequacy of the Durban Platform for addressing climate change, the ambiguous terms of a path forward, and the inequitable impacts that will result, full consensus among all country delegates was reached. Rather than the UNFCCC completely falling apart as some had anticipated, the climate regime succeeded at maintaining political stability, and not a single low-income state objected.

### Waiting in Doha, Warsaw, and Lima

Consensus among states was achieved again the following years in Doha in 2012, Warsaw in 2013, and Lima in 2014. Doha represented the end of the fast-start finance period, and NGOs such as Oxfam pointed to a "climate fiscal cliff," with few new financial commitments being made to support vulnerable countries. The negotiations in Doha took place immediately following massive Typhoon Bopha in the Philippines, which caused more than a thousand deaths. The lead delegate of the Philippines, Naderev Saño, gave a powerful speech to broad applause, calling for more ambitious action on climate change, arguing, "If not us, then who? If not now, then when?"[66]

In this context, Doha saw the emergence of a strong coalition of civil society and low-income states making demands for a "loss and damage mechanism"[67] to provide compensation for extreme and slow-onset climate disasters that could not be readily adapted to. After achieving language in the agreements in Doha for the creation of such a mechanism and with very limited progress being made on the issue of mitigation, low-income states and numerous civil society organizations in the Climate Action Network made this issue a primary focus in the negotiations in Warsaw. Wealthy countries demonstrated very limited consideration of scaling up finance for adaptation, the voluntary pledge and review approach to mitigation became increasingly institutionalized (with the newly termed "intended nationally determined commitments"— INDCs),[68] and the Philippines experienced yet another severe climate-related event just prior to the negotiations. This time Typhoon Haiyan killed more than six thousand people.

In this context, the LDCs, AOSIS, and other coalitions such as the Central American Integration System made explicit statements that they would not consent to any agreement that did not provide a new loss and damage mechanism.[69] However, the United States, Canada, and especially Australia pushed back strongly against the demand for a separate mechanism for loss and damage, arguing that such issues could be dealt with under existing adaptation institutional arrangements. In response to overall weak progress in the negotiations, many civil society groups staged a walkout on the second to last day of the negotiations.

In the end, a deal was unanimously agreed to in Warsaw. Although it did little to move the negotiations forward on the issue of mitigation, it did establish a new international loss and damage mechanism under the Cancun Adaptation Framework. However, the text is ambiguously worded with no indication that developed countries should be liable to pay into this mechanism in relation to their historical emissions.[70]

The main negotiations moved next to Lima, where a draft mitigation framework was sought in time to generate the necessary foundation for a treaty in Paris in 2015. The penultimate draft presented on the final Saturday morning in Lima contained an extremely weak set of positions on what nations would even *report* to the UN about their planning for reducing emissions.[71] The room was split sharply, largely along old North–South divisions, with developing countries primarily being the actors that expressed immense frustration as they rejected the agreement as utterly unacceptable.[72] Only the AILAC group of Latin American countries (Independent Association of Latin America and the Caribbean) and a few island nations broke ranks from the Southern bloc, saying that it was too important to leave Lima without an agreement.[73]

Just before midnight Saturday, a day after the UN climate negotiations were supposed to have ended, the Peruvian president of the meeting, Manuel Pulgar-Vidal, announced the new final draft was up on the web, which he gave Parties an hour to review. Several new paragraphs and clauses had appeared in the text, a few of which had been clamored for by developing states. Several other highly desired paragraphs were still missing or omitted.[74] All 194 states present accepted the final text in what is officially named the Lima Call for Climate Action—though many did so with lukewarm support.[75]

The new text clarified that the work under the new agreement would be guided by the long-agreed principles of the Convention (presumably, including those associated with equity, precautionary action, and fair burden sharing).[76] Importantly, the new text also added urgency, noting with grave concern the significant gap between the aggregate emission pledges by Parties and what is needed to have a likely chance of holding the increase in global average temperatures below 2 degrees Celsius or 1.5 degrees Celsius. The new text also clarified that LDCs and small island developing states have special circumstances with regard to mitigation commitments.

The agreement is new because it confirmed that all nations will submit reports on their "Intended Nationally Determined Contributions," or INDCs. In the past, only the historically wealthy OECD nations had responsibilities to reduce emissions. These INDCs will cover many issues, such as reducing emissions, but they also can report how a country will adapt to climate impacts, and what they could do if they had more funding and technical support.

There are many weak areas of the text. Perhaps most importantly, there is a general lack of commitment during the pre-2020 period, when scaling up mitigation action is essential. Moreover, despite language that "urges developed country Parties to provide and mobilize support for ambitious mitigation and adaptation actions, especially to Parties that are particularly vulnerable to the adverse effects of climate change," there was no inclusion of concrete pledges in public funding support for developing nations to reduce their emissions and to prepare for future climate-related disasters. There is no indication of how that funding will be generated, and who will deliver it. Rather, the fight for adequate funding was kicked down the road to the pivotal negotiations in Paris in 2015.

In addition, there is nothing in the Lima Accord (what is officially named the Lima Call for Climate Action) to advance the loss and damage mechanism so desperately desired by developing countries (the words "loss and damage" were in fact completely deleted in the final text). This is another issue that will arise in the negotiations in Paris. Most crucially, the whole process is now about voluntary pledges by nations, not any rational sharing of the remaining available "atmospheric space" before we irreparably destabilize the climate. In February of 2015, another round of negotiations in Geneva produced agreement on an eighty-six-page draft text to serve as the basis for negotiation in Paris later in the year. There remained hundreds of brackets throughout the text indicating different options yet to be agreed upon.[77]

### Choice on the Periphery

A 2011 report found that developing country mitigation commitments coming out of Cancun far exceeded that of developed countries.[78] This raises the question of why low-income and other developing states that had been opposed to the Copenhagen Accord shifted their position to embrace a new deal with essentially the same approach to mitigating climate change in Cancun, Durban and Lima. Why would small island states like Tuvalu and the Maldives and LDCs like Sudan and the Gambia embrace a framework that could leave their countries ravaged by rising seas, drought, and other climate change impacts? The popular explanation is that consensus resulted from expert diplomacy carried out by the presidency and secretariat of the negotiations. This, and particularly Mexican diplomat de Alba's globe-trotting meetings in the year leading up to the conference in Cancun, indeed played an important role in achieving norm alignment in the negotiations. However, our analysis more broadly points to three distinct processes that were pivotal in the negotiation of consent.

First, low-income state consent has been contingent on the provision of strategic material and institutional concessions by wealthy states framed as rightful governance. The Copenhagen Accord, which represented a dramatic weakening of existing international climate change regulation, included numerous promises for material and institutional support for states particularly vulnerable to the impacts of a changing climate. Consenting to the terms of the Accord presented the promise of new and perhaps unprecedented financial benefits for low-income states. This reflects a core contribution of Gramscian political theory: hegemony is dependent on redistributive sacrifices of an economic corporate kind that promote social cohesion in a class-divided society.

Second, we have shown that such concessions have functioned as more than merely top-down bribes—what realist theorists call "side payments." While side payments, bribes, and other forms of coercion have been prevalent, the effectiveness of promises of climate finance in building consensus has largely depended on the legitimacy of the concession of climate finance within the political community. As such, the negotiations have been marked by a contentious process of norm alignment, through what Gramsci calls a war of position, by which states with competing class interests and group identities have come to some agreement on what are deemed legitimate terms of consent. The role of charismatic leaders such as Mexico's de Alba have played an important role in coordinating different actors to align their normative expectations and the corresponding concessions to be delivered.

In this case, concessions have been a direct strategic response to the long-term collective organizing and framing of demands by low-income-state coalitions in coordination with civil society. As shown in table 4.2, climate finance was presented in the Copenhagen Accord and then in the Cancun Agreements in terms that closely mirrored the demands of low-income-state coalitions, particularly those that had effectively developed an identity of vulnerability.

Moreover, the subsequent organizing and framing by various coalitions, and in the diplomacy efforts of Mexico, have been essential to the legitimation of such concessions. After a rejection of the Copenhagen Accord by key states, the concessions in the agreement the following year in Cancun were no longer portrayed by low-income state delegates as a bribe but rather a legitimate program for international financial support for adaptation. While many low-income state delegates continued to prioritize and express great frustration on issues of mitigation, adaptation increasingly became a core focus, as did eventually loss and damage.[79] In the end, the concessions of adaptation finance and, later, the loss and damage mechanism, despite ambiguous terms, were embraced by weak and strong states alike as core areas of progress in the negotiations.

A process of constructive ambiguity, we argue, has served to enable norm alignment. For wealthy states, the ambiguous terms of the agreement text on both mitigation and finance issues have provided a means to enhance the legitimacy of the post-Copenhagen regime, while providing plenty of room to shirk future finance and mitigation responsibilities. This process was articulated well by journalist Nitin Sethi after the highly ambiguous outcome of the Lima Agreements: "...the carefully crafted legal ambiguity in the document left room for even countries with conflicting and nonnegotiable issues to claim their respective interests had been safeguarded at Lima."[77] Low-income states would have preferred more concrete promises; nevertheless, the ambiguous terms provided a foundation for making future claims around rightful practices. Thus, the biggest conflicts were pushed ahead to be resolved in Paris in 2015 and beyond.

Overall, despite the promises made in Copenhagen and Cancun, low-income states have since been only moderately successful in influencing donor countries to adopt their interpretation of the key phrases related to climate finance.[80] While at first glance it appears as if wealthy countries came close to meeting their dollar figure promises during the fast-start period as they claimed the next year in Doha, Qatar, the fine print, represented in table 4.2, tells a different story. Most prominent, only about a third of the financing has actually been new and additional to existing development finance as promised.[81] This means that most of the money has simply been diverted from other pressing development needs.[82] Wealthy countries have also not fully delivered on promises related to climate finance such as transparency, predictability, adequacy, and balance (between mitigation and adaptation). Nor have they met promises on mitigation. We return to these issues in the next chapter.

Importantly, the Cancun Adaptation Framework, agreed on in 2010, with the purpose of pushing action on adaptation, was later a stepping-stone for the adoption of a decision in Doha in 2012 to establish international arrangements to address what was called loss and damage. If fulfilled, this has the potential to establish an international mechanism to require financial compensation and institutional support for countries that suffer from both extreme weather events and slow-onset events such as the rise in sea level.

Overall, it is clear that low-income states have had some influence. By leveraging the identity of vulnerability, delegates along with allies in civil society have influenced how materially rich countries have framed, and in some cases acted on, material and institutional concessions. The fact that wealthy countries are continually compelled to justify their behavior in the negotiations in these terms speaks to the influence that low-income states and civil society has had on the process.

Third, we argue that the negotiation of low-income-state consent has been intimately tied to structural conditioning and strategic coercion shaped by historical conditions of the broader world order. As we have discussed, vocal opposition to the Accord in the months following Copenhagen came with the potential consequence of cuts in international aid. Conversely, consenting to the terms of the Accord presented the promise of new financial benefits, perhaps particularly persuasive from the perspectives of heads of state and treasury ministers back home in debt-ridden states who did not yet view climate change as a primary concern at the level of economic development.

In terms of economic power, low-income states are heavily dependent on trade with the North.[83] They have been hard hit by the global economic recession[84] and suffer from a double exposure to economic harm and climate change disasters.[85] Trade dependency, vulnerability to external shock, and a heightened need for development assistance due to a global recession left low-income states particularly vulnerable to diplomatic pressure in Cancun.

In addition, a fragmentation of the "South" identity in global politics inhibited the possibility of strong and unified class-based demands, such as those advanced during the 1970s as part of the New International Economic Order. In particular, the G77 negotiating bloc, while still active, has increasingly come to represent subgroups with divergent class interests.[86]

While there were clear divisions growing for years, the unity of the G77 was further compromised in Copenhagen when its largest actors, the BASIC countries, joined with the United States in drafting the Copenhagen Accord. The Accord thus came to represent not merely a US or Northern betrayal of the Kyoto Protocol but, rather, the fragmentation of a South–South agreement on a path forward. This not only weakened the practical task for low-income states to cohesively organize against this new arrangement of interests, but also further disrupted the long-held identity in climate politics of South solidarity in relation to the North. An additional wedge in solidarity among low-income states was fostered by competition over designations of vulnerability deemed critical for accessing scarce adaptation funds, a topic we return to in the next chapter.[87]

A related structural constraint has been inequality in capacity between states in the formal negotiating process. Despite having consensus-based voting structures in the UNFCCC, low-income states are overwhelmingly outmatched in terms of financial resources, political influence, and negotiating capacity.[88] And since Copenhagen, exclusive meetings that bring together select negotiators who wield particular influence have been used, exacerbating divisions in decision-making power.

Ironically, another structural weakness relates to low levels of climate pollution emitted by low-income states. The forty-eight LDCs, for example, account for less than 1 percent of cumulative global carbon dioxide emissions.[89] Because these countries are insignificant contributors to climate change, they have low "polluter power"[90] to either leave the regime without dire consequence or address the problem on their own. This weakens their ability to withhold consent and has led to their exclusion in certain crucial negotiations concerning mitigation action.

Given that there is very little that low-emitting countries can do on their own to mitigate climate change yet are disproportionately affected, it is not surprising that they were reluctant in Cancun to obstruct or walk away from the UNFCCC process. With the first commitment period of the Kyoto Protocol set to expire in 2012, low-income states, lacking key resources and forms of political leverage, were presented with a choice of deciding between a new and inadequate take-it-or-leave-it mitigation framework, and no international mitigation framework at all. Rejecting the agreements in Cancun and Durban would have been tantamount to accepting an international system without rules governing behavior on climate change.

## Consent in a Warming World

This chapter has examined why low-income states consented in Cancun in 2010 and in the negotiations after to a new climate change framework that is inadequate for preventing catastrophic events in the poorest and most vulnerable countries. We have drawn on neo-Gramscian theory discussed in chapter 2 to employ a strategic view of power in the processes of negotiated consent. This approach makes an important contribution to cooperation theory, which has failed to adequately account for the interaction between agency and structure in international governance and the specific mechanisms of low-income state consent. We have identified three mechanisms in the strategic negotiation of consent: material concessions, norm alignment, and structural conditioning.

The eventual consent of low-income states reflects the reality that for weak actors, "bad rules that are universally acknowledged are better than no rules."[91] However, we have shown that the politics of consent and cooperation also necessitates empirical attention to the specific forms of low-income state agency. Importantly, a strategic view of power reveals that the terms of consent have been largely influenced and legitimized by low-income-state coalitions, wielding identities related to vulnerability, and these states have achieved important gains in the process, often in collaboration with transnational civil society networks.

Understanding how the politics of consent unfolds empirically in multilateral contexts also offers a window into how more radical confrontations to power are diverted, co-opted, fragmented, and accommodated. As this case reveals, low-income states have been strategically compelled by wealthy states to accept ambiguous promises for concessions in return for their consent to a largely gutted mitigation framework. As Robert Cox argues, "Hegemony is like a pillow: it absorbs blows and sooner or later the would-be assailant will find it comfortable to rest on."[92]

Overall, this analysis suggests that the process of international cooperation is dynamic, contentious, co-constitutive, and historically contingent.

As David Levy and Daniel Egan argue, "Capital's international hegemony is not uncontested in the international sphere; rather, it secures legitimacy and consent through a process of compromise and accommodation that reflects specific historical conditions."[93] This supports Gramsci's core insight that hegemonic resilience rests not in its rigid and unresponsive structures of domination, but rather, in the adaptability of lead actors to make certain accommodations to those they profess to lead.

What will this mean as we head toward the pivotal negotiations in Paris in 2015 to negotiate a new mitigation framework? We argue that our analysis suggests that low-income-state coalitions must establish stronger links of solidarity and ties to powerful actors and strong civil society networks in order to withstand future attempts of wealthy states to coerce and negotiate their consent to inequitable climate policy. While such coalition solidarity alone would not likely shift the intransigent positions of the United States and China, it will be instrumental in creating the conditions where a more just and sustainable path forward is at least possible. This is a monumental task given the immense forms of structural inequality and compounded vulnerabilities that low-income states face in today's global economy. The draw of promises of near-term finance for low-income states might be too great to resist.

As China becomes ever more dominant and enmeshed in African, Asian, and Latin American development in its search for natural resources to feed its booming economy, it may also prove increasingly challenging for low-income states to publicly challenge its interests.[94] Will low-income states continue to partner with the EU in the negotiations, pushing the rest of the world to follow their lead? Or will this alliance formed in Durban prove more symbolic than substantive, as the EU suffers its own internal fragmentation on the issues of climate change and economic strife?

In the next chapter, we argue that maintaining strong unity among states that are particularly vulnerable to climate change impacts will prove increasingly challenging as the world warms and climate disasters escalate when there are scarce adaptation funds to go around. We are, in fact, already seeing a wedge forming between highly vulnerable states as they compete for scarce adaptation resources. With these dynamics in mind, we now turn to the international politics of adaptation finance.

# 5

# The Politics of Adaptation

## Not Enough for Coffins

A meeting of about one hundred African negotiators, civil society members, and legislators was hastily called at the chaotic Copenhagen climate negotiations in 2009. It was the first week of the meeting, and the divisive Danish text had just been leaked. The lead G77 negotiator, Lumumba Di-Aping from Sudan, turned on his microphone, tears running down his face.[1] "Ten billion dollars for climate change [promised to developing countries by the European Union] may be an inducement for some countries," he said. "It is not enough to buy coffins for everyone who will die because of climate change in Africa. I would rather burn myself than accept these peanuts."[2]

Di-Aping's words call attention to a dramatic shift that has recently taken place in the global climate change negotiations: financing for developing countries to adapt to the adverse impacts of climate change has emerged as an issue at the top of the agenda alongside mitigation of greenhouse gas emissions. Adaptation to (human-induced) climate change has been defined by the Intergovernmental Panel on Climate Change (IPCC) as "initiatives and measures to reduce the vulnerability of natural and human systems against actual or expected climate change effects."[3] At the core of the politics of adaptation funding is the reality that poor developing countries have contributed very little to causing climate change, yet they are experiencing the impacts worst and first.[4]

This raises some basic questions. With very little progress having been achieved in two decades of negotiations on mitigation issues and with some major climate change impacts now inevitable, what would a just approach to adaptation finance look like? What precedent is there for a just approach in the negotiating texts, and what obstacles stand in the way in practice? Relatively little scholarly literature exists on adaptation politics and the role of funding. What does exist often portrays an oversimplified North–South split, ignoring divisions between countries on both sides of the global divide.

In the previous two chapters, we explored the shifting political alliances within the climate negotiations and how a process of "negotiated consent" resulted in a near-consensus agreement to an inadequate and inequitable mitigation framework. Here we shift our attention to the politics of international adaptation. The chapter is organized into two main parts. First, we look more closely at the complex strategic politics that have emerged around finance for climate change adaptation within the UN process. We assess why the issue of adaptation was long neglected in UN politics and how adaptation rose to the top of the agenda in recent negotiations. Drawing on the theory of justice of Amartya Sen (discussed in chapter 1), we develop a definition of adaptation finance justice based on decisions in the texts of the UN Framework Convention on Climate Change (UNFCCC) and its subsidiary bodies.[5] We apply our definition of adaptation finance justice to identify and assess three main points of contention between countries on both sides of the North–South divide: we call these conflicts the Gap in raising the funds, the Wedge in who is prioritized to receive funds, and the Dodge from using just governance institutions.

We then shift our attention outside the UN process, and beyond climate finance in particular, to consider more broadly the governance challenges and geopolitical dynamics that are likely to emerge as drastic climate change impacts become inevitable. Here we consider the potential of 4 degrees Celsius of average global warming, or worse. We explore related issues of climate-induced migration, climate-related security issues, disappearance of states under rising sea levels, fragmented intergovernmental structures for disaster management, geopolitical conflicts over the thawing Arctic, and the role of insurance companies and private actors in adaptation. We also discuss the perplexing governance, political, and social challenges related to large-scale technological attempts to engineer the climate.

To conclude, we connect the issue of adaptation politics back to our theoretical framework of strategic power in the book. We also lay out some concrete initial steps of action in order to enable just and lasting solutions to the climate crisis for vulnerable developing countries, the countries whose interests are most crucial on this issue and the most unlikely to be met. With Sen, we ground our analysis in existing political conditions, in the hope that by doing so, a more just solution can be reached. We also introduce the next set of chapters, which shift our attention from state actors to the private sector and civil society. We begin with a look back to when discussing climate adaptation was widely considered taboo.

## Long Neglected

The core focus of climate negotiations for their first decade was preventing global warming (i.e., mitigation), initially through voluntary and then through legally binding greenhouse gas emission reduction commitments (under the UNFCCC and under the Kyoto Protocol, respectively). The cardinal principle in the UNFCCC of Parties taking actions based on their "common but differentiated responsibilities and respective capabilities" was employed as a rationale for developed countries to take action on mitigation issues but was generally not discussed in terms of adaptation. The low profile of adaptation early on was evident from the fact that as late as 2006, only six of the forty-four proposals for a post-Kyoto regime dealt with adaptation as a policy issue.[6] This was in spite of the fact that there was, from its beginning in 1992, language in the Convention on measures concerning adaptation to the impacts of climate change.

We highlight five reasons that adaptation and, by extension, adaptation finance was not a core issue during the first decade of the UNFCCC. First, there was scientific uncertainty in the initial years about impacts of human-induced climate change. Although the Second Assessment Report of the IPCC came out in 1995, attribution of climate change to human activities was not very strong, so there was uncertainty over the extent to which climate change would occur.[7] Inclusion of adaptation issues in the Second Assessment Report was nominal: "Of the 728 pages of substantive text, about two thirds are devoted to impacts, one third to mitigation and only 32 pages to Adaptation [4 percent]."[8]

A related issue is that the UNFCCC addresses only anthropogenic climate change and does not extend to climate variability.[9] This compounds the difficulty of distinguishing vulnerability from human-caused climate change from existing vulnerability to natural climate variability. As a result, adaptation actions were viewed to be limited to changes that are proven to be anthropogenic and distinct from climate variability.[10] This proved a methodological barrier to advancing adaptation work under the UNFCCC because some wealthy nations were unwilling to commit to funding projects in developing countries to deal with weather variability that might not be human caused.[11]

Third, there was reluctance among many actors in both developing and developed countries and civil society to engage the issues of adaptation because of fears that it would distract from efforts to achieve an adequate mitigation framework.[12] Attention to adaptation was given in terms of how much mitigation was needed rather than focusing on adaptation measures in their own right.[13] Many developing countries thought that discussing adaptation at this stage might encourage developed countries to avoid mitigation of greenhouse gases.[14] In the early days, *adaptation* was sometimes seen as a dirty word. Anthony Anderson put it plainly: "Adaptation to a changing climate will be unavoidable. But it is a subject that carries a heavy ideological freight, for many people in the environmental movement suspect that any discussion of adaptation can only distract attention from the efforts to cut emissions."[15] In terms of timescale, some saw adaptation as a longer-term strategy than mitigation because climate change would be more evident with time.[16]

Discussing the funding of adaptation was also politically charged in a different way during this time. The wealthy countries initially resisted any attempt for fund provisions, perhaps understanding that a focus on adaptation might be an acknowledgment of their responsibility and liability, since they were the main cause of the accumulation of greenhouse gases in the atmosphere.[17] Some key developed countries avoided adaptation issues because acknowledging culpability on this issue might fuel demand for solutions to other global problems, such as poverty, health, and human rights violations.[18]

Fourth, mitigation projects in developing countries bring clear benefits to developed countries (reducing global greenhouse gas concentrations); however, the far-reaching benefits of adaptation projects are less obvious. Debates over who should pay the costs of adaptation projects ran afoul of this problem, especially since so much early funding was controlled by the Global Environment Facility (GEF), managed by the World Bank. The GEF's mission statement demanded that the agency spend its funds only on global environmental public goods, not local issues.

Finally, adaptation as a strategy was also held back by intra-G77 disunity related to Article 4(8) of the Convention. This Article considers actions related to funding, insurance, and the transfer of technology in response to the adverse effects of climate change on developing countries. However, it also refers to the "impact of the implementation of response measures," especially on "countries whose economies are highly dependent on income generated from the production, processing and export, and/or on consumption of fossil fuels and associated energy-intensive products." Accordingly, Saudi Arabia and some other OPEC countries demanded compensation to help them diversify their oil-dependent economies, which they argued was an adaptation strategy. The EU and other developed states have fervently rejected this demand for compensation.[19] This standoff delayed the negotiation process over adaptation for many years.[20]

However, not all actors avoided the issue of adaptation in the early years. The forty-three-member Alliance of Small Island States (AOSIS) was active from the beginning of the process in putting adaptation high on their agenda at the talks. This was due to their extreme vulnerability, particularly from typhoons and sea level rise, which was already measurably and perceptibly occurring. As early as 1990, AOSIS proposed insurance as an instrument of adaptation, including the creation of an international insurance pool.[21]

### Risk Rising

Adaptation would not remain off the agenda. The IPCC Third Assessment Report, published in 2001, concluded with more confidence that climate change could be attributed to human activities. This strengthened the position of AOSIS and the newly formed Least Developed Countries (LDC) negotiating group, such as on the need for redoubled efforts on adaptation under the treaties. In the negotiations in Marrakesh in 2001, adaptation emerged for the first time as a major strategy to address climate change impacts. Article 4.9 drafted there focused on the LDCs, with an agreement that they would benefit from relatively uniform national planning efforts directed at climate vulnerability and measures to adapt.[22] The National Adaptation Programmes of Action were supposed to address the most immediate and urgent needs in the LDCs.

In response to demands from developing nations, three funds were agreed to in Marrakesh: the LDC Fund, the Special Climate Change Fund, and the Adaptation Fund. The first two would include all countries, since they would operate under the Convention, but the third would include only countries that had ratified the Kyoto Protocol. The Adaptation Fund, placed under the Kyoto Protocol, was to be financed by a 2 percent levy from selling carbon permits through the new Clean Development Mechanism.[23] This gave the fund the potential for substantial revenues, given the explosive growth of the carbon markets in the mid- to late 2000s.

Surrounded by the poverty and growth pains of Delhi, India, the UN-FCCC conference in 2002 there saw a new form of public protest—what could be called one of the first large actions of an emerging climate justice movement.[24] The new coalition marching in the streets of Delhi consisted of fishers from Kerala and West Bengal representing the National Fishworkers' Forum, farmers from the Agricultural Workers and Marginal Farmers Union, and a delegation of Indigenous peoples threatened by the massive Narmada dam and from mining-impacted areas of Orissa. Delegates of NGOs from twenty other countries participated.[25] "This is the human face of the rising movement for Climate Justice," the movement declared.[26] The protesters affirmed that "climate change is a human rights issue" affecting "our livelihoods, our health, our children and our natural resources." They declared, "We will build alliances across states and borders to oppose climate change inducing patterns and advocate for and practice sustainable development. We reject the market-based principles that guide the current negotiations to solve the climate crisis: Our World is Not for Sale!"[27]

Inside the Delhi conference hall, the negotiating situation was tense. Fearing that limits on their carbon emissions would lead to economic stagnation, the G77 group of poorer nations, led by the host nation India, pushed for Kyoto to focus on sustainable development. The Delhi Ministerial Declaration on Climate Change and Sustainable Development that came from the conference declared that "economic and social development and poverty eradication are the first and overriding priorities" of developing countries, that "climate change and its adverse effects should be addressed while meeting the requirements of sustainable development."

What was known as the Green Group, a coalition of the EU and developing countries that had been working together since Berlin, were at odds,[28] but at Delhi, the developing countries signaled they had effectively taken control of this element of the Kyoto process, staking out their own right to development. The Delhi Declaration stated plainly that "Parties have a right to, and should, promote sustainable development. Policies and measures to protect the climate system against human-induced change should be appropriate for the specific conditions of each Party and should be integrated with national development programmes, taking into account that economic development is essential for adopting measures to address climate change."

As discussed in the previous chapter, at about this time, the concept of ecological debt or climate debt emerged and began to be demanded vociferously at the negotiations.[29] The G77 and a coalition of more than thirty

Western NGOs, policy institutes, and think tanks (many of them instrumental in changing the international debt regime) began to push more aggressively for some remuneration of the ecological and climate debts.

The issue of adaptation advanced year on year, rising steadily up the agenda, even as mitigation disputes continued. The Africa COP (Conference of Parties) held in 2006 in Nairobi was regarded as a milestone in the adaptation agenda under the UNFCCC. UN Secretary-General Kofi Annan reinforced this feeling during his address to delegates as he called for renewed urgency on adaptation for the poor.[30] Some developing country delegates also called for adoption of a separate adaptation protocol.[31] A five-year Program of Work on Impacts, Vulnerability and Adaptation to Climate Change was adopted with a list of nine activities drawn up for implementation. This program, dubbed the Nairobi Work Program, continues to be developed and extended.[32]

Finally, at the pivotal conference held in Indonesia in 2007, the Bali Action Plan was adopted, which placed adaptation as one of four pillars, together with mitigation, technology transfer, and finance. The issue of response measures needed to help diversify oil-dependent economies was transferred to the mitigation track, which satisfied OPEC and finally allowed adaptation to move significantly forward on the agenda. The Adaptation Fund was operationalized, with the GEF working as the trustee, as it was also for the other two funds: the LDC Fund and the Special Climate Change Fund. Notably, the Bali Action Plan called for "improved access to adequate, predictable and sustainable financial resources and financial and technical support, and the provision of new and additional resources" for both mitigation and adaptation.[33]

And in 2009, the Copenhagen Accord for the first time put forward a concrete dollar pledge to be provided by rich countries to developing countries for adaptation and mitigation measures. The Accord promised $30 billion fast-start finance between 2010 and 2012, with balanced allocation between adaptation and mitigation, and scaling up to $100 billion a year by 2020.[34] In the following year, with adoption of the Cancun Adaptation Framework, the agreements anchored the finance pledge into the UNFCCC treaty and affirmed that "adaptation must be addressed with the same level of priority as mitigation."[35] The following two years were consumed with struggles by Parties and observer groups over representation in the governance of the Green Climate Fund and the Adaptation Framework Committee.[36]

Three factors drove the shift of the issue of adaptation to center stage in the negotiations. First was the increase in climate-related disasters in recent years.[37] The latest IPCC Report on Extreme Events (2012) aligns these developments, particularly extreme precipitation and flooding, with anthropogenic climate change. Moreover, the economic losses due to the climate events are found to be higher in developing countries. This has helped sharpen the cognitive frame in which to understand and assign blame for these losses. The climate justice cognitive frame advanced by civil society organizations, academics and many developing country parties tied emissions in the North to suffering in the South, as seen in the Delhi protests.[38] Beyond the UNFCCC negotiations, even at the UN General Assembly special session on climate change organized by the UN secretary general in September, 2009, in New York, nation after nation recounted horrible disasters that appeared tied to climate change.

Second, in the years following the adoption of the Kyoto Protocol in 1997, it has become increasingly clear that the nations most responsible were not taking adequate action to prevent "dangerous anthropogenic interference with the climate system" as they had agreed under the 1992 Convention (Article 2). Then, in 2009 in the Copenhagen Accord, there was a dramatic shift away from the legally binding greenhouse gas limits for most developed countries, established in the Kyoto Protocol, to a system of bottom-up emissions reduction pledges. Designed by Brazil, India, China, and South Africa (BASIC) and the United States in closed-door sessions, in this approach there was neither an agreed-on aggregate figure for emissions reduction nor any system to ensure that the pledges made are deep enough to meet emissions reductions that the body of scientific evidence suggests are needed.[39]

Achieving gains on adaptation in this context was widely seen as a more winnable fight by developing countries.[40] And to mollify the G77 group sentiments on the lack of progress on mitigation, industrial countries for the first time made concrete financial pledges in the area of adaptation. Thus, the negotiations since Copenhagen simultaneously produced the conditions for increased attention of developing countries to the issue of adaptation and increased willingness among developed countries to address this issue. The language was still loose with plenty of room for flexibility,[41] but for the first time, there was a promise to respond to adaptation needs, particularly for the most vulnerable countries, in a way that appeared to be potentially more balanced with mitigation efforts.

The $100 billion per year in climate finance promised by wealthy countries by 2020 is having some complex impacts on the negotiations. Support for adaptation in developing countries, for example, may contribute to increasing their perception of fairness in international climate policy.[42] And the willingness of the G77 to demand adaptation finance to address the needs of the most vulnerable countries, along with a loss and damage mechanism, may be one of the only threads that still hold the group together.

Third, together with increasing extreme weather events, the failure of a mitigation regime to stop them, and the new language of "ecological debt" and "climate justice," a new agenda item has added a new dimension to international adaptation policy: that of loss and damage to address the impacts beyond what can be adapted to.[43] This took center stage in the negotiations in Doha and Warsaw in 2012 and 2013, where developed countries were pressured to agree to an international mechanism. Loss and damage also played a prominent role in the negotiations in Lima in 2014. However, the extent to which the institutional mechanism created will provide real benefits to vulnerable countries remains uncertain, and there is the risk that the negotiations on this issue will eclipse attention to the pressing issue of new commitments of adaptation finance, not to mention the need to dramatically ramp up action on mitigation.

**Just Finance?**

Issues of justice and ethics in efforts to mitigate climate change have received a fair amount of attention through the years.[44] Analyses often focus on how the burden for reducing global greenhouse gas emissions to a safe level should be shared in a just way between different countries. Substantially less attention has been directed to conceiving of justice as it relates to who should pay for the task of adapting, and how.[45] While there is indeed overlap between these two issues, adaptation finance raises some entirely new justice issues. Some of these relate to the fact that the distributive questions that are posed by the ethics of adaptation are "not only between burden-takers (i.e., those who take adaptive or mitigating action) but also between the recipient of benefits."[46] Perhaps the most notable attempt to define adaptation finance justice is by Italian geographer Marco Grasso, who calls for a "fair process, that involves all relevant parties, of raising adaptation funds according to the responsibility for climate impacts, and of allocating raised funds [by] putting the most vulnerable first."[47]

As far as we know, there has yet to be an attempt to construct a definition of justice for adaptation finance comprehensively based on what is in the text of the 1992 Convention and subsequent decisions of its subsidiary bodies. Such a definition offers the advantage that it reflects terms that have already been agreed upon by Parties, even if they have been agreed on through an imperfect and in many ways unequal political process.

In developing such a definition of justice in adaptation finance, we draw on Amartya Sen's "realization-focused comparison" theory of justice, which we outlined in chapter 1. For Sen, justice means sustaining people's capability to have and safeguard what they value and have reason to attach importance to.[48] His approach to justice focuses on how practical reasoning should enable society to reduce injustice and advance justice rather than focusing on the abstract components of a perfectly just society. Sen focuses on the actual behavior of actors and not just the arrangements of institutions and how they are supposed to work. And his perspective recognizes that there are often ambiguities between divergent approaches to organizing society that are all reasoned as just.

We believe these perspectives on justice provide a useful theoretical entry point for developing a definition of *adaptation finance justice* in relation to the particular context of the UNFCCC negotiations, assessing the political realities of adaptation finance in relation to this definition, and discussing strategies for achieving justice in adaptation finance.

How does this approach to justice relate to the text regarding adaptation finance in the Convention, subsequent UNFCCC decisions, and the Kyoto Protocol? A central problem is that these texts often focus on planning rather than action on adaptation. The Convention uses language such as "prepare for"[49] rather than "implement," and "take climate change considerations ... to the extent feasible" into consideration[50], rather than giving them highest priority. The Kyoto Protocol has similarly weak language in this area, and provided no entity to coordinate adaptation issues.[51]

However, at the insistence of developing countries, particularly by AOSIS and the LDCs, there have been changes to address these weaknesses. The Cancun Adaptation Framework was adopted in 2010 in order to bring all the adaptation concerns into a coherent package. Then in Doha in 2012, a new program of National Adaptation Plans to address mid- and long-term goals was adopted for the LDCs and other developing states. This sought to complement the existing NAPA program, which addressed the most urgent short-term adaptation needs in LDCs. In addition, the Adaptation Committee, designed to promote the enhanced action on adaptation in a coherent manner, was operationalized with equal representation from developed and developing countries.

Still, it can be argued that compared to mitigation, the legal basis for key elements of adaptation finance under the Convention is weak. And the tough practical questions of how to pay for the work of adaptation, and who should be prioritized, have been put off and have hardly been addressed to date.

Despite this, as demonstrated by table 5.1, the overall intent of the UNFCCC decisions related to adaptation finance is relatively straightforward. Most prominent is the commitment of states to take action based on "common but differentiated responsibilities and respective capabilities."[52] While this principle is theoretically relevant to issues of adaptation finance, it has been mostly referenced in relation to mitigation.

More explicit are Articles 4.3 and 4.4 of the Convention. Article 4.3 says that developed countries will provide "new and additional financial resources to meet the agreed full costs" and "full incremental costs" of actions taken by developing country parties. The qualifying word *agreed* poses a problem, as agreement is ever elusive. However, the last sentence of this article says that they shall take into consideration the "need for adequacy and predictability in the flow of funds, and the importance of appropriate burden sharing among the developed country Parties." But Article 4.4 is explicit in terms of adaptation. It says that countries included in Annex 2 (which includes all of the OECD countries but not former Soviet Union states) "shall also assist the developing country parties that are particularly vulnerable to the adverse effects of climate change in meeting costs of adaptation to those adverse effects." Subsequent agreements offer further clarity on the raising of "adequate," "scaled-up," "predictable," "balanced," "new," and "additional" funds.

Based on this analysis of the UNFCCC texts in terms of what has been agreed upon in relation to adaptation finance (table 5.1), we present the following summary definition for adaptation finance justice:

*Adaptation finance justice* requires that developed country Parties take measures to assist developing countries to adapt to the adverse effects of climate change by providing adequate, predictable, and balanced finance that is new and additional to earlier levels of foreign assistance, with priority in allocation of funds to countries particularly vulnerable. Action should be taken on the basis of equity and in accordance with the principles of common but differentiated responsibilities and respective capabilities, and should follow a country-driven, gender-sensitive, participatory, and fully transparent approach, taking into consideration vulnerable groups, communities, ecosystems, and indigenous knowledge.

We recognize that this definition still contains a great deal of ambiguity in the details of how it should be carried out in practice. This definition is representative of current agreement in the UNFCCC texts, but it will have to be amended over time as the agreement among Parties evolves.

**Table 5.1**
Principles of Adaptation Funding Justice in UNFCCC Texts versus Current Practice

| Principle of adaptation finance justice | Related articles and decisions of the Conference of Parties | Political reality of adaptation finance |
| --- | --- | --- |
| 1. Balance: "Affirms that … adaptation must be addressed with the same priority as mitigation and requires appropriate institutional arrangements to enhance adaptation action and support." | CP.16 2 (b)* CP. 16 IV (a) 95 | *Imbalance*: Finance to developing countries for mitigation has been far greater than that for adaptation. (the Gap) |
| 2. Fair burden sharing: "The Parties should protect the climate system for the benefit of present and future generations of humankind, on the basis of equity and in accordance with their common but differentiated responsibilities and respective capabilities." | 3.1* CP.16 I.1 CP.16 II (14) | *No agreement on fair burden sharing*: No consistent or transparent formula for developed country adaptation finance commitments; developed countries have been unwilling to discuss adaptation finance in terms of responsibility or capability. (the Gap) |
| 3. Precaution: "The Parties should take precautionary measures to anticipate, prevent or minimize the causes of climate change and mitigate its adverse effects." | 3.3* 4.4 4.8 | *Not precautionary*: Finance pledges represent movement toward a precautionary approach; however, levels of funding are inadequate to meet developing country needs. (the Gap) |
| 4. Predictable, adequate, new, and additional: "*Decides* that, in accordance with the relevant provisions of the Convention, scaled-up, new and additional, predictable and adequate funding shall be provided to developing country Parties, taking into account the urgent and immediate needs of developing countries that are particularly vulnerable to the adverse effects of climate change." | CP.13 1. (e) CP.13 1. (i) CP.16 2. (d) CP.16 II (18) CP.16 IV (a) 97* | *Not predictable, adequate, or clearly new or additional*: Pledges made by developed countries represent a step toward scaling up climate finance; likely that adaptation finance is not new or additional to existing official development assistance; adaptation finance is not adequate for meeting basic developing country needs related to climate change; due to a lack of transparency and uncertainty about future adaptation finance provisions, funding levels are highly unpredictable. (the Gap) |

Table 5.1 (continued)

| Principle of adaptation finance justice | Related articles and decisions of the Conference of Parties | Political reality of adaptation finance |
|---|---|---|
| 5. Needs-based targeting of funds: Parties shall be guided by "the specific needs and special circumstances of developing country Parties, especially those that are particularly vulnerable to the adverse effects of climate change." | 3.2*<br>4.4<br>4.8<br>4.9<br>CP.13 1. (i)<br>CP.16 II (11)<br>CP. 16 IV (a) 95 | *No agreed allocation protocol*: Least Developed Countries, small island developing states, and African countries are considered the "particularly vulnerable Parties"; however, there is increasing controversy about which parties should be included in this group and how vulnerability should be assessed for the allocation of adaptation funds. (the Wedge) |
| 6. Transparent, recipient, and science-led allocation and governance considers especially vulnerable groups: "*Affirms* that enhanced action on adaptation should be undertaken in accordance with the Convention, should follow a country-driven, gender-sensitive, participatory and fully transparent approach taking into consideration vulnerable groups, communities and ecosystems, and should be based on and guided by the best available science and, as appropriate, traditional and indigenous knowledge." | CP.16 II (12)*<br>CP.16 II (20) a<br>CP. 16 IV (a)<br>100, 103, | *Not country driven*: The National Adaptation Programmes of Action (NAPA) for Least Developed Countries represent an attempt at a country-driven approach to adaptation planning and funding; however, less than a quarter of NAPA projects have been funded; COP and Kyoto funds created with the intent of facilitating a country-driven and participatory approach have received about 1 percent of climate funds; no evidence that adaptation finance has been sensitive to the particular needs of women, indigenous peoples or other marginalized groups; limited transparency in adaptation finance—inconsistent reporting disallows summing and comparison. (the Dodge) |

a. See Ciplet, Fields, Madden, et al. 2012.

## The Gap

The conflicts in adaptation finance politics fall in three categories: those having to do with supplying, allocating, and governing the funds. We begin our discussion with the conflict related to supplying adaptation funds. What we call the Gap refers to the inconsistency between the promises that have been made for adaptation and the funding that has actually been delivered. The convention or the Kyoto Protocol did not specify the level of financing to address climate change issues. However, it does imply the need to strike a balance between necessity and availability, with identification of sources and their predictability.[53]

In reality, even after two decades, wide differences exist among groups of countries in the climate negotiations concerning their positions on how climate finance should be mobilized.[54] There has been a yawning gap in the amount of adaptation funds available to developing nations compared to assessment of need. Recent estimates of need are particularly worrisome. For example, the United Nations Environment Programme's 2014 *Adaptation Gap Report*, estimates the cost of adapting to climate change in developing countries as likely to reach at a minimum two to three times the previous estimates of $70–100 billion per year by 2050.[55] And in the near-term, the report estimates adaptation costs and the residual damage in the LDCs alone is likely to reach $50 billion a year by 2025/2030 and possibly double this value ($100 billion a year) by 2100.[56] The report estimates plausible costs for all developing countries at $150 billion a year by 2025/2030.[57]

Indeed, estimates of the financial need for adaptation in developing countries are ever changing. This is largely for three reasons. First, estimates have shifted as the science has become more robust in terms of the anticipated impacts experienced at distinct temperature changes. Second, these estimates are dependent on anticipated future temperature rise in a given context, which is itself highly dependent on the outcomes of international and national political and social processes on climate change mitigation. Finally, estimates of damages and harm depend upon how much human society anticipates the problems and take adaptive actions.

In order to assess the degree of justice achieved thus far on this issue, attention to the numbers themselves in the Copenhagen Accord and Cancun Agreements are crucial: $30 billion in short-term fast-start finance for 2010 to 2012 and scaling up to $100 billion a year by 2020. However, the true meaning of these numbers depends on the interpretation of key phrases in the text, many of them loosely defined or not defined at all.

First, the texts promise "adequate funding," yet developed countries have fallen short in this area. Donor countries have not been at all transparent about how they have determined their financial contributions for adaptation. Why has $100 billion a year been allowed to count as "adequate" funding? This is clearly a political decision, based on perceptions of what was feasible to the key players.

It is also not clear what proportion of the funding will be in the form of pure grants, partial grants, or purely market rate loans. It is difficult to see how vulnerable countries could respond to the requirement to repay loans for adaptation. The Copenhagen Accord also says, "This funding will come from a wide variety of sources, public and private."[58] In spite of repeated complaints about this mixing of two very different types of finance, there is no improved clarity in the agreements from Cancun, Durban, Doha, Warsaw, or Lima concerning what proportion of funding should or must be publicly raised.

Predictably, contributor countries are focusing on mobilizing private rather than public sources of funding and attempting to quantify what that might mean for meeting their responsibility.[59] The Accord says the funds will come from "bilateral and multilateral, including alternative sources of finance." The first part of this clause makes clear that contributor nations are protecting their right to channel climate finance through their own bilateral agencies (and not only through UN or World Bank funds). The second part suggests that Parties are willing to take on board so-called innovative financial mechanisms such as those put forward by the High-Level Panel on Climate Finance (i.e., airline and bunker fuel levies). However, it is entirely unclear which, if any, of these mechanisms Parties will accept, and high-level panels have failed to arrive at ones acceptable to all the key players.

Second, the Copenhagen and Cancun texts promise "predictable" funds, which means clearer and better-met targets. Predictability is essential for developing countries to establish their own budgets and to plan for adaptation responsibly. But predictability has not increased since Copenhagen; Northern governments have struggled to muster the will or the political support for sources of public funding for increased climate finance under the current economic circumstances.

How can funding be predictable, which is desperately needed for planning anticipatory adaptation, without agreement on some sort of autogeneration mechanism? Some quite developed proposals have been advanced to levy international air passengers a small flat fee or to tax bunker fuels used in international shipping, placing a tiny tax on international financial transactions, a tax on carbon, or even a tax on arms trade. At the time of this writing, none of these proposals are moving forward and in the most recent rounds of negotiations, they have almost completely fallen off the political agenda. Climate finance remains voluntary, depending on the political expediency in the wealthy countries. The mocking irony is that with continued global financial crisis and no major interstate war right now, world military spending has stayed constant at more than $1.7 trillion a year.[60]

Another issue having an impact on the predictability of funding is the fragmentation of aid. Foreign assistance has grown increasingly fragmented, with more than seven dozen channels of funding, both governmental and multilateral, and now also private foundations active.[61] With so many funding channels and very little transparency about what is being funded, it is difficult for both contributors and recipients to adequately assess where money is going.

Third, the phrase "scaled-up" is another aspect that has not been adequately addressed. After years of the wealthy nations putting only token amounts of voluntary funding into UN climate funds (see table 5.1), developing nations pushed for scaled-up funding at Copenhagen. This phrase has come to stand for the period from 2013 (after fast-start finance ends) to 2020, when the Cancun Agreements specify a tenfold larger scale of funding per year. Yet no language in UNFCCC decisions indicates a plan for this crucial scaling-up period. The G77 and China, including AOSIS and the LDCs, has demanded a road map, along with $60 billion a year to reach by 2015, but the Doha, Warsaw and Lima agreements cite no number at all.[62] A fiscal cliff looms for mid- and long-term climate change mitigation and adaptation finance.

Fourth, the Copenhagen and Cancun texts also promise "new and additional" funding, which suggests that it would be above conventional official development assistance. These words have been much debated since Copenhagen, and their meaning is not at all clear.[63] Most countries have provided no explanation of their baseline at all.[64] There is serious concern that developed countries are recycling their official development assistance toward climate finance (and thus taking money away from other pressing development needs such as health and education) or renaming past pledges as commitments to fast-start finance. Oxfam estimated in November 2012 that only 33 percent of fast-start finance was actually new.[65] These approaches undermine the credibility of financial pledges made at the international level and damage trust in the political process.[66]

Finally, both the Copenhagen Accord and Cancun Agreements promised "balanced allocation between adaptation and mitigation." Our analysis of the fast-start reports provided by the key Parties through November 2012 showed, however, that somewhere between 75 and 80 percent of the funding actually focused on mitigation.[67] This ratio can hardly be called balanced.

So how much funding overall has actually been delivered to developing countries to meet adaptation costs? By reviewing the reports of wealthy countries, we estimated that as little as $1.5 billion a year was pledged by wealthy countries to developing countries for adaptation during the fast-start period from 2010–2012.[68] Another analysis by the OECD's Development Assistant Committee found that in 2013 $3.4 billion in official development assistance was committed from wealthy to developing countries with a principle objective of addressing adaptation.[69]

Overall, while low-income states have made some gains, they face major challenges in their efforts to scale up adaptation finance, and it is not clear if the loss and damage agenda will amount to rebalancing this power dynamic. The Gap raises critical issues of whether adequate funds are being allocated where they are needed and disbursed when they are needed for appropriate interventions.

We now move to the Wedge issue of how to allocate the scarce funds that have been provided for adaptation.

**The Wedge**

While the supply side of adaptation finance concerns the question of where funds are to come from, the demand side deals with the questions of who gets access to available adaptation funds and based on what criteria. In other words, if $50 billion in funds per year is to be available for adaptation in 2020 ("balanced" funding to adaptation under Copenhagen and Cancun language), how are these funds to be allocated based on a fair set of criteria? If there is a shortfall from promised funding, and totals seem likely at this writing to be perhaps one-tenth of that figure, then the pressure to allocate funds fairly is even more critical and tense. Thus, this puzzle relates to providing funds with priority to those parties that are particularly vulnerable.

We call this puzzle the Wedge: if current patterns of financing continue, the issue of how to divide too few funds among too many actors stands as a potential wedge to disrupt solidarity between developing country actors in both the mitigation and adaptation negotiations. There is also the issue that funds may not be allocated to prioritize the most vulnerable groups; instead, some funding allocation approaches reflect donor country interests more than the needs of vulnerable actors.[70] The Wedge puzzle raises the question: How can fair funding allocation criteria be developed without disrupting developing country solidarity?

Back at the Berlin negotiations held in 1995, the COP adopted a "stages for adaptation activities and funding" plan, that specified early work "to identify particularly vulnerable countries or regions and policy options."[71] Twenty years on, perhaps due to the sensitive nature of the issues, explicit criteria to determine how vulnerability should be assessed in order to allocate adaptation funds have yet to be fully developed. Little has been decided, except that "particularly vulnerable" or "most vulnerable" developing nations are to be prioritized.

In the formative text of the Convention, LDCs were identified to have specific needs and special situations concerning funding and transfer of technology.[72] Small island developing states (SIDS) and African countries were subsequently included alongside LDCs in the Bali Action Plan as "particularly vulnerable to the adverse effects of climate change."[73] The Bali Action Plan also used the expression "most vulnerable"[74] in relation to these groups of countries for the first time in the UNFCCC process. Then, in the Copenhagen Accord, the language distinguishing these groups became even more explicit with the phrase "most vulnerable developing countries."[75] Some countries of the G77, including Pakistan, opposed this category. Most recently, developing countries, including Guatemala, Colombia, Bolivia, and Pakistan, that are not LDCs, SIDS, or African countries called for inclusion of language in the Cancun Agreements that would designate them as "highly vulnerable countries" in order to ensure that they would be in a position to receive adaptation funds. The civil society newsletter *ECO* referred to this effort negatively as a "beauty contest" over which country is most vulnerable.[76]

Although the proposal for highly vulnerable countries was rejected by the G77 and China bloc, it indicates the perceived benefits that gaining specific vulnerability status can have for countries in the UNFCCC. This process also indicates the risk that concessions based on special status can have on disrupting the solidarity among developing countries. Given that there are 134 countries in the G77 and China bloc with very diverse characteristics and interests, it is notable that this group has stayed relatively united in its demands and actions for nearly two decades of climate negotiations.[77] Clearly, much of the negotiating leverage of developing countries depends on their ability to maintain common positions and solidarity among actors. However, as discussions in chapters 3 and 4 show, this unity has weakened.

As our definition of justice in adaptation finance indicates, according to UNFCCC parties, allocation for adaptation funds must prioritize the most vulnerable. This approach runs counter to prior schemes such as the GEF resource allocation framework, which is not based on the vulnerability of a country but on its performance and its ability to contribute to environmental benefits beyond its borders. This approach has resulted in an inequitable distribution of funds; the numbers show that "in the climate change focal area, 36 countries share $674 million, with $236 million left to 124 countries."[78] The GEF's activity profile shows an overwhelming bias toward mitigation. For example, between 1991 and June 2011, out of 914 projects carried out in 156 countries with a total budget of $3.84 billion, 755 were mitigation projects, with a value of $3.4 billion; the remaining limited funds were dedicated to adaptation and enabling activities [11 percent].[79]

What would a just allocation approach based on relative vulnerability look like? The concept of vulnerability is used as an analytical tool for describing states of susceptibility to harm, powerlessness, and marginality of both physical and social systems and for guiding normative analysis of actions to enhance well-being through reduction of risk.[80] Thus, it is essential to note that not only are certain groups more or less geographically exposed to physical environmental threats such as sea-level rise, droughts, floods, and disease, but various characteristics of a group and their context make them more or less likely to be able to prepare for, cope with, or adapt to such impacts. If the criteria focus only on geographical factors, then the SIDS, with some of them only a few meters above sea level, are probably the most vulnerable. In terms of low socioeconomic indicators, however, the LDCs are the most vulnerable. From a justice perspective, allocating funds based on the assessment of vulnerability is a process fraught with ambiguity.

The first place where we have seen any effort to plan the allocation of adaptation funds with attention to vulnerability is in the Adaptation Fund.[81] The Fund's current prioritization formula for selecting projects and programs includes the level of vulnerability, the level of urgency and risks arising from delay, and ensuring access to the Fund in a balanced and equitable manner, among others. In addition, the Adaptation Fund Board has considered three protocols for fund distribution among eligible Parties: a cap per eligible country, an allocation per region, and criteria to prioritize among specific eligible projects.[82]

Although the attributes in the Adaptation Fund frameworks are theoretically sound, there is a lack of clarity in the criteria. Such metrics thus
prove difficult to quantify and assess and allow much discretion or interpretation in prioritization. As a result, using these criteria to allocate
funding is potentially problematic. Questions facing the Adaptation Fund
Board and other funding entities on how best to allocate scarce resources
are numerous. What time frame should be prioritized in terms of allocating funds? Should the funding entity allocate funds primarily to individual projects, or should national programs that take a more comprehensive
approach receive priority? At what scale should vulnerability be assessed
in the allocation of funds? Should countries that already have plans in
place or those with less institutional capacity be prioritized? Should loss
of culture be considered more or less important than economic loss or
loss in human lives?

Unfortunately, science alone cannot rescue us from the politics of adaptation funding allocation and steer us toward a more just solution. Many
models assess national vulnerability,[83] all with strengths and weaknesses,
depending on what criteria are prioritized in assessment. An important
reflection of the tensions that lie ahead is the weight given to specific
indicators of vulnerability. For example, Bangladesh argues for giving
more weight to loss of lives and livelihoods, while some G77 members,
including AOSIS, that are small in population size argue for an approach
that considers their geographical and cultural integrity. This conflict is
indicative of the practical and political decisions that are necessary in any
allocation model that assesses national vulnerability.

## The Dodge

The third puzzle of adaptation finance, related to governance of funds, we
call the Dodge. Thus far, very limited funds have been allocated through
institutions that meet the demands of developing states and adequately
address the justice dimensions in our definition. This puzzle raises the
question: What can be done to ensure that funding institutions established with much deliberation under the supervision of the UNFCCC are
not dodged by donors in favor of more donor-friendly institutions?

Developing countries have been united in their demands for adaptation and mitigation funds to be administered by the UNFCCC and parties to the Kyoto Protocol. Chief among demands has been for direct
access to funds, in which accredited national or regional institutions
in recipient countries assume the role of directly administering funds,

thereby enhancing national and local level autonomy to carry out adaptation projects and programmes.[84] Similarly, there has been a strong push to have majority representation from developing countries on the boards that oversee funding decisions. These two measures are part of a larger platform to shift donor assistance from the micromanagement of funds at the point of disbursement to establishing more democratic global funding mechanisms and greater national ownership and autonomy in making decisions about funding priorities in recipient countries.[85] Developing countries also advocate greater control over funds to provide a more streamlined process for accessing those funds, given that it has been slow to reach recipients.

During the second decade of climate negotiations, the focus of larger developing countries such as the BASIC countries remained more on mitigation than adaptation. As discussed in chapter 3, feeling they were not getting adequate attention in the negotiations, the LDCs created their own caucus in 2001. And their organizing paid off: in Marrakesh in 2001, three funds were established: the Least Developed Countries Fund and the Special Climate Change Fund under the Convention and the Adaptation Fund under the Kyoto Protocol.[86]

There were major struggles over who should oversee these funds and how they should be structured. Developing countries pushed for the COP to oversee the funds, with the understanding that this would enable them greater decision-making power.[87] In contrast, developed countries preferred their funds be overseen by the GEF, an institution established in 1991 by the World Bank and administered together with the UN Development Programme and UN Environment Programme. Since major donors have near veto power at the World Bank, developing countries objected to the GEF having administrative power over UN funds. Further controversy was added by the GEF's use of a resource allocation framework,[88] which is based on two criteria: global benefits from some activity and country performance in executing aid projects. Even with a new criterion to prioritize states with low gross domestic product, LDCs see the criterion of global benefit as a way to divert most of the GEF resources for greenhouse gas mitigation, while leaving almost nothing for adaptation. Despite developing country opposition, the LDC Fund and Special Climate Change Fund continue to be administered by the GEF.

The core paradox with the Dodge is the amount of attention that governing UNFCCC climate finance is requiring for how few resources are being channeled through those institutions. Because the financial architecture of the regime is extremely fragmented, a new high-level twenty-member standing committee on finance, with direct accountability to the COP, has been tasked with bringing coherence, accountability, and transparency to climate finance. However, despite careful design and establishment with equal representation from developed and developing countries, it is not at all clear that the standing committee will have any impact on core decision making.

The experience of the Adaptation Fund is especially informative for the issue of the Dodge. The unique structure of the Fund enabled developing countries and civil society to have leverage in their fight to achieve the governance practices that they sought. In particular, in Nairobi in 2006, it was agreed that the Adaptation Fund would be filled by a novel international revenue-generating source, a 2 percent levy on carbon emissions permits in the Clean Development Mechanism. Countries like Brazil, China, and India, which host the overwhelming share of Clean Development Mechanism projects, see the levy on this mechanism as a solidarity payment from larger developing countries to the LDCs.[89] Since this funding source is not attached to national budgets of Northern governments, this gave developing countries more control over the fund's operation.

A final factor that made it possible to govern the Adaptation Fund differently was that it is operated under the Kyoto Protocol, a treaty to which the United States is not a party. As a result, the Adaptation Fund represents a very different balance of power in comparison with nearly all other international funding agencies. However, it is suffering severely from a funding deficit, since the price of the Clean Development Mechanism–certified emissions reductions has crashed largely because of the uncertainty around the future of the Kyoto Protocol.

Most recently, an umbrella funding institution, the Green Climate Fund, was introduced in the Copenhagen Accord, and key elements of its structure were agreed to in Durban in 2011. Notably, parties agreed that the fund is to be overseen by a body under the United Nations, as advocated by developing countries, rather than the GEF, which was advocated by the United States and the European Union. Furthermore, the Green Climate Fund, finally established in Seoul, Korea, is administered by a twenty-four-member board, with equal representation from the developed and developing countries.[90]

Devoting tremendous attention to the design and establishment of these funds, developing country delegates and civil society campaigners from both the North and South have essentially pursued a strategy of "if you build it, funds will come." However, despite the effort in establishing and refining these funds, only about 2 percent of fast-start climate funds have been channeled through UNFCCC and Kyoto Protocol funding institutions, with limited consistent or predictable sources of revenue.[91] For example, the Adaptation Fund garnered very few fast-start funding pledges[92] despite strong calls to donors from civil society to do so.[93] Perhaps more promising, the Green Climate Fund has received pledges for $10 billion over its initial four-year period,[94] including a $3 billion pledge by the United States,[95] with a balanced share designated to support adaptation in vulnerable countries. Many states and members of civil society have celebrated this as an important accomplishment. However, this means that well under $1.5 billion a year—a relatively small amount when compared to developing country need—is likely to be delivered by the Fund to vulnerable states for adaptation, and there is no established mechanism for scaling up money moving forward. To date, most Northern donors have simply dodged these funds, preferring to pursue other channels.

And while the Clean Development Mechanism for a time provided a relatively steady stream of funding to the Adaptation Fund, the overall amount is still small, and there is great uncertainty about its future viability (see table 5.2). Similarly, of the estimated $5 billion needed to fully fund National Adaptation Programmes of Action in the LDCs,[96] donors have paid a mere $832 million into the LDC Fund. Meanwhile, more than $10 billion has been directed to the World Bank, particularly to its Climate Investment Fund and other agencies controlled mainly by the North.[97]

**Table 5.2**
2014 Status of UNFCCC/Kyoto Protocol Funding (in millions $)

| Funding source | Pledged | Paid | Disbursed |
| --- | --- | --- | --- |
| LDC Fund[a] | $879.8 | $831.5 | $726.25 |
| Special Climate Change Fund[b] | $331.1 | $299.1 | 242.26 |
| Adaptation Fund[c] | $223.57 | $205.53 | $69.1 |
| Green Climate Fund[d] | $10,000 | NA | NA |
| Total | $11, 434.47 | $1,336.13 | $1,037.61 |

a. As of March 31, 2014. Global Environment Facility 2014a.
b. As of March 31, 2014. Global Environment Facility 2014b.
c. Climate Funds Update 2014.
d. UN and Climate Change 2014b.

History shows that many such funds have been created in various multilateral processes, only to be abandoned by Northern donors.[98] While the steps toward the establishment of the Green Climate Fund have been widely celebrated as a victory, important questions remain unanswered about how much funding it will govern in the future. If indeed only 1 or 2 percent of climate funds continue to be channeled through UNFCCC and Kyoto Protocol funds, the creation of funding structures that reflect principles of adaptation finance justice may be largely hollow victories. In such cases, efforts to establish more just institutional funding frameworks may come at the cost of diverting attention from other goals, such as improving funding practices in institutions like the World Bank, which deliver the majority of climate funds. In the case of the Green Climate Fund, it is still an open question as to the value of these efforts.

Several civil society campaigners we have talked with hope that at a minimum, the Green Climate Fund will serve as a model for more justice-oriented funding practices for others to follow. Even the highly regarded Adaptation Fund has done little to address the justice issue of gender sensitivity, such as developing funding practices to account for the unique vulnerability of women to climate change impacts. Nor has this fund taken measures to ensure that the most vulnerable populations within a given nation are able to effectively participate in adaptation planning and funding decisions. These are areas for which the Green Climate Fund could serve as a model for best practices.

**Runaway Warming**

In considering the issue of international adaptation politics, we believe it is important to consider high-risk scenarios of the IPCC,[99] of uncertain likelihood but where temperature rise may exceed important ecological tipping points, and thousands or millions of people become "climate refugees." The institutions built over the last two decades to raise and spend climate finance are based on the presumption that adaptation will be a process that can be managed. While we spin out a series of six scenarios in chapter 9, we pause here to consider what the impact of a world that is 4 degrees Celsius warmer would have on the effectiveness and equity of various forms of governance. This section looks beyond the UN climate regime, to consider other international and national bodies that will be confronted with issues related to vulnerability and climate change in the years to come. Are these governance bodies equipped to handle such a context? How will such dynamics affect relations between various peoples and states?

These are the questions that global communities are now just beginning to grapple with, since all assessments suggest that the target fixed by the UNFCCC community for a maximum allowable increase of up to 2 degrees Celsius temperature rise is quickly becoming unachievable. The latest trends for a number of indicators, such as warming, rising sea levels, and extreme weather have far exceeded the Intergovernmental Panel on Climate Change (IPCC) predictions made in 2007.[100] The 2012 IPCC special report on extreme weather events warns that "a changing climate leads to changes in the frequency, intensity, spatial extent, duration and timing of extreme weather and climate events."[101] Even normally staid institutions like the World Bank and the International Energy Agency in recent reports have delivered warnings of runaway climate change, such as an increase of 4 degrees Celsius to 6 degrees Celsius and its dire consequences.[102]

How soon could we see a world that is 4 degrees Celsius warmer? The 2014 White House Report on Climate Change links greenhouse gas emissions to unambiguous climate changes in the United States.[103] The World Bank asserts that even with the current mitigation pledges fully implemented, there is roughly a 20 percent chance of exceeding 4 degrees Celsius by 2100; if those commitments are not carried out, the Bank report says, a warming of 4 degrees Celsius could occur as early as the 2060s. The report explains that "such a warming level and associated sea-level rise of 0.5 to 1 meter, or more, by 2100 would not be the end point: a further warming to levels over 6°C, with several meters of sea-level rise, would likely occur over the following centuries."[104]

The Bank report warns of the consequences: "No nation will be immune to the impacts of climate change. However, the distribution of impacts is likely to be inherently unequal and tilted against many of the world's poorest regions, which have the least economic, institutional, scientific, and technical capacity to cope and adapt. ... It is likely that the poor will suffer most and the global community could become more fractured, and unequal than today."[105] Though the Bank report emphatically urges that such a grim reality must not be allowed to happen, given the absence of political will, it may come to pass. In such a world, how are global communities likely to adapt? Can they do so?

In any case, there are likely to be disruptive migrations.[106] Sub-Saharan Africa and South Asia, with the highest concentrations of poverty, are likely to be affected most in such a warming world. Inequality within and between countries, already on the increase, will intensify, presenting a grimmer world from regional and global security points of view. There may be as many as 200 million people around the world displaced as climate refugees by 2050.[107] Whatever the number of displaced people, because of a causal chain in climate change impacts, there will be substantial declines in agricultural productivity, intensification of water scarcity, political instability, and social tensions. As a result, people will be forced to leave some places. The UNFCCC agreed-on text at Doha in 2012 urges further understanding of the "patterns of migration, displacement and human mobility."[108]

Opinions differ about the impacts of such migrations: some argue that it is not good for the origin countries because they lose precious human resources and brain power.[109] Others argue that such migration can take the pressure off the limited or stressed resources of the sending region. Furthermore, securitization of migration and its handling by stricter immigration laws and militarization of borders are not likely to serve well the issue of migration as an adaptation strategy.

Importantly, the existing Refugee Convention does not accommodate the undefined status of climate refugee. The principle of nonrefoulement in human rights law, that is, prohibition of forcible return of climate refugees to their place of origin, does not yet have the power of enforcement, though it is evolving.[110] Neither of the two UN Conventions on Statelessness, the 1954 Convention Relating to the Status of Stateless Persons and the 1961 Convention on the Reduction of Statelessness, can provide protection for climate refugees.[111] The problem is that the status of climate refugees has not yet been clarified in international law, so analysts argue for expanding the existing Refugee Convention or creating a new one specifically for the purpose.[112]

The likelihood is that with intensification of climate change impacts, the UN Security Council could take up the issue of linkage between climate change and security. It has already hosted three debates on the issue since 2007, but there is great disunity among the UN member states about its role. Of the five permanent members, France, the United Kingdom, and the United States support the involvement of the Security Council in handling climate change issues, while China and Russia, supported by other major developing countries, oppose it. Those opposed make the argument that climate change issues are disruptive of sustainable development efforts, and hence they should be discussed in the UN General Assembly, the UNFCCC, or the Economic and Social Council, where developing countries have a majority voice.

Another issue of a runaway warming world is the plight of low-lying island states, many of which are likely to disappear under rising seas. This is the reason the island nation members of the negotiating group AOSIS have been active both within and outside the UNFCCC process to pursue their case for relocation with rights in some new locations.

In the eventuality of the disappearance of states such as the Maldives, Kiribati, and Tuvalu, novel problems, such as displacement of whole countries, their relocations, statehood without territory, or the management of their vast exclusive economic zones (EEZs) are likely to shake up the legal world. This poses fundamental challenges to the zealously guarded sovereignty-based world system of states.

For example, the twenty-two Pacific island small states have a land area of about 90,000 square kilometers, but they command an EEZ of over 27 million square kilometers.[113] The tiny state of Kiribati, with about 100,000 people and a land area of 811 square kilometers, commands an EEZ of 3.5 million square kilometers. If those states sink, what will be the status of those vast EEZs? Will the host states, which give refuge to those stateless people, have command over these economic zones? Or can the refugees living in another state command sovereignty over the EEZs?

Three additional issues need mentioning when we consider a world with runaway warming. First, existing bureaucracies are focused on the global response to humanitarian crises and natural disasters and are increasingly focused on disaster risk reduction. For example, the United Nations works in parallel with the World Bank, such as in the UN's Office for Disaster Risk Reduction and the Bank's Global Facility for Disaster Risk Reduction. Thus, the work of coordinating and carrying out climate adaptation is being spread across UN agencies, multilateral and bilateral agencies, and at the levels of nations, subnational states, local governments, nonprofit agencies, and NGOs. Yet there is no central coordination strong enough to ensure that the greatest needs are being met.

Second, private actors such as insurance companies are stepping in to govern climate change risk in new ways. Some are excluding coverage for areas of high risk, such as coastal and riverine floodplains. Others are taking the approach of requiring policyholders to meet technical standards in their facilities' construction. Some in the area of agriculture are requiring farmers to take certain measures to reduce their risk if they want access to insurance.[114] Government agencies are taking up some of these standards to reduce their risks of extending coverage to citizens or companies building in high-risk zones.

And third, there are geopolitical tensions among members of the Arctic Council with declining ice sheets over the Arctic.[115] Governance of the now-accessible region and its resources is contested. There are just five nations with direct shoreline on the Arctic Ocean. However, the Arctic Council has three other member countries and several other countries have designs on the resources of the region. The full world community would like to be part of the decision-making in the Arctic region, so these groups are considered exclusive.

**Plan B?**

If there is a failure to stop global warming through the UN process, another darker issue arises. Here we turn to an issue that has made remarkable headway in the academic and policy debate during the past few years: geoengineering. Geoengineering (or climate engineering) comes in two forms: carbon dioxide reduction (CDR) and solar radiation management (SLM).[116] Carbon dioxide reduction includes enhanced weathering of rocks, afforestation, liming oceans, large-scale production of algae, ocean fertilization, and direct capture of carbon dioxide from the air. Solar radiation management techniques include enhancing tropospheric clouds, reducing cirrus clouds, increasing the albedo of marine or terrestrial surfaces, shooting sulfate particles into the stratosphere, and deploying reflective mirrors in space to cool the earth.[117] We have little doubt that rafts of other ideas are being developed by individuals, companies, think tanks, military research groups, and substantial scientific research groups. The question is how to govern them.

As with any other kind of new large technological fixes, solar radiation management geoengineering is giving rise to a raging debate among the ethics community.[118] Supporters argue for its deployment, saying it is efficient, cost-effective, or necessary. Opponents argue the potential lock-in with a dangerous technology, with risks of no return, and ultimate failure in addressing the main issue, reduction of greenhouse gas emissions. Their rationale is that in effect, geoengineering is a means to give preference to transient or lesser values, such as consumption of fossil fuels over the need to come to terms with natural and atmospheric dynamics.

However, apart from the technological side of geoengineering, there are dire geopolitical and social consequences. Some ethics scholars argue that some kinds of CDR, such as afforestation, may go with nature, while SRM, such as reflecting back sunlight, is antinature and more dangerous.[119] While the carbon capture and storage (CCS) technology involves actions within a country, there are potential spillovers across boundaries because geoengineering involves actions over global commons. These actions are likely to bring benefits to specific countries or regions (for example reliable rainfall for growing crops), leaving others powerless or even inflicting more harm and more inequality. The poor and vulnerable countries and communities, as well as future generations, are likely to suffer more and bear more costs under a scenario of deployment of such technologies.[120] Geoengineering may lead to further uncontrolled emissions, requiring the continuous application of sun-blocking technologies.

Overall, this discussion shows that solving a policy problem as diabolical as climate change, where one witnesses a temporal and spatial separation of causes and effects, will be exceedingly difficult to mobilize, both inside and beyond the forum of the UN climate regime. Tremendous geopolitical challenges are likely to unfold in a runaway warming world. Both climate refugees and geoengineering have failed to be met with an adequate response by UN treaties, and private and national efforts so far are piecemeal and incomplete.

## Small Change

Building on our analysis in the previous two chapters, here we have argued that low-income and other particularly vulnerable states have been able to win certain concessions on adaptation finance from wealthy states. By leveraging identities of vulnerability and strategically mobilizing in coalitions to negotiate the conditions of their consent, these countries have effectively pushed the issue of adaptation to the center of the negotiations. However, we find that, taken together, the story of the Gap, the Wedge, and the Dodge reveals that thus far, the critical elements for financing for adaptation have not reflected the basic tenets of justice agreed on by parties in the UNFCCC decisions.

The story is rather simple and disheartening: most of those who are most responsible for climate change and capable of supporting adaptation actions have fallen far short of their obligations. Despite clear language in the Convention in which wealthy countries agreed they have a responsibility to provide adequate funding to developing countries to adapt to climate change, there remains an ever-widening chasm between funds that are needed and what has been promised and delivered. Wealthy countries have repeatedly pointed to the global economic crisis as justification for the inadequacy of funds. Yet they have continued to pour massive investments into conventional priorities, including military enhancement and, ironically, fossil fuel subsidies.

The problem is more insidious since this inadequacy in funding levels has contributed to tensions between developing countries over which ones should have priority to access scarce resources. This threatens to disrupt decades of solidarity among actors in the global South, particularly among low-income developing countries. That solidarity has historically been essential to the poorest countries' finding any leverage at all in the negotiations. And the widely celebrated initiative to establish a Green Climate Fund will do little to promote a country-driven, gender-sensitive, participatory, and fully transparent approach to adaptation finance, if, like other UNFCCC and Kyoto funds, adequate and predictable resources are not deposited in its vaults. A justly governed fund without sufficient money won't do enough to promote justice for the millions of people already experiencing adverse impacts of climate change.

In addition, we have argued that harmful levels of warming are already inevitable, and unless we take bold mitigation action, 4 to 6 degrees Celsius global average warming is well within the realm of possibility within this century. We face a potential flood of climate refugees, nations without territories above water, and a set of dangerous technical solutions that are currently ungoverned or are governed in highly unequal ways. The current fragmented governance system is hardly capable of effectively and equitably managing issues such as climate-induced migration, climate-related security issues, disappearance of states under rising sea levels, fragmented intergovernmental structures for disaster management, geopolitical conflicts over the thawing Arctic, and the role of insurance companies and private actors in adaptation. And there exist gaping governance, political, and social challenges related to large-scale technological attempts to engineer the climate.

To connect this back to our main argument in the book, the emerging world order that low-income countries face—including a rapidly warming climate, a historic global recession, the fragmentation of interests in the global South, a recalcitrant global hegemon, and a highly fragmented global governance system—has created both new opportunities and threats for these countries to adapt. Overall, in chapters 3 to 5, we have shown that given the particular structural conditions of the contemporary world order, states alone show little promise of arriving at a sustainable, effective, and equitable international climate treaty, or international governance of climate-related issues, more broadly. States experiencing the worst impacts of climate change are also the ones that have the least power in the negotiations and are particularly vulnerable to co-optation in the current historical context. However, these same states have strategically mobilized within this context to achieve significant, yet still insufficient, gains.

This concludes our focus on state coalitions and conflicts within the UN negotiations. In the next three chapters, we turn to nonstate actors, including business, civil society, NGOs, and social movements, in shaping international climate politics. We seek to understand the role of these actors in influencing international climate change policy in recent years, as well as their potential to become a positive catalyst for dramatic change moving forward.

# 6

## The Staying Power of Big Fossil

### Energy Lives Here

Welcome to Copenhagen. Imagine you are a delegate from Canada. You gather your agenda for the day, pick up a civil society–produced "*ECO*" newsletter about the negotiations, and head through the Bella Conference Center. Walking through the main doors, you are thrust into a room with a maze of booths of nongovernmental organizations.

As you rush to attend your first meeting of the day, you walk quickly past dozens of messages on signs calling for "climate justice," "systems change," "gender justice," "Indigenous rights," and "Repay your climate debt!" At other times, you encounter Indigenous peoples and allies running through the halls with linked arms shouting out, "No Rights, No REDD," referencing the need for rights protections in the newly proposed policy to combat deforestation.

You also pass by a number of display booths from renewable energy associations promoting next-generation technology. After your meeting, you have lunch next to a huge wall-size mural depicting society's shift to a green future. As you leave the negotiations in the late afternoon, you encounter the civil society organization CAN-International presenting the Fossil of the Day award to the state actor deemed most obstructive to progress in the negotiations. Today the award is gifted to your country, Canada, for arguing that the base year for future cuts in emissions should be changed from 1990 to 2006, thus weakening overall mitigation targets.[1]

But there is also something notable that you didn't encounter today as you walked around the Bella Center: public messaging by fossil fuel associations about the benefits of doubling down on carbon-intensive fuels. At face value, the Bella Center appears to be a small city seized from the hands of the dominant fossil fuel party, claimed by a revolutionary "climate justice" force.

However, mistaking the visual absence of the fossil fuel industry or the diversity of green business and carbon trading groups making their rounds as a shift in power would be a mistake. As you find out at your after-hours cocktail party, the fossil fuel industry has indeed bought its tickets to Copenhagen. Its lobbyists have diligently raised their core issues to you. A few of them have even invited you to dinner the next evening. More important, fossil fuel lobbyists had already done their legwork prior to the negotiations, at home, in the comfort of their familiar stomping grounds. The business and industry NGOs (BINGOs) are present at the negotiations, but their voice is more subdued than the groups calling for climate justice. They advocate for continuity, for "going slow," for preserving a stable and predictable business climate in order to avoid disruption to national and global economies. As a delegate, you already know what positions your bosses are willing to take. Sure, there might be a bit of flexibility in the details, but your job description was pretty clear prior to arriving in Copenhagen.

You also know that it would be a mistake to naively think the unprecedented visual presence of climate justice messaging at the negotiations is now a powerful force to be reckoned with. However, this does not mean that this development is without impact or that it doesn't have any influence on your thinking and that of your fellow delegates from the North on important issues. But your actions are constrained.

In the previous two chapters, we considered whether low-income states have the potential to dramatically shift the trajectory of the negotiations. We argued that short of the emergence of a far more powerful state solidarity movement rooted in the ideas of vulnerability and equity, this is unlikely; they are more likely to win certain small concessions in exchange for their consent to the status quo. In this and the next two chapters, we shift our attention to nonstate actors in the negotiations. We seek to understand the potential of business and civil society actors to transform international climate policy. We broadly discuss business and environmental nonstate actor engagement in the climate regime, how their efforts have changed over time, and insights from the literature in this area. The goal is to understand how they each influence humanity's response to climate change, through the UN process and back home in their countries of origin, and the potential for them to be a catalyst for more ambitious and equitable climate action.

In this chapter, we argue that even with a recent diversification of business interests in the negotiations and the emergence of transnational carbon trading coalitions,[2] old school obstructionist fossil fuel companies still hold the most sway over states in the negotiations. This conclusion will not surprise many people. We begin this chapter with a discussion of the scholarship on business influence in the climate negotiations. A major thread of this literature points to the importance of the diversification and fragmentation of business actors in the negotiations, which, the argument goes, has diluted the influence of obstructionist fossil fuel companies. Some have gone so far as to argue that a new reality has emerged based on the logic of profitability that is inexorably driving companies to implement practices that are more responsible to people and the planet.[3]

Next we discuss the history of business groups in the UN negotiations during the first decade and then examine the fragmentation of business interests in the late 1990s and the rise of carbon trading. We consider the impact of carbon trading and the greening of business in the negotiations to challenge the dominance of fossil fuel industries, the primary contributors to climate change. Our argument is that carbon trading has not meaningfully weakened fossil fuel interests. Despite important gains for renewable industries in the past decade, fossil fuel industries still receive overwhelmingly higher levels of state subsidies and overall far higher investment than renewables while consistently achieving record profits. This is all despite a struggling global economy and governments everywhere searching for revenues to plug yawning gaps in their budgets. Moreover, the global price of carbon and the carbon markets themselves have floundered, indicating little willingness of governments to aggressively regulate greenhouse gas emissions from fossil fuels. There are, of course, notable exceptions, but they are few.

We identify and discuss three main processes by which fossil fuel interests have remained dominant in this context despite national and international efforts at mitigation policy. First is the carbon trading diversion, a powerful myth implicitly propagated by carbon coalitions that mitigation is possible without aggressive near-term behavior change in industrialized countries away from fossil fuel development. This has enabled a debate focused on sustainable *projects*, while leaving powerful *interests* largely untouched. Second, tacit power points to the fact that the structural advantage and political capital of powerful and established industries make it challenging to uproot them in domestic contexts. Fossil fuel lobbies have been extremely aggressive and effective in obstructing domestic climate policy and shaping it in their interests. Third, the business boomerang points to how fossil fuel lobbyists engage at the UN Framework Convention on Climate Change (UNFCCC), as a hub for decision makers around the world, to build far-reaching social capital needed to influence domestic policy and maintain stable markets for their core products. This process is then instrumental in shaping states' international climate positions. This is especially important, we argue, in an era when new fossil fuel extraction technologies such as hydraulic fracturing and horizontal drilling are rapidly shifting the geopolitical energy landscape.

In the chapter conclusion, we argue that while a transnational capitalist class that defines its interests in putting a price on carbon is emerging, this class remains weak and will always be unsuccessful in mitigating climate change unless it is willing and able to aggressively challenge fossil fuel interests. Simply put, carbon markets mean little in terms of mitigation unless they translate into trillions of dollars in lost profits for fossil fuel companies.

**A New Business?**

Nonstate actors were instrumental in founding the UNFCCC, and they have been active since its inception in 1992. During the early years of the negotiations, these actors were grouped into two main constituencies: business and industry NGOs (BINGOs) on the one hand and environmental NGOs (ENGOs) on the other. The rules of the COP established accredited NGOs as observers; by invitation of the president of the COP, they have the right to participate in proceedings that are deemed relevant to them,[4] as well as some private meetings.[5] Unlike state representatives (called Parties), observers do not have the right to vote and cannot gain access to all parts of the negotiations.

Irja Vormedal identifies three main determinants of influence of business groups in the negotiations.[6] First, a key variable is the access that these groups have to national delegations and informal consultations with influential actors such as the UNFCCC secretariat. Second, resources that actors bring to the negotiations are critical. These include intellectual capital, technological and technical expertise, administrative assets to navigate the complicated UNFCCC processes, and financial assets to organize membership and manufacture networking opportunities such as cocktail receptions. A third factor is the strategic engagement of groups in leveraging their resources and access. Activities include coordinating constituent interests in daily meetings, facilitating events to raise attention to their issues, engaging in the drafting of legal texts for contact groups, and networking effectively with state delegations.

In the case of the coal industry, Vormedal argues that the industries' technological ownership, innovation capacities, and know-how—what he calls technological power—have been particularly important for BINGOs in this area to influence the negotiations in their favor.[7] In addition, industry groups have indirect influence through national policymaking—what has been termed "tacit power" in the negotiations. This means that state government representatives, who are structurally dependent on private sector profitability, may anticipate resistance from powerful business and related interests at home to initiatives that threaten established industries.[8] They may self-censor or temper their personal desire to act on climate change for fear of igniting a firestorm back home, or they may define their interests narrowly based on a mutually constituted understanding of the problem at hand.

It's generally understood that business NGOs have a structural advantages over environmental NGOs in the negotiations. For example, in terms of financial resources, in 2011, the International Emissions Trading Association, an industry lobby pushing for market mechanisms in the treaty, had a lobbying budget roughly two to four times the size of the largest civil society network, the Climate Action Network (CAN).[9] However, one could argue that overall, the ENGOs have greater legitimacy because they are perceived as being driven by moral and collective concerns rather than financial interests.[10] It is thus not surprising that business interests that have teamed up with ENGOs in "Baptist-and-bootlegger" coalitions,[11] such as the case of coalitions for carbon trading, have been particularly influential.[12]

Simone Pulver argues that in the contemporary period, there is "a lack of coherence, consensus and unity in business positions, due to the conflict between those companies that see climate change action as a business opportunity and those that regard it as a liability."[13] She points to the fact that due to this rift, there is an inherent challenge in organizing business as a united front on the issue. She explains that the group that sees climate change as a business opportunity includes industries that anticipate a climate-driven shift away from coal and oil and toward the natural gas, wind, solar and energy efficiency industries.[14] Pulver also points to the challenges of industries even in the same sector, such as oil, to agree on a common political strategy. Thus, the argument goes, industry is inhibited in achieving maximum influence in the negotiations by severe disunity.

Others have posited that we are seeing the emergence of a new transnational capitalist class that identifies their interests with mitigating climate change or establishing carbon markets. For some, the emergence of such a business class is still at an "embryotic" or "patchy" phase.[15] They acknowledge that responding to climate change effectively requires decarbonizing the global capitalist economy "to restructure or dismantle huge economic sectors on which the whole of global development has been built."[16] Others take a rosier view of the change that is needed and suggest that it is already well underway, driven by smart businesses looking to protect their bottom line. As Hunter Lovins and Boyd Cohen rather optimistically celebrate, "A new reality, now recognized as 'the sustainability imperative' is inexorably driving companies to implement practices that are more responsible to people and the planet because they are more profitable."[17]

A broader group of scholars have observed or posited the emergence and dominance of pro-environment behavior in business. The theory of ecological modernization, which argues that market actors in modern societies are meaningfully incorporating ecological rationality into their decision making, has been debated for decades.[18]

Thus, it is worth asking: Has a transnational class that views climate change as a business opportunity emerged as expected? And does its engagement have the potential to shift the climate negotiations in a more promising direction? In other words, how meaningful is the new "multiculturalism" of business interests in the negotiations in terms of pushing us toward ambitious action to mitigate climate change? Are scholars like Lovins and Cohen correct that a sustainability imperative is driving businesses to dramatically shift their practices, or do fossil fuel interests still reign supreme? Moreover, does the literature adequately account for business influence, and that of fossil fuel industries in particular, in the contemporary context? We begin with a discussion of the early years of business engagement in climate politics.

## Early Days

Through the years, there has been a wide range of collective responses by companies and their officials to efforts to reduce emissions. After the near collapse of global environmental governance in the UN system after the 1982 Nairobi conference on the human environment (when business and environmental advocates butted heads as the oil crises and related economic concerns had taken the wind from the sails of the green movement), the Brundtland Commission in 1987 released its landmark report, *Our Common Future*.[19] Sustainable development meant many different things to different actors, but for the Brundtland group, it meant "development which meets the needs of current generations without compromising the ability of future generations to meet their own needs." Norwegian prime minister Gro Brundtland's commission proposed that sustainable development was possible and that business could be, and in fact had to be, part of the solution.

In the early years of the UNFCCC, including the first stage of negotiations of the Kyoto Protocol, the dominant form of business engagement was that fossil fuel companies and their representatives advocated against requirements for emissions reductions.[20] Scholars have argued that business actors had an early advantage in the negotiations over civil society with the appointment of Maurice Strong, head of a Canadian electric company and seen as pro-business, as the secretary general of the 1992 Rio Earth Summit.[21] During its first five years, a range of business groups participated in the negotiations; they represented various interests, from the relatively proregulatory Business Council for Sustainable Development, to the staunchly antiregulatory Global Climate Coalition, a group representing US and some European oil, coal, automobile, and chemical companies.[22]

Meanwhile, back in their home nations and between the big UN meetings, rather than let environmentalists form expectations for their behavior for them, firms banded together in industry groups that sought to shape the discourse about how they should be expected to act in the face of climate change. A number of trade groups and industry organizations created environmental stewardship programs, such as ISO 14000 and the chemical industry's "Responsible Care" initiative.[23] This was an effort with internal and external goals. After the terrible disaster at Union Carbide in Bhopal, India, the chemical industry learned that its public image was only as strong as its weakest link. Externally, the initiative included programs to regularize reporting and process documentation and, especially, to handle outreach to communities where environmental exposures were a concern. After the wave of environmentalism in the 1980s largely focused on the products and by-products of big chemical and manufacturing companies, these kinds of corporate social responsibility (CSR) programs were in place by the time concerns about climate change arose in the early 1990s.

Individual firms also produced CSR reports, which increasingly included information on total greenhouse gas emissions reduced and initiatives taken to improve energy efficiency. The message of most of these initiatives was to tell publics and regulators that a new wave of climate-related regulations was not necessary because individual firms and industry organizations were capable of self-regulation. In some cases, these reports downplayed environmental risks and far overstated firms' improvements in performance on the basis of emissions of climate pollutants. Critics called these kinds of reporting and lobbying efforts "greenwashing."[24]

To paint all firms and industries in all countries with that brush, however, appears unproductive. Some places and firms seemed to be taking environmental considerations into their core decision making, and some of these firms were profiting and gaining enthusiasm from workers and their communities and regulators for sharply reducing waste and risks to workers and communities. In Holland, environmental sociologists Gert Spaargaren and Arthur Mol called this "ecological modernization" and sought to describe the conditions under which it was occurring.[25] Debate has swirled around whether the conditions for ecological modernization are limited to Northern Europe and certain industries, or whether they represent a transformation occurring in capitalism globally.[26] The case of climate change and fossil fuels must be a central one in that debate.

Still, when it came to international climate policy, the primary role of the fossil fuel industry in the early 1990s was one of obstruction. The Global Climate Coalition, the biggest coalition in the negotiations representing many of the most powerful corporations in the world, along with the International Chamber of Commerce and International Petroleum Environmental Conservation Association, largely oriented its engagement to obstruct and stall progress in the negotiations.[27] Other industry groups, such as the International Climate Change Partnership, advocated a more moderate position on greenhouse gas mitigation. As an industry group representing a broad range of manufacturing, some of these companies focused on manufacturing insulation, electronic control equipment, and other energy-efficiency technologies because they believed that they could benefit from a high price on carbon emissions.[28]

However, despite the plethora of industry associations at the time, the intransigent Global Climate Coalition held considerable leverage, in part due to its unwavering position.[29] In these early years, this coalition, representing mostly US-based companies, engaged primarily to exert its influence through the JUSCANNZ bloc (Japan, United States, Canada, Australia, Norway and New Zealand).[30] But as the pivotal negotiations in Kyoto approached, the unity of the Global Climate Coalition began to fracture.

## Fracturing Interests?

Beginning in 1996, the position of the Global Climate Coalition divided in the international negotiations when industry actors, including British Petroleum, DuPont, and others in the International Climate Change Partnership, teamed up with the environmental organization Environmental Defense to advocate market-based regulatory strategies in the Kyoto Protocol.[31] In forming this alliance, Environmental Defense broke from the long-held commitment to a regulatory approach, described by critics as "command and control," held by the civil society group the Climate Action Network. As a result, at the time, Environmental Defense informally withdrew from the Climate Action Network and no longer participated in its daily strategy meetings.[32] Later, the Climate Action Network's position would shift to working to improve carbon trading, while still advocating for strong internationally binding targets for nations. Environmental Defense has since reengaged in the Climate Action Network's processes.

The business-environmental NGO alliance of Environmental Defense and the International Climate Change Partnership, along with the Pew Center, had an instrumental role in ensuring that the Kyoto Protocol top-down commitments were complemented with a market mechanism for offsetting emissions in industrialized countries.[33] The commitment to free market principles advocated by this alliance was embraced by President Bill Clinton as a middle ground between hard-liner and polarized environmental and business camps and was pivotal in ultimately shifting the EU's position to compromise on a pro-trading regime.[34] As Jonas Meckling argues, "Business was not able to prevent mandatory emission controls—the initial preference of the corporate mainstream—but ... it was able to influence the regulatory style of climate politics by building momentum for carbon trading."[35] The new NGO-business coalition effectively did the job of finding a compromise position that policymakers quickly adopted.

In this way, the negotiations in Kyoto in 1997 produced a central market mechanism, the Clean Development Mechanism (CDM). The CDM was established to allow wealthy states a means to achieve emissions reductions at a lower cost. They did this by purchasing carbon credits, certified by the CDM, from project developers in developing nations. The CDM was also intended to encourage the participation and investment of private sector actors in developing countries.

As the UN-approved CDM market grew, a whole market in voluntary emissions reductions (VERs, or carbon offsets) also arose, in response to demand from customers that companies address their carbon footprint. The market grew explosively on both the business side for entrepreneurs and young professionals looking for careers and for customers and firms looking to offset their emissions. It seemed that every plane flight and conference offered customers and attendees the opportunity to "offset" one's impact on the climate by making a small contribution to companies that promised to plant trees, distribute improved cook stoves or invest in renewable energy in developing countries.

Thus, in the post–Kyoto period, there was a shift in business representation "towards the green(er) end of the spectrum, reflecting a more general shift in corporate strategies from opposition towards more accommodative and constructive approaches to climate change mitigation."[36] This included a broad spectrum of industries, ranging from business groups promoting "clean coal," to carbon markets, to renewable energy and efficiency. Reflecting the institutionalization of a market-based strategy to reduce emissions—an approach that has become hegemonic in the climate regime—the largest represented industry group in the negotiations for over a decade has been the International Emissions Trading Association, launched in 1999. However, efforts of fossil industries to undermine confidence in climate science and policy continued, carried forward by conservative think tanks, especially in Washington, DC.[37]

In the early 2000s, Benito Müller of the Oxford Institute for Energy Studies argued that the insertion of carbon trading into the Kyoto Protocol was building a "new class," in whose interest it was to have a strong set of binding targets for nations to meet in reducing their carbon emissions.[38] The markets, he argued, required strong downward pressure on the number of total tons of emissions, to drive and sustain prices at viable levels. But to what extent has the emergence of carbon coalitions and carbon trading, and the diversification of business actors involved in domestic climate change politics, facilitated the emergence of a powerful class that identifies their interests in mitigating climate change, and challenging fossil fuel interests? In the next section we make the case that fossil fuel interests still reign supreme despite the fracturing and diversification of business interests within the negotiations.

## Marketing Diversion

While it is true that in the last decade of the Convention, we have witnessed a notable diversification of business engagement and positions in the negotiations and the fracturing of the obstructionist Global Climate

Coalition, the impact of the diversity of business interests in relation to climate change policy has mostly been overstated. In particular, the emphasis on the fragmentation and diversification of business interests by leading observers usually glosses over the reality that obstructionist fossil fuel companies still hold the most sway over states, within and, especially, outside the negotiations.

The evidence we can muster for this claim is fourfold. First, there continue to be overwhelmingly disproportionate state subsidies, and overall investment, to fossil fuel industries compared to green industries. In 2012, for example, the International Energy Agency estimated that fossil fuel industries (oil, gas, and to a lesser extent coal) received 500 percent larger subsidies than renewable energy did.[39] In the same year, the chief economist of the International Energy Agency, far from a radical organization, described fossil fuel subsidies as "a hand brake as we drive along the road to a sustainable energy future." And fairly incredibly, he went on to say, "Removing them would take us half way to a trajectory that would hold us to 2 degrees C [in global average temperature change]."[40]

The International Energy Agency says that just thirty-seven governments spent $523 billion on artificially lowering the price of fossil fuels in 2011. The NGO Oil Change International points to an overall figure of at least $775 billion, and potentially over $1 trillion, in fossil fuel subsidies in 2012.[41] So our estimates of fossil subsides vary by almost an order of magnitude, from $500 billion to $1 trillion, but the point is clear: this industry is benefiting wildly from government support around the world. We need to understand these subsidies and how the fossil fuel industry keeps the subsidies coming.

What's stopping competitor businesses from achieving a level playing field? Clearly if a transnational business class existed that was powerful enough to lead us to bold action on climate change, they would at least be raising the issue of disproportionate state subsidies in favor of their competitors. But by and large, they're not. Rather, it has been mostly more radical NGOs such as Friends of the Earth International, Sierra Club, and Greenpeace that have called for the removal of these subsidies. Recently international organizations such as the International Energy Agency, the International Monetary Fund, and the World Bank have also made statements about the need to remove these subsidies.

There are indeed signs of progress on renewable energy. From 2004 to 2014, despite the global recession, overall global investments in renewable energy (both public and private dollars) increased fivefold. However, this trend masks the fact that 2012 and 2013 saw significant dips in global renewables investment. The largest area of growth has occurred in developing countries, which now have nearly as much investment in renewable technologies as countries in the North do. Most of this growth is in China, which in 2013 accounted for 60 percent of developing country renewable energy investments. Total world investments in 2013 in renewable energies were $214 billion.[42] While this is promising, especially combined with dramatically lower prices for solar and wind energy production, it is still outmatched by fossil fuel investments, which were $270 billion in 2013.[43] Sociologist Richard York recently found that even in countries with major growth in renewable energy, fossil fuel use is continuing to grow.[44] York's sobering conclusion is that renewables only supplement rather than displace fossil energy.

Meanwhile, fossil fuel companies continue to reap enormous profits, with limited regulations on their extracting and bringing to market the very product that is upsetting the stability of the global climate system. Between 2001 and 2013, the top five oil companies combined made well over $1 trillion in profit.[45] (Notably, there were dips in their profits in 2012 and 2013 since the banner year of 2011.) Still, this level of profit ($93 billion in 2013 alone) is remarkable given that during this period, the Intergovernmental Panel on Climate Change released four subsequent reports linking continued fossil fuel use to increasingly catastrophic and scientifically robust climate change scenarios.[46] This period also included the Deepwater Horizon oil spill disaster by the company British Petroleum in the Gulf of Mexico in 2010. There is a sharp disconnect between knowledge and the behavior of global society as a whole, which needs to be understood.

Even the coal industry, which has experienced a series of notable blows in the last few years, continues to grow rapidly. As discussed in chapter 2, there have been notable developments with both private and public investors being less willing to finance the development of new coal-fired power plants. Major international development banks such as the World Bank, the European Bank for Reconstruction and Development, and the European Investment Bank have all recently adopted policies to severely limit their coal lending. The US and UK governments have adopted similar policies for their development agencies and operations abroad. Universities such as Stanford, at the behest of student activists, have recently taken up policies to limit their investments in coal.[47] Finally, there are indications that private banks such as Wells Fargo and JP Morgan Chase are cutting ties with the coal industry.[48] Nevertheless, a 2014 International Energy Agency report anticipated that coal demand will grow at an average rate of 2.1 percent per year through 2019,[49] assuming that there won't be massive public push-back in the coming years against the burning of coal.

Third, there has been the general floundering (some would say collapse) of global carbon markets. After peaking in 2011, the value of global carbon market transactions plunged 36 percent in 2012.[50] The plummet has been so bad that the World Bank has now stopped doing public analysis to provide an overall figure reflecting the state of the market.[51] Without a clear future for the Kyoto Protocol, the only global trading mechanism, the CDM, was reported in dire need of rescue by the assessment of a UN panel due to a more than fivefold collapse in the price of credits.[52] The much-anticipated UN carbon market for the purpose of forest conservation (REDD+) has not developed into a functioning market. The EU Emissions Trading System is the largest carbon trading market and the only one with a foreseeable future demand, but it too has seen the price of carbon permits plummet. In January 2013, carbon permits dropped to a record low in price when the European Parliament rejected a plan to prop up prices. Permits in this system have lost 85 percent of their value since mid-2011.[53]

The collapse of the price of carbon is only one indication of the failure of carbon markets. There have been other critiques leveled at the effectiveness and impacts of the carbon trading approach to mitigating climate change. Some critiques are related to disagreements on the likely climate benefits of carbon offsets, particularly around the measurement of what's called "additionality." To verify that a project is "additional" and thus eligible for financing, the CDM requires a counterfactual—that pollution would have occurred without CDM financing. The concept of additionally poses an important question: Were real emissions avoided by the funding of this project? In fact, critics argue that many projects would have likely taken place without carbon credits.[54]

There is also much debate about the ability of consumers to judge the difference between real and bogus carbon offsets. One researcher recently discovered a project developer that simply cut and pasted large chunks of documentation from one project to another.[55] Many observers have questioned whether unscrupulous project developers could sell the same carbon "reduction" to several buyers. A WikiLeaks cable from Indian government officials said that none of the CDM projects in India can be claimed as reducing emissions from what would have been the case without such credits.[56] Others have argued that in the case of a highly potent greenhouse gas HCF-23 (a waste byproduct of the refrigerant gas

HCFC-22), the CDM has produced perverse incentives in China for industry actors to generate *more* of the gas, in order to then receive CDM funds for its disposal.[57]

The difficulty of proving that CDM projects actually result in emissions reductions (as compared to a hypothetical case without such spending) has left developers—whether searching for places to start tree plantations, cogeneration plants, microhydro, or to install equipment on chemical factories around the world to capture emissions—facing comparisons with the Wild West as "carbon cowboys."[58] This question of additionality leaves the program open to powerful critiques on the validity and value of carbon offsets from a scientific perspective, conflicts of interest inherent in the certification process, corruption, and evidence of manipulation of the data.[59]

Recently the UN's Environment Programme released a report directly acknowledging the risks to carbon markets from corruption and seeking a middle ground of better accountability, transparency, and regulation as a way to protect those markets.[60] In addition, green-labeling groups have formed to help consumers judge the value of their carbon emissions reductions. For example, as confidence began to erode in the measurable benefits of carbon offset projects, the World Wildlife Fund and over eighty civil society groups including Care International and Mercy Corps joined with governments and corporations such as H&M, DHL, Nokia, and Virgin Atlantic to create the "gold standard" designation for voluntary emissions reductions.[61]

In addition to questions of the effectiveness of carbon trading, there is also a somewhat uncomfortable but fundamental moral question of relying on markets to handle such a core part of the task of reducing emissions. Ethical issues raised by stakeholders include critiques of allowing the wealthy to evade their responsibilities by buying "indulgences," the morality of commodifying the natural environment, and distributional justice issues associated with emissions trading.[62] A related critique is that carbon offset projects have resulted in the displacement of communities or livelihoods of vulnerable peoples, an issue that we return to in chapter 8.[63] One reason for this is that despite having an objective of promoting sustainable development, the CDM only rewards a project's ability to mitigate greenhouse gas emissions, but does not consider other positive or negative environmental and social impacts.[64] The piecemeal approach focused on generating offsets in isolated projects has also been critiqued as incapable of making a significant contribution to sustainable development.[65]

Others argue that in addition to being scientifically unsound, carbon offset markets actually encourage more greenhouse gas emissions by cultivating a sense of complacency. In a viewpoint published in *Nature*, scholar Kevin Anderson of the Tyndall Centre writes, "Offsetting, on all scales weakens drivers for change and reduces innovation towards a lower-carbon future. It militates against market signals to improve low-carbon travel and video-conference technologies, while encouraging investment in capital-intensive airports and new aircraft, along with roads, ports and fossil-fuel power stations."[66] Although such a critique is certainly debatable, it raises a crucial question: Having chosen carbon markets as the primary mechanism for mitigating climate change, what have been the opportunity costs of not pursuing other political or technical strategies given the major failures of this strategy to date?

The failures of carbon markets are often chalked up to a series of technical failures: the initial overallocation of permits, a global recession that decreased demand (and thus led to an oversupply, driving down the price), or a series of loopholes or glitches in the design of the program (i.e., the failure to negotiate a strong second period of the Kyoto Protocol in Copenhagen). However, carbon markets, and carbon offsets in particular, have no chance at becoming a viable and adequate mitigation strategy when they are not part of a comprehensive system that takes action to put firm limits on fossil fuel use, which have never been set.

The carbon market approach to mitigation has largely ignored the evidence recently brought to public consciousness by environmental writer and activist Bill McKibben, who has argued that addressing climate change necessitates preventing the extraction of most of the fossil fuel reserves already on the books of major corporations. McKibben, citing work by Malte Meinshausen and colleagues in the prestigious journal *Nature* in 2009,[67] argues that in order to stay within the safe level of emissions, we have to leave four-fifths of known fossil reserves in the ground.[68] Notably, major international institutional bodies such as the International Energy Agency have made claims that are only moderately more conservative.[69] Another important study, by Richard Heede found that only ninety investor-owned and state owned companies produced nearly two-thirds of all greenhouse gas emissions since the beginning of the industrial age.[70] This has brightened the spotlight on the major fossil fuel companies in their role in contributing to climate change.

However, keeping these reserves in the ground would come at a tremendous cost to fossil fuel companies, their shareholders and the states where the largest share of proven reserves reside. In their book *The*

*Burning Question* (2014), Berners-Lee and Clark estimate that at a price of $100 a barrel, currently proven reserves of oil add up to more than $170 trillion, or the equivalent of more than two years of global domestic product. And this figure does not include reserves in natural gas and coal, as well as promising-looking unconventional reserves, which they argue might double or triple this figure.[71] Perhaps not surprisingly the states with the largest proven carbon reserves underground are the United States, Russia, and China,[72] three countries that have played a largely obstructionist role in international climate negotiations for more than two decades.

The collapse of prices in carbon markets can largely be attributed to the unwillingness of the major emitting countries of the world to adopt obligations to cut emissions that would harm development of fossil fuels.[73] This failure is indicative of the unwavering influence of the fossil fuel industry to ensure that states adopt the position that there is no real alternative to reliance on fossil fuels for economic development. Overall, this suggests that while there have been some positive trends, such as the global growth in investments in renewable energy, Lovins and Cohen overstate the proposition that a "sustainability imperative" is inexorably driving companies to adopt behaviors that protect people and the planet. There is little to no evidence to suggest that the influence of fossil fuel industries has been weakened in the negotiations or elsewhere. But what explains fossil fuels industries' continued dominance?

**Staying Power**

Building on the literature about business influence in the negotiations discussed above, we argue that three processes have been key to maintaining fossil fuel dominance in international climate policy. First is the carbon trading diversion, which refers to the fact that while perhaps not necessarily an inherently flawed approach, carbon trading has been leveraged as a delaying tool by fossil fuel companies and intransigent states to distract attention from more substantive measures to curb emissions that would involve their losing business in the short term.

At the international level, the focus on carbon trading and offsets has shifted attention to the types of projects that are capable of reducing emissions individually and away from robust caps and other measures that directly challenge fossil fuel interests in maintaining existing practices. An enormous amount of time and energy have been spent designing, debating, implementing, and fine-tuning emissions trading regimes such as the EU Emissions Trading System, the CDM and REDD+, and national programs such as the Regional Greenhouse Gas Initiative and the California cap-and-trade program in the United States.

In practice, the focus on trading mechanisms and offsets has upstaged the focus on a cap or ceiling to emissions. A truly adequate cap would put us on a trajectory of deep reductions to stay below a 1.5 degrees Celsius global average temperature rise. However, we are nowhere near achieving even the less ambitious 2 degrees Celsius target. Moreover, the focus on offsets and trading regimes has shifted attention to reducing emissions in the global South, where reductions are cheaper, while often offering a relatively free pass in the North, where such reductions are more costly.

To be fair, certain NGO actors in carbon coalitions have been very concerned about securing a cap on global emissions. However, the overall debate on carbon trading has conveniently obscured the fact that even a highly active carbon market does not necessarily mean a net reduction from business-as-usual emissions, particularly from fossil fuel sectors, without robust regulations driving emissions down sharply in the immediate and medium terms.

This is not surprising, given that carbon trading coalitions have included energy giants such as British Petroleum and the utility Duke Energy that have an enormous financial stake in preserving fossil fuels development, whether it be oil, coal, or natural gas. Carbon trading coalitions often assume the strange position of lobbying for climate change policy that doesn't do too much to adversely harm the industries that are primarily responsible for causing climate change. For example, at the national level in the United States, the 2009 Waxman-Markey cap-and-trade bill, supported by the business-NGO alliance the Climate Action Partnership, contained large loopholes and giveaways for the coal and oil industries and key states with large coal use. Concessions in the bill to the coal industry included free emissions permits, billions in subsidies for carbon capture-and-storage technology development, and a decade of delayed action without sharp emissions requirements.[74] The bill would have stripped some of the Environmental Protection Agency's authority to regulate climate pollution under the Clean Air Act. It is also not surprising that despite a less obstructionist approach of fossil fuel industries based in the EU to climate policy,[75] the EU emissions trading system is coming under increasing scrutiny for failing to penalize polluters.[76]

Second, the unrivaled welfare for the fossil fuel industry, and its enduring dominance, is no doubt connected to its lobbying presence in powerful nations and the perceived need by these states for investment in fossil fuels to sustain their economic growth. This is what Irja Vormedal refers to as "tacit or "structural power."[77] For example, Nick Campbell of the International Chamber of Commerce recently told a reporter with the Center for Public Integrity, "The only way you really get leverage [in the UNFCCC] is if you can convince a delegation at home that it's in their interest to have their instructions say this or that."[78] Not surprisingly, prior to the negotiations in Copenhagen, more than 1,150 companies and advocacy groups deployed 2,810 climate lobbyists in Washington (five lobbyists for every member of Congress) to fight federal cap-and-trade climate legislation—an increase of 400 percent from six years prior.[79]

As discussed in chapter 2, fossil fuel industries have strong ties with the UN Saudi Arabian delegation, as well as that of other OPEC and fossil trade–dependent states, which have had a large impact in obstructing progress in the negotiations for over two decades. The influence of fossil fuel interests over lead states such as the United States, Canada, and China is also considerable.

Third, what we call the *business boomerang* highlights that the concept of tacit power is perhaps oversimplified. This concept fails to fully account for the indirect and multilevel nature of fossil fuel influence in the negotiations. In particular, as a hub for decision makers around the world, the international climate negotiations are a key site for fossil fuel executives and operators to build far-reaching social capital needed to influence domestic policy and maintain stable markets. While many scholars have often portrayed domestic and international influence as separate processes, we see them as highly connected. Tacit power is a dynamic process through which business lobbyists engage in the UNFCCC negotiations in order to make stronger connections to use later for lobbying in national contexts, which in turn influences the international position on states. This is particularly important in the age of the global corporation, where industry actors can gain advantage by having strong relationships with decision makers in various key states.

Campbell and other industry lobbyists point to the amazing access to national delegates at the negotiations and the ability to form useful new relationships. One industry representative called it "loitering with intent."[80] As Exxon Mobil's Brian Flannery said, "You form contacts all over the world, people who you know will answer the phone because they know you. To me, that's tremendously valuable to be able to discuss."[81] Similarly, David Hone from Royal Dutch Shell explained, "Talking to the delegates opens doors for people back in Shell Brazil, who may then go and have a follow-up conversation." Thus, fossil fuel industry lobbying can also be understood to have a boomerang effect[82]: lobbyists engage at the UNFCCC to build the relationships needed to influence domestic policy, which is then instrumental in shaping nations' international climate positions.

Moreover, as we discussed in chapter 1, the global energy landscape is shifting dramatically. Countries that have never had access to large energy reserves are now sitting on top of huge newly accessible shale oil and gas deposits.[83] For the hungry oil company, forming new relationships with decision makers in unfamiliar contexts where there are new reserves to access is likely more important than ever before. In this light, the UNFCCC is an ideal networking opportunity, where shared interests in fossil fuel extraction can be discovered.

### Still Marketing

Where does all of this leave us? As climate change science has become increasingly difficult to dispute, fossil fuel associations has become less visible in the international negotiations. However, while there has been fragmentation and diversification in the approaches of different business actors in international climate politics, we do not see significant evidence that the obstructionist forces of fossil fuel lobbies have waned in power.

Rather, fossil fuel companies have not seen the need to mobilize the same resources since Copenhagen. This is because they have essentially succeeded in forestalling any threatening action by states at the national or international levels. The sharpest obstructionist climate change public relations firms and think tanks in Washington, for example, have seen their donations for climate work rise and fall depending on the level of threat they perceive within statehouses and UN negotiation halls.[84]

In the current context, renewable energy industries have made significant strides, but evidence suggests that fossil fuel industries remain unequivocally dominant. Fossil fuel industries still compete on a highly unequal playing field, subsidized by the very governments that negotiate international climate treaties. The shift of some in the industry to a carbon market approach, once admonished by some of the more obstructionist fossil fuel companies, has not proven a real threat to fossil fuel interests, and the biggest actors have continued to post record profits with no sign of slowing down.

Nor do we see evidence of an emerging and powerful transnational capitalist class with the ability to respond effectively to the climate crisis.

At least in its current state, these actors lack the might or will to push for policies that will keep fossil fuels in the ground. Thus, despite the diversification of business interests represented in the negotiations, fossil fuel interests still dominate.

Moreover, we have argued that the continued dominance of fossil fuel industries in international climate politics has been made possible by three processes: the carbon trading diversion, tacit power, and the business boomerang. Indeed, the change that is needed will not simply be designed by business coalitions with their eye on new markets. If financial interests drive business actions (and we believe that they largely do), then the existing investments made by fossil fuel interests are simply too great, their profits too astronomical, their instrumental and discursive power too substantial, and the diversity of their product too limited for them to undermine their own ability to hold on to this status quo. Very strong sentiments would be needed to overcome interests that great.

Notably, while the benefits of a carbon trading approach have come under intensified critique, there continues to be political energy spent introducing new carbon markets in places such as California, Australia, Japan, and Canada, as well as developing countries such as Brazil, China, Mexico, Thailand, Vietnam, and South Africa. We argue that the focus of these policies must shift from carbon trading markets to robust caps on greenhouse gas emissions, structures of accountability that limit loopholes and corruption, and dramatic shifts in industry and consumer behavior in industrialized countries, particularly in the United States and China where emissions must decline rapidly. In other words, the focus on trading must not come at the cost of diverting attention from regulating powerful industries.

However, businesses also need to be involved in combating climate change. New markets for technologies and practices that reduce greenhouse gas emissions are essential, and profits can be made by those able to produce them and get them to market. But perhaps never before has there been an industry with so much power and so much to lose through domestic and international policymaking. As Bill McKibben recently explained, effective climate policy will require conflict. There is too much profit at stake for it to be any other way.[85]

How does this connect to the main argument of this book? Climate change will not be solved by isolated groups of actors of states, business, or civil society. Rather, a neo-Gramscian strategic approach to power relations points us to look for the possibilities of a broad-based and diverse countermovement coalescing against fossil fuel interests, with the legitimacy and force capable of shifting the balance of forces on this issue. Moreover, such a coalition will have to leverage major tensions in the broader world order. Thus far, we have argued that such a power shift will not transpire through state-level politics, institutional design, or market-level incentives alone.

This brings us to the role and trajectory of transnational civil society inside and outside the negotiations, which we believe to be critical. But has civil society engagement during the two decades of the UNFCCC negotiations demonstrated any hope for it as an emancipatory force? In particular, has civil society undergone changes that have made it more of a force to meaningfully challenge fossil fuel interests and push states in a direction to implement policies that dramatically curb carbon emissions?

In the next two chapters, we argue that the defined interests and representation of civil society in the climate negotiations have largely fragmented in recent years. There are now three main sets of actors: professionalized NGOs and academics who continue to collaborate with states and market actors in the negotiations to work for pragmatic and incremental change, NGO and social movement actors that converge in the negotiations to push a more radical agenda but lack direct links to key powerful actors, and actors that have given up on participation in the negotiations (or never participated in the first place), choosing more local or grassroots strategies for combating climate change.

Overall we see limited coordination among these divergent approaches. Resources and links to power still rest overwhelmingly in the hands of professionalized NGOs taking a more reformist approach. As a result of the largely fragmented and unequal condition of climate civil society, we argue that those on the inside, while relatively flush in resources, have limited leverage to put pressure on states that would cause the transformational change needed, nor a commitment to a political project that fundamentally challenges existing power relations. Rather, they serve as narrowly conceived norm entrepreneurs, without real power for shifting the balance of forces in the negotiations, even as the world warms.

# 7

## Society Too Civil?

### A Cold March on Washington

On a cold February day in 2013 in Washington DC, 35,000 activists marched through the streets, calling on the Obama administration to reject a major pipeline and boldly take up climate action in his second term. Organized by the blossoming 350.org network led by writer Bill McKibben, the march was called "the largest climate protest in U.S. history."[1] (This was prior to the People's Climate March in New York in 2014, with more than 400,000 marchers). The Keystone XL pipeline would bring extremely high-carbon oil extracted from tar sands in Alberta, Canada, through the American heartland to refineries on the Louisiana and Texas coast.[2] Not unnoticed was the fact that President Obama wasn't even in Washington at the time: he was golfing in Florida with a pair of key oil, gas, and pipeline players from Texas.[3]

Senator Sheldon Whitehouse from Rhode Island spoke to the raucous crowd and afterward commented on the Weather Channel that "my experience, up the street in Congress, is that polluters own the place." He said that in order to address climate change, "You've got to create the political environment; that becomes not only important, but necessary."[4]

The march on Washington is only one of many climate protests worldwide in recent years. In Copenhagen at the tumultuous international negotiations, a major protest wound through the cold city streets with a crowd estimated at from 30,000 to 100,000 protesters, ending with hundreds arrested when police heard of plans for looting by anarchists at the center of the march.[5] Two smaller protests took place later in the week, when it became clearer that talks were not heading toward serious action on the issue. The next year, in April 2010, the World People's Conference on Climate Change and the Rights of Mother Earth was held in Cochabamba, Bolivia. Bolivian President Evo Morales was the catalyst for the event, which brought together what organizers estimated as 30,000 people to develop radical alternatives to the failures of Copenhagen.[6] In October, 350.org organized a day of climate action called a "Global Work Party," which included 7,000 events held in 188 countries worldwide.[7] At these events, mostly small groups of people took photos of creative symbolic actions calling attention to the climate crisis and pushing for the solidarity of people to take action worldwide. Then in September 2014, the People's Climate March took place in New York to coincide with UN Secretary General Ban Ki Moon's UN Climate Summit. The historic event, which brought together more than 400,000 marchers, coincided with hundreds of smaller marches around the world.

These are significant numbers of people, yet they are relatively small and widely dispersed numbers when compared, for example, with civil rights or anti–Vietnam War protests or, more recently, the protests across Europe and the world against the war in Iraq (estimated at between 6 and 10 million protesters in sixty countries on the weekend of February 14 and 15, 2003).[8] On the environment, the obvious comparison would be with the first Earth Day protests on April 22, 1970, when an estimated 20 million people were out in the streets at protests across the United States. Even twenty years later in 1990, the Earth Day Network reported that "200 million people in 141 countries" protested to raise the issue globally.[9] In comparison to these, mass mobilizations on climate change have been downright tiny.

So if not massive protest, what types of movements from civil society have we seen to combat climate change? What have been the strengths and weaknesses of different forms of activism? Where do we see the most potential for a catalyst for the change that is needed to take on this vast issue, particularly for realizing a fair and sustainable international treaty in the face of powerful fossil fuel interests seeking to stall or stop it?

We begin this chapter with a broad look at civil society responses to climate change. We identify and discuss three main approaches: environmental justice movements, "big green" advocacy, and corporate responsibility activism. We discuss the strategies of these approaches and the obstacles that prevent each from finding greater influence.

We then shift our attention to the role of civil society in the international negotiations. We seek to understand the changes that civil society has undergone in the negotiations in recent years and the potential for it to transform international climate policy moving forward. We discuss what scholars in this area have identified as the main conditions for civil society to find influence in this context. We then turn our attention to the role of civil society in the first decade and a half of the negotiations and discuss the subsequent fragmentation and diversification of civil society with the growth of the climate justice movement.

This history shows that the defined interests and representation of civil society in the climate negotiations have largely fragmented in recent years. There are now three main sets of actors: professionalized NGOs and academics who continue to collaborate with states in the negotiations to work for pragmatic and incremental change; social movement and NGO actors who converge on the negotiations pushing a more radical agenda, but largely without links to power; and actors who have defected from participation in the negotiations (or never participated in the first place), choosing more local or grassroots strategies for combating climate change. There is only limited coordination between these divergent approaches.

The literature on civil society in international climate change politics, we argue, has not fully explained why civil society has failed to influence mitigation action. We highlight three main deficits. First, despite the diversification of actors involved in the negotiations, resources and links to power still rest overwhelmingly in the hands of professionalized NGOs that take a reformist approach. Those on the inside, while mustering greater resources, have limited leverage to put the type of pressure on states that would cause the transformational change needed. Nor have they demonstrated a commitment to advancing a political project that fundamentally challenges existing power relations. Rather, they serve primarily as norm entrepreneurs, without real power to get truly transformational norms adopted. Aware of their limitations and the scope of their ability to make change, their tendency to make weak demands is not surprising.

Second, rather than a linear path of declining civil society relevance in the climate regime over time, as some authors have suggested, we argue that there have been hinge moments when there are opportunities for bold change. In the contemporary period, civil society has failed to develop a coordinated and viable strategy for building strength to realize influence at these key moments. Finally, in addition to the UN, various other international governance frameworks are relevant to action on climate change, including international trade regimes, financial institutions, scientific bodies, and a host of other fragmented and layered governance systems. Thus far, civil society has primarily devoted its attention at the international level to the UN climate processes while largely neglecting some other relevant venues and bodies.

## Divided We Stand

We discussed in chapter 1 that we believe that effectively combating climate change necessitates realizing a binding international treaty. However, when it comes to civil society, most of the activity around climate change happens in local and national contexts. Various scholars have argued that achieving an adequate international treaty requires shifting behavior in national contexts.[10] Thus, before engaging with the role of civil society in the UN regime, we begin our analysis with the role of civil society in countering climate change broadly at the local, national, and transnational levels. Here we focus on what we see as three main types of civil society responses to climate change: environmental justice movements, big green advocacy, and corporate responsibility campaigns.[11] We outline the strengths and weaknesses of each of these approaches and discuss the extent to which they collectively represent an adequate response to the problem at hand.

### Environmentalism of the Poor and Environmental Justice Movements

Numerous environmental justice struggles around the world are engaged in actions relevant to climate change. These movements often challenge false stereotypes that poor people and communities of color don't care about protecting the environment.[12] Some actions have emerged in isolated locations and focus on fighting particular polluting industrial projects such as coal mining, oil extraction, waste incineration, and industrial agriculture. For example, since 2012, protests have emerged in northeastern Colombia in response to pollution and the dislocation of communities resulting from mining operations in the area. These protests may have contributed to the national environmental licensing authority's denying several multinational mining companies new licenses to mine for coal in this region.[13]

Another example is an anti–land grab coalition between peasant organizations and civil society that has emerged in the Senegal River Valley in West Africa.[14] Some activists here are identifying their struggle as part of the global climate justice movement. Likewise, for decades, the U'Wa Indigenous peoples in Colombia and Ecuador have mobilized to keep oil companies from drilling on their land.[15] In Nigeria, a combination of violent and nonviolent actions has been instrumental in halting oil production. Even in China, where protest is readily suppressed, sustained protests have led to the cancellation of at least three proposed waste incinerators in Beijing, Guangzhou, and Wujiang.[16]

There are also environmental justice struggles that focus on achieving positive social and environmental alternatives to polluting projects. Such examples include activist struggles for zero waste composting, recycling, and waste reduction strategies in Buenos Aires; a coalition for cleaner shipping ports in Los Angeles; and a bus riders' union for improved public transportation in Boston. These are merely examples; protests regularly occur in all parts of the world, led by marginalized communities in response to environmental concerns. Many of these have implications for climate change, and an increasing number of them identify themselves as climate justice activists.

Other environmental justice movements have emerged as broad-based transnational struggles uniting various organizations and communities facing similar issues. For example, in 2013, an alliance of ten Canadian and US Aboriginal groups united to oppose tar sands oil extraction and transport. The issue is highly relevant for climate change, since tar sand extraction and refining is highly energy intensive. They have vowed to block three multibillion-dollar pipelines that are planned to transport tar sands oil.[17] All of the members of the alliance live near the Alberta tar sands or the proposed pipeline routes. This action builds from the "Idle No More" movement in which Indigenous groups blockaded rail and roads to protest poor living conditions and legal violations that their peoples have endured.

Another broad-based transnational movement, La Via Campesina, the world's largest federation of peasant and smallholder farmers, convened a march of thousands of peasant farmers and climate justice activists at the UN climate negotiations in Cancun in 2009, calling for climate justice. They argued that small farmers and villagers around the world are most affected by climate change, and the strategies that governments and transnational corporations have presented to the climate crisis, such as biofuels and transgenic crops, are "false solutions." As we'll discuss in chapter 8, the urban poor have also begun to organize in transnational coalitions to demand climate change solutions that value the work of waste pickers (informal recyclers) and protect their rights to their livelihood.

Environmental justice movements often raise the political and economic costs of projects, making it increasingly risky for corporations and states to invest in activities that further destabilize the climate. Their power resides in being able to physically halt or delay activities, leverage national and international laws, and challenge the perceived rightfulness of project activities and goals through engaging the media. In successful movements, there are often high numbers of participants, links with resourced NGOs, a clear goal, unity and resistance to being co-opted, willingness to endure trauma, longevity, and a communications strategy that shifts social perceptions of the project being challenged. It also may be helpful when groups claim identities that offer international recognition or assume legal status as groups with standing as affected parties in damage lawsuits.

These movements are often the only entities that are directly and meaningfully confronting fossil fuel interests and the drivers of deforestation. While it is impossible to know the rate of success of these movements in halting projects, it is safe to say that there are many failures in this regard. Radical grassroots movements often come at a high social cost. Due to the marginal status of movement participants, environmental justice struggles sometimes result in the death, displacement, or disenfranchisement of participants, particularly in places with limited legal or political recourse.

Local environmental justice struggles are also often complex, with tensions raised when some residents are dependent on employment offered by polluting industries, or identify with associated development possibilities. Some movements take NIMBY (Not In My Back Yard) forms and focus on preventing pollution or harm in one location, while others take a broader view of protecting the global environment (sometimes described as NIABY—Not in Anyone's Back Yard—or NOPE—Not on Planet Earth) and preventing climate change in particular. Transnationally linked environmental justice movements perhaps offer the best hope in this area as a force capable of shifting the political calculus of states and corporations rather than merely shifting the burden of pollution elsewhere. While there has been increasing coordination of social movements under the broad banner of climate justice, it would be premature to say that a global climate justice movement of the poor has emerged or that it is well coordinated and resourced.

Importantly, a limited number of linkages exist between well-resourced environmental and development-focused NGOs and grassroots movements. The vast majority of foundation dollars still concentrate on high-level policy advocacy, with very limited funding supporting efforts at change on the ground, where the impacts are experienced.[18] Without financial resources, and access to decision makers, the media, and legal support, environmental justice movements are often too easily marginalized, victimized, or ignored by powerful fossil fuel interests. However, such linkages, when not well grounded in principles and relationships of solidarity, can also serve to co-opt the radical goals of grassroots movements, as larger groups sometimes make compromises for local groups without their consent.[19] Unfortunately, decades of conflicts between NGOs and grassroots movements have led to some deficits of trust and few examples of effective collaborative movement building.

## Big Green Advocacy

In the United States, the majority of foundation dollars dedicated to fighting climate change has gone to supporting legislative efforts of the biggest environmental organizations in Washington DC.[20] Beginning in 2006, big green groups such as the Environmental Defense Fund, Natural Resource Defense Council (NRDC), and the Nature Conservancy teamed up with major fossil interests such as Shell, British Petroleum, Dow Chemical Company, and Duke Energy in the US Climate Action Partnership (USCAP). Groups such as the Environmental Defense Fund sought to repeat achievements that they had in the late 1980s of limiting sulfur dioxide emissions, by advocating for a cap-and-trade regulatory approach to address climate change. The groups believed that the most promising strategy was to team up in a coalition with the biggest polluters in order to agree on a compromise legislative solution. In this way, USCAP was successful, for the first time in the United States, in bringing forward a coalition of powerful actors in favor of climate change policy.[21]

From the beginning, environmental justice activists fiercely criticized this effort as detached from real movements for change and an example of the co-opting of the environmental movement by big business.[22] The environmental justice movement in the United States largely opposed efforts for cap-and-trade policy, arguing that it creates new forms of exploitation and environmental inequality, creates profits for polluters, and locks us into an inadequate framework for addressing climate change.[23] They highlighted a series of local impacts that cap and trade allows to remain in place, including failing to reduce greenhouse gas co-pollutants such as fine particulate matter, mercury, and sulfur dioxide and the possibility of allowing emission increases in some locations that would create enduring pollution hot spots. They also criticized the efforts of mainstream groups in Washington, DC, for largely excluding representatives of affected communities from their leadership and consultations.

In 2013, sociologist Theda Skocpol, while not acknowledging environmental justice concerns specifically, echoed critiques of the efforts of USCAP as based on "a mistaken assumption of how U.S. politics works." In particular, she argued that this movement underestimated the ideological divide in Congress with the emergence of the Tea Party and overestimated the ability to strike a deal without there first being a formidable grassroots mobilization. We agree that to achieve the substantive change necessary to address climate change, "reformers will have to build organizational networks across the country, and they will need to orchestrate sustained political efforts that stretch far beyond friendly Congressional offices, comfy board rooms, and posh retreats." Importantly, this will also require responding adequately to the concerns expressed by constituents. It will inevitably necessitate a willingness to take bolder and more confrontational positions in Washington, DC, that may not fit neatly into reformist and technocratic politics.

As the *Washington Post* pointed out in 2013, the big greens continue to lack racial diversity, even though communities of color and low-income communities are those most adversely affected by environmental pollution. As high-profile environmental activist Van Jones explains, "We essentially have a racially segregated environmental movement. …We're too polite to say that. Instead we say that we have an environmental justice and a mainstream movement."[24]

While still a largely white organization, the Sierra Club has recently made efforts to redefine its activities and budget priorities to support grassroots mobilizations to stop coal production with its Beyond Coal campaign. Other big green groups like NRDC have also complemented their often reformist-oriented Washington advocacy efforts with strong legal support for communities fighting polluting industries. Organizations such as Earthjustice (formerly the Sierra Club Legal Defense Fund) have pursued primarily legal approaches to addressing environmental and climate issues. Greenpeace has often served as a more aggressive voice in policymaking and has focused much of its strategy on communications and direct action to draw attention to its issues. Of the big greens, perhaps Friends of the Earth International has been the most connected to local struggles, seeking to bridge movement building with aggressive environmental policy.

Importantly, the predominantly "elitist" approach of the big greens hasn't been confined to the US context. For example, US conservation organizations such as Conservation International and World Wildlife Fund have long been criticized for not taking the needs of local communities into account in their work in the global South. In Europe, mainstream environmental organizations have been more effective in shaping and passing environmental and climate policy than in the US context. Examples include the chemicals policy REACH platform and the EU's relatively strong emissions reduction targets and its emissions trading scheme (ETS). After initial resistance, as with their US counterparts, mainstream environmental groups in the EU moved to advocating market-based approaches to curbing emissions.[25]

However, as discussed in chapter 6, the impact to date of the ETS on reducing greenhouse gas emissions is debatable, and the price of carbon permits has recently bottomed out, leading some to speculate on whether the system will collapse completely.[26] This approach has come under attack from more radical environmental groups in the EU and elsewhere. They have argued that the framework has left important loopholes for powerful polluters and has failed to establish a robust cap on emissions. Others have pointed to various forms of corruption in the emissions trading system.[27]

As a recent statement of over ninety civil society organizations states, "It is time to stop fixating on 'price' as a driver for change. We need to scrap the ETS and implement effective and fair climate policies by making the necessary transition away from fossil fuel dependency."[28] The floundering of the EU's ETS has important lessons for policy efforts moving forward in the EU and places such as California, Australia, Japan, and Canada, as well as developing countries such as Brazil, China, Mexico, Thailand, Vietnam, and South Africa.[29] The decades of work on this approach by mainstream groups in many nations raise questions about the viability of big green advocacy to adequately rein in global warming.

### Corporate Responsibility Activism

Various environmental organizations have focused on shaping the policies of corporations and their financiers, including banks, universities, and city governments. Their main strategies include communications campaigns targeting company branding, shareholder activism, and legal challenges.

Most prominent in this area has been the Rainforest Action Network (RAN), which has worked to push corporations such as Home Depot and Victoria's Secret to adopt more environmentally sustainable policies and practices. Most recently, RAN and other organizations such as Pacific Environment, and International Policy Studies, have organized to push major financiers such as Bank of America, the World Bank, the US Export-Import Bank, and the Overseas Private Investment Corporation to adopt certain safeguards or divest from fossil fuel development. For example, RAN has targeted Bank of America with a campaign demanding that it completely divest from bankrolling coal mining, infrastructure investments, and coal plants. It has organized protests at shareholder meetings; engaged in public relations smearing activities such as banner drops and posting signs on the bank's ATM machine screens; conducted civil disobedience resulting in arrests; and organized a coalition of investor and legal and environmental leaders to urge the bank in a letter to restrict its investments in coal.[30]

In another example of corporate responsibility activism, Friends of the Earth and other environmental organizations filed a lawsuit against the US Export-Import Bank in relation to its fossil fuel investments. The 2009 lawsuit settlement required the bank to develop a carbon policy. However, these groups deemed the policy that was developed to be completely inadequate for curbing fossil fuel investments in gas, oil, and coal.[31]

Other organizations have sought to hold corporations accountable for their pollution in the global South. For years, the organization Amazon Watch has engaged in a multipronged strategy to get the oil giant Chevron to pay up for its multibillion-dollar oil contamination of the Ecuadorean Amazon. It has helped to garner media attention, including *Vanity Fair* magazine and the news program *Sixty Minutes*, to the issue and the plight of local communities; organized company shareholders and brought affected individuals to speak at the meetings; and offered support to the communities in their legal challenge in Ecuador. The legal challenge resulted in a ruling in an Ecuadorean court that the company owes $18 billion in damages,[32] but the company has yet to pay.

A number of other organizations have targeted universities and cities as part of divestment or social responsibility activism. Most prominent perhaps is the recent campaign of Bill McKibben's 350.org to push colleges and cities to divest from the biggest US fossil fuel companies. This encouraged Stanford University to adopt a divestment strategy related to coal in 2014.[33] In the same year, the World Council of Churches, a fellowship of 300 churches, claiming to represent some 590 million people in 150 countries, agreed to phase out its fossil fuel investments and encourage its members to do the same.[34] Shorty after, heirs to the oilman John D. Rockefeller, announced that their $860 million philanthropic organization, the Rockefeller Brothers Fund, had divested completely from coal and tar sands investments, with indications that they might divest from other fossil fuels moving forward.[35] The social investment consultancy, Arabella Advisors, claims that all together, 181 institutions and local governments and 656 individuals representing over $50 billion in assets have pledged to divest from fossil fuels.[36]

Overall, corporate responsibility activism has the promise of shifting corporate and institutional practices in more sustainable directions. This is particularly the case when there is the possibility of financially motivating companies to do something in a more sustainable or socially just way. These campaigns have sometimes resulted in important precedent-setting shifts in corporations. However, monitoring and enforcing new standards can be problematic; it is difficult to verify that companies have met their promised changes in practice. In addition, one can only have low expectations that a major oil or coal company will profoundly improve its environmental performance when its profits are generated from extracting fossil fuels. At the end of the day, financial return is the bottom line of publicly traded companies; it is their fiduciary obligation to their shareholders to maximize it.

Focusing on the financiers of fossil fuel companies, such as banks, universities, and cities rather than fossil fuel companies themselves, makes sense, but to have a big impact, these campaigns have to be able to make fossil fuel extraction unprofitable, or at least undermine the perceived value of fossil fuel assets. This is a tall order and as many corporate responsibility campaigners recognize, such activism needs to be complemented with broader social movements that strategically employ multiple strategies.[37]

Looking across these three types of social movement organizations, we see fragments of what can be understood as a strong climate change movement. Transnational environmental justice movements (e.g., Idle No More and Via Campesina) that are making serious efforts to challenge the most egregious excesses of global capitalism at their source have remained largely detached from far better resourced environmentalist efforts that have focused on policy reform at the national and international levels. One of the exciting aspects of the movement mobilizing against the Keystone XL pipeline has been that it has brought together a wide spectrum of actors, from Indigenous leaders to college students to Sierra Club members to labor representatives.[38] Still, most of the major environmental organizations with substantial funding remain largely detached from such types of organizing.

In the next section we ask: How has the climate movement taken shape within the corridors of the UN negotiations? Have there been promising signs in recent years of a more powerful and bold movement to push states to adopt a robust mitigation treaty? What accounts for the failure of civil society to find success in this context on the big issues? We begin with a discussion of the factors that scholars have identified as the main determinants of civil society influence in this context.

## Access and Sway

Historically, there is some evidence that business interests have been more effective at influencing the negotiations than have civil society groups. One could look at the lack of an adequately ambitious treaty after twenty years of negotiations to confirm this. In the case of comparing environmental NGO and business influence on the Clean Development Mechanism, Lund found that the structural influence, including institutional and discursive factors, seems to have favored business NGOs (BINGOs) as compared to environmental NGOs (ENGOs).[39] Others have argued that the stage of the negotiations matters in determining whether ENGOs or BINGOs have had more influence.[40] For example, Burgiel demonstrates that in the case of the Cartegena Protocol on Biosafety, ENGOs had the greater influence during the agenda-setting stage of the negotiations, when ideas and public pressure were important, but during the middle and late implementation stages, industrial groups were more effective at using economic arguments against a far-reaching agreement.[41] Sociologist Brian Gareau documents how the later phases of the Montreal Protocol for the protection of the ozone layer have seen a rise in the effectiveness of business groups fighting to protect their profits and rolling back precautionary approaches to environmental protection.[42]

The influence that one can observe of civil society groups in the climate regime has been attributed to factors including access to the UNFCCC Secretariat and state delegates from the Northern and Southern governments. It is also due to significant resources, including technical and specialized knowledge and the ability to bring legitimacy to the process and link with public opinion.[43] However, during the first decade of the negotiations, civil society influence was limited by its relative lack of financial resources, by being excluded from closed meetings, and by the inability to circulate on the floor during plenary sessions. In this way, Corell and Betsill argue that the UNFCCC offered relatively weak access to ENGOs as compared to its sister Convention on Desertification.[44]

In terms of the role of civil society in global politics and climate change politics in particular, the scholarship is divided. In one camp, proponents argue that civil society involvement increases transparency, strengthens representation of marginalized constituents, and provides knowledge and expertise capacity.[45] Civil society involvement also has the potential to expand the negotiation agenda and create space for marginalized state actors to make strong demands. For example, when asked about the role of civil society activism in Copenhagen, the lead delegate of Tuvalu, Ian Fry, whose fiery quote opened this book, explained, "It helped. Certainly the impromptu demonstration outside the plenary hall was very helpful, I think, in highlighting the fact that our concerns couldn't just be swept under the carpet."[46] However, Fry also acknowledged that Tuvalu was unable to make progress on its demands.

In another camp, scholars argue that civil society itself is not necessarily a democratic space,[47] that it may represent narrow interests,[48] that it decreases in impact with increasing levels of democracy (the democracy-civil society paradox),[49] and that civil society reproduces certain inequalities.[50] For example, in relation to reproducing inequality, Marceau and Pedersen argue that creating more spaces for civil society in the negotiations means that "some groups get two bites of the apple": the best-financed groups have favorable access in wealthy countries at the national context and then are granted a second opportunity to influence other states at the international level.[51]

We take the perspective that civil society is a complex and highly unequal sphere, and sometimes it is financially influenced by foundation funders; however, achieving legitimate and more just international policy necessitates civil society's active and heightened empowerment—particularly as a check on the influence of private sector actors like fossil fuel companies in the negotiations. We argue that the literature has not adequately accounted for the failure of civil society to find leverage on the issue of climate change and realize an effective treaty during the contemporary period. We offer our view on the weakness of civil society in this context below. First, though, we turn to a discussion of the history of civil society in the negotiations leading to a major fragmentation in the climate movement in 2007.

### Early Days

The civil society umbrella organization with the most capacity and longest history in the negotiations is the Climate Action Network (CAN). Founded in 1989, the network has been active in transnational advocacy with a complex organizational structure of national coordination groups, regional CAN organizations (e.g., CAN-Asia, CAN-Australia), and one umbrella group, CAN-International. The network was the only officially recognized umbrella group with the right to speak for civil society for over a decade.

The network operates in a series of working groups that are open to member organization staff and volunteers. These groups focus on a broad series of issues in the negotiations, from mitigating greenhouse gases and improving climate finance to taxing and regulating bunker fuels and paying poor nations for loss and damage from climate change. Text and policy ideas flow up from these working groups to a smaller drafting group, which pulls together a CAN position for the negotiations, with input welcomed over the larger CAN-talk electronic listserv and at the daily CAN assembly meetings during the negotiations. While truly global, CAN is dominated by the Northern large environmental organizations with the greatest capacity, especially those with strong European and US chapters with professional staff: World Wildlife Fund, E3G, Sierra Club, Germanwatch, Union of Concerned Scientists, Oxfam, and so on.

It is not surprising that CAN has received most of the attention in scholarship in this area.[52] In addition to its unmatched resources, it is also unmatched in its institutional influence among civil society groups. It is largely credited with paving the way for the establishment of the Kyoto Protocol in 1997.[53] Humphreys argues that NGOs that became involved in the negotiations in earlier years are more likely to find influence.[54]

CAN's relative influence among civil society groups in the negotiations might also be attributable to its ability and willingness to have policy proposals that fit within the dominant norm discourse of the negotiations.[55] In other words, groups with more radical proposals that counter ideas such as the power of the market, technology-driven change, and the benefits of economic growth, may be less likely to find traction or a seat at the table where decisions are being made. Scholars have found that this has been particularly true in latter stages of negotiations, when institutional constructs have been developed in relation to particular discourses and positions become more rigid and polarized.[56]

During the formative Rio Earth Summit negotiations in 1992 and the first half-decade of the convention, civil society organizations often took bold positions and mobilized various constituents. However, from the establishment of the Kyoto Protocol in 1997 to 2005, civil society became largely professionalized and oriented toward technical solutions.[57] There were so many technically complex details to be worked out on monitoring emissions, how to enforce the terms of the treaty, and assessing what counts in carbon trading schemes (among dozens of other issues) that one had to first understand the issues on their own terms. The alternative was to be written off entirely by negotiators as naive or uninformed.

During this period, some organizations that focused on issues of social equity, development, and justice—including Oxfam International, Christian Aid and Bread for the World—largely disengaged from the process.[58] At the same time, more radical components of civil society largely ignored the UNFCCC process and focused instead on other targets, such as militarization and corporate-led globalization, as epitomized by the newly articulated global justice movement that successfully shut down the World Trade Organization negotiations in Seattle in 2009. Important exceptions to this include the first large-scale climate justice summits that were organized as alternatives to the COPs. These included the alternative summit in the Hague in 2000, which led to the emergence of a new climate justice network called Rising Tide,[59] and a diverse mobilization of climate justice activists at the alternative summit in Delhi in 2002. In addition, an international coalition of justice organizations convened in Johannesburg in 2002 to address climate justice concerns at the Earth Summit. There they released a formative statement for the climate justice movement called the "Bali Principles of Climate Justice."[60] At the time, CAN did little to incorporate environmental justice activism into its platform and organizing.

### System Change?

The absence of global justice organizations in the negotiations began to change in 2000 and 2002. At this time, social justice NGOs, including ones that rejected the UN process as elitist and co-opted, organized climate justice summits parallel to the negotiations in the Hague and New Delhi, respectively.[61] Beginning in about 2005, more social justice–oriented and radical segments of civil society began to reengage in the UN climate negotiations. This was driven in part by the new scientific evidence pointing to increasingly pessimistic scenarios related to the timing, severity, and feedback loops of climate change. It was also a result of growing recognition by segments of civil society that the existing approach to climate change through UN negotiations was neither adequate nor equitable.

Groups began to develop vocal opposition to carbon trading, "false solutions" such as biofuels and industrial tree plantations, the commodification of nature, violations of Indigenous peoples' rights, and the disproportionate climate impacts on women (for example, natural disasters lower the life expectancy of women more than that of men),[62] as well as their own marginalization in climate politics. In 2004, various global justice activists and organizations gathered in Durban and drafted the Durban Declaration on Carbon Trading, which rejected the claim that carbon trading would halt the climate crisis. The document states, "We denounce the further delays in ending fossil fuel extraction that are being caused by corporate, government and United Nations' attempts to construct a 'carbon market,' including a market trading in 'carbon sinks.'"

In Bali in 2007, several activists were excluded from CAN's daily meeting during the COP there, lacking the CAN sticker on their credentials that was required for entry. Sitting outside the meeting, these activists shared common frustrations with CAN's process, its strategies, and its effectiveness. The activists quickly mobilized to found Climate Justice Now! (CJN!), "a network of organizations and movements from across the globe committed to the fight for social, ecological and gender justice."[63] The positions and tactics of CJN! countered CAN's more reformist and technocratic orientation. CJN!'s founding press release argued that the calls of affected communities, Indigenous peoples, women, and peasant farmers for solutions to the climate problem has "failed to capture the attention of political leaders." Their demands included:

- Reduced consumption
- Huge financial transfers from North to South based on historical responsibility for climate change and ecological debt for adaptation and mitigation costs, paid for by redirecting military budgets, innovative taxes, and debt cancellation
- Leaving fossil fuels in the ground and investing in appropriate energy efficiency and safe, clean, and community-led renewable energy
- Rights-based resource conservation that enforces Indigenous land rights and promotes sovereignty of the people over energy, forests, land, and water
- Sustainable family farming and food sovereignty[64]

This movement played an important role in bringing the climate justice frame into the negotiations in Bali in 2007 when the influential Bali Action Plan was adopted. Bali ended up being one of the most hopeful moments in the negotiations, with a road map put in place that addressed many developing countries' demands. The CJN! movement gained momentum with a climate justice conference organized by the NGO Focus on the Global South in Bangkok in 2008. This conference brought together 170 activists, including fishers and farmers, forest and Indigenous peoples, women, youth, workers, researchers, and campaigners from thirty countries.[65]

At the same time, the Third World Network, an NGO, established itself as a strong voice for justice within the formal negotiations, making interventions and documenting the proceedings from what had been a largely absent perspective. By 2009, CJN! emerged with a major presence in the climate negotiations; the UNFCCC offered CJN! members an equal number of display booths to that of CAN in the main hall of the Bella Conference Center. Whether a result of being pushed or inspired by climate justice organizations, several CAN member organizations also adopted many of these same messages in their advocacy. For example, the TckTckTck! Campaign,[66] with big green environmental partners including World Wildlife Fund and the NRDC, adopted the radical slogan, "System change, not climate change!"

Since 2005, as interest in climate change expanded, constituency groups showing up at the negotiations multiplied to include local governments and municipal authorities, Indigenous peoples, more research and independent NGOs and trade unions, farmers and agricultural nongovernmental organizations, women and gender nongovernmental organizations, and youth and nongovernmental organizations.[67] Other intergovernmental organizations like the UN World Meteorological Organization have also emerged in the negotiations; despite not having voting rights, they do have speaking intervention rights equivalent to that of states.[68]

## Collision

Copenhagen in 2009 was a key moment with a blossoming of transnational advocacy networks. The participation of civil society members swelled as the meeting exploded to 30,000 delegates and observers. The Bella Conference Centre in Copenhagen was overwhelmed with people who had traveled across the world to observe, influence, or protest the key negotiations there. Various groups and agendas were represented: besides environmental NGOs (ENGOs) there were ranks of representatives from RINGOs (research institution NGOs), BINGOs (business and industry NGOs), InGOs (Indigenous NGOs), and YUNGOs (youth NGOs), among others. Virtually every one of these categories had within it factions with different goals, strategies, and styles of action. Others, as part of the climate justice movement, came mainly with the intention of protesting what they saw as a process devoid of democracy and parroting the interests of global capital.

Despite representing the pinnacle of civil society engagement within the negotiations, Copenhagen also represented the moment when the doors closed on civil society. With most of the 30,000 civil society observers scheduled to arrive the second week of the negotiations, the secretariat made a last-minute decision to limit access to those who could acquire one of only 7,000 secondary registration cards.[69] These cards were rationed according to an unclear process. Furthermore, corresponding with the arrival of several heads of state from various countries, civil society access was denied completely the final Wednesday morning of the negotiations. Then on Thursday, only 1,000 observers and Friday only 90 observers were provided access.[70] Thus, many people who traveled to Copenhagen for the negotiations were never allowed to set foot in the conference center. Several more NGOs were also shut out completely from the negotiations during the second week for their protesting behavior earlier in the meetings, including Friends of the Earth International, Avaaz, and TckTckTck.[71]

Sociologist Dana Fisher identifies three factors leading to civil society being shut out of these negotiations: increased registration numbers, poor planning, and the merging of movements. In particular, the merging of movements perspective highlights that many climate justice organizations and individuals traveled to the negotiations in Copenhagen with an expressed purpose to protest its process.[72] This took form within the negotiations with daily outbreaks of protest. Officially, protests required permitting by the UNFCCC in advance, but as the negotiations progressed and it became clear that things were not going to end well, they began to erupt with increasing spontaneity and volume.[73] At times, NGO representatives had their badges confiscated when they were deemed to not be complying with protocol and they were unable to return to the conference center.

Civil society protest of the negotiations culminated with a call from CJN! and the European-based network Climate Justice Action to "take over the conference for one day and transform it into a People's Assembly."[74] Due to a clampdown on security at the center, activists were largely unable to find access and disrupt the negotiations. The police response was swift, aggressive, and in some cases preemptive, with many reports of physical force being used to quell protest outside.[75]

Demonstrations erupted all over the city. These actions came on the coattails of the major protest of 30,000 to 100,000 people in the streets, and hundreds were detained by police, including many that were merely walking peacefully as part of the protest.[76] Fisher draws from this lesson from the Copenhagen experience, "Ironically, the more civil society actors try to participate—and the diversity of the perspectives represented by the civil society actors involved—the less access they are likely to have."[77] However, diversity of opinion and nonviolent protest does not have to equate to losing the right to participate. We observed that at least a few groups simultaneously pursued insider and outsider strategies for change; this view conflicts with the common portrayal of clear and definitive divisions among the roles of actors.

## Wonks Remain

Since Copenhagen, civil society interest and in some cases access in the UNFCCC process has waned. In Cancun the following year, civil society delegations were often limited to badges for half or fewer of the number of people they had brought in previous years. Organizers of the conference also isolated NGO booths and side events in a separate building that was a fifteen-minute bus ride from the actual negotiations. In addition, accessing both the convention center and side event area required traveling by bus or private vehicle past several checkpoints and dozens of armed military personnel, some with their guns pointed uneasily at the passing drivers and passengers.

While various social movement groups, including a large caravan of the transnational farmers' rights group Via Campesina, had converged on Cancun for the negotiations, their activities remained far removed from the negotiations. These strategies had the impact of creating separation between civil society (especially its more radical contingents) and negotiators, and many observers expressed feeling marginalized from the process.

The conference organizers were largely successful in buffering the negotiations from any sense of disruption or notable protest. Enforced order was the overwhelming tone set by the Mexican government hosts. Beginning in Cancun, civil society participation swung gradually back toward technocratic reform and away from civil disobedience. At the same time, without a clear international target and with a fairly loose structure and less resourced member groups, CJN! began to lose steam as a coherent organizing body for the climate justice movement.

The following year in Durban, civil society numbers had dwindled even more. This can be explained mainly by three factors: expensive

airfare travel from many parts of the world, declining interest in the negotiations due to plunging confidence in the utility of the UNFCCC to
address climate change, and declining numbers of credentials for entry
extended by the secretariat. Convention organizers in Durban didn't deploy the same level of military presence or the infrastructure separation
strategies found in Cancun. However, while climate justice protests and
events were organized throughout the city, the climate justice community
remained largely outside the convention.

Only on the final Friday of the negotiations did the presence of the climate justice community become highly visible inside the convention center. This is when a group of about a hundred people, organized through
e-mails and word-of-mouth, held a protest just yards from the official negotiations, in the corridors outside the conference ballroom. They chanted "Occupy the COP," "The people united will never be defeated," and
"Climate justice now!" in response to what had been a largely inadequate
series of international negotiations over the course of the two weeks.[78]

After being asked to disperse, some of the organizers were told by the
UNFCCC security that they had three options: move outside to demonstrate far removed from the delegates; disperse and end their demonstration; or if they stayed, their credentials would be confiscated, they would
be escorted out of the convention center, and they would be permanently
banned from taking part in a COP in the future.[79] This final measure was
unprecedented, and some expressed being unsure of whether the security
officials had the authority to carry it out.[80]

After much deliberation among the group, the mobilization largely
splintered, with some participants refusing to budge, others symbolically
offering their credentials to the security officers, and still others simply
dispersing. In the end, security was largely effective in relatively quietly
bringing the protests to an end. The fragmented response among participants left some disappointed or angered that they did not hold their
ground collectively and pointed fingers at what they considered to be
weak or even complicit movement leadership. Some individuals later
wielded barbed accusations of collusion with authority by some climate
justice activists from Greenpeace and 350.org.[81]

The atmosphere the following year in Doha felt as if the UNFCCC
had once again almost fully gravitated back to becoming a space mainly
for negotiators and a limited number of professional NGOs. This included mainly actors that had largely accepted (and had helped to establish) the guiding norms of international climate governance and worked
within the confines of its institutions, as compared to a space for social

movements, climate justice activists, and affected peoples. The corridors of the cavernous brand-new convention center on the edge of the desert city were nearly empty. A further setback for civil society was a new paperless policy, which banned distributing information on tables and walls in the main halls of the negotiations and thus hindered their efforts to communicate with delegates. The next year in Warsaw, a large contingent of civil society staged a walkout at the end of the second week of the negotiations, a symbolic gesture to protest what they viewed as an utter failure of the parties to make progress on core issues in the negotiations. Many also protested what they viewed as the increasing corporatization of the UN process, including official corporate sponsors of the negotiations such as BMW, General Motors, and the Polish Energy Group.

What we have seen in the contemporary period is a gradual weeding out of more radical and diverse voices from the UNFCCC negotiations and a narrowing of approaches to change being advocated. At the same time, although CAN-International and its member organizations remain capable of forming and following positions on a large number of issues within the talks, they have been unable to achieve progress on their main collective demand: a binding international treaty that is scientifically adequate in preventing catastrophic climate change. In the next section, we explore why civil society has failed at achieving this central goal.

## Reasonable beyond Reason

The scholarship on civil society influence in the negotiations gives us several clues about the ineffectiveness of civil society in the contemporary period. Civil society has been largely outmatched in terms of financial resources and political capital of the still powerful fossil fuel interests (as discussed in chapter 6). It has also faced institutional hurdles. For example, decisions increasingly have been made in closed meetings at the negotiations, and even plenaries involving all state delegates have often become inaccessible to NGOs and members of civil society in recent years.

Most critical, civil society has encountered a problem in that climate change is perhaps the most complex and entrenched issue in the history of humanity. Addressing climate change requires profound changes in society's central institutions, and major reductions in consumptive behavior are needed. Virtually every facet of our global economy, daily practices, and physical infrastructure contributes to the problem. There is no simple fix, and thus it may seem a bit unfair to blame civil society for not being able to turn this around. (And to be clear, as we point fingers, we count ourselves among those in civil society who have failed at our central task!) However, we believe that there are three further strategic issues at the heart of civil society ineffectiveness in this context.

First, big greens control the vast majority of resources available to environmental groups and have been largely unwilling to share them. Private philanthropic foundations bear much of the blame here: they have been almost completely one-sided in favoring technocratic, insider, market-based, and compromise solutions for solving the climate problem. Unyielding positions, arguably required to push the political system to keep us below dangerous concentrations of greenhouse gases in the atmosphere, were never rewarded with support from foundations and other NGOs.

Very few resources (relatively speaking) have been devoted to building a climate movement. Here lies the biggest problem: there is very little connection between efforts for justice where climate impacts are being felt and the corridors of national capitals and international forums where mainstream environmental groups have concentrated their efforts. As a result, peoples' movements have had very limited resources to mobilize strategically, and the big green environmental groups have limited leverage at the community level to demand real change, shift public consciousness, and serve as a wrench in the gears of economic systems when necessary.

To be clear, there are more than coordination problems; there are tensions that have deep historical roots, a lack of trust between mainstream and grassroots groups, and sometimes a basic incongruence of worldviews.[82] Big green groups still by and large understand and work for change in fundamentally different ways than do activists who are organizing what can be understood as movements.

This reformist approach is largely responsible for advancing the carbon trading diversion discussed in chapter 6. They have been far too willing to compromise, knowing that they have very little leverage to make more ambitious and meaningful demands without a mass movement on their side. The unfortunate political reality of climate inaction by governments has often been accepted at face value, with very limited willingness, resolve, or strategy by a sufficient number of activists to establish the political conditions in which civil society might be in a position to push for a new and different reality. And as we have discussed, in some cases, civil society organizations such as Environmental Defense and the Pew Center have been initiators in shifting climate policy to embrace more market-friendly approaches.

Second, and related, we reject the idea suggested in the literature that the negotiations are a linear process, whereas civil society has potential influence only in the founding years, before path dependency takes over or only technical expertise is valued. Rather, there have been sporadic moments in the UNFCCC when civil society has at least partially transcended merely technical and professionalized interventions to shape the agenda and norms of the negotiations.

This happened during the founding of the Convention, during designing the Kyoto Protocol, and in the making of the Bali Action Plan. Rather than a linear progression from early agenda-setting to later implementation stages in the regime, the UNFCCC has been characterized by various and regular hinge moments when both regressive and progressive bold action seems possible. These hinge moments become more numerous when we consider important decisions that are being made nearly every year in various regime institutions, such as the Clean Development Mechanism, the Adaptation Fund, and the Green Climate Fund. These are the moments when it is unclear which way major policies will swing.

In the years between the major hinge moments in the negotiations, we have seen a waning, professionalizing, and homogenizing of civil society engagement as attention has shifted from bold ideas and design to implementation. After some raucous moments in the negotiations in Bali in 2007, the civil society presence in Posnań in 2008 was utterly forgettable. At the national level, the United States elected a leader who had publically committed to pursue action on climate change. But at the same time, there was essentially no broad-based movement to hold him accountable for his campaign promises on this issue. Copenhagen was a disaster not because of what happened in the negotiating halls, but because of what didn't happen prior to Copenhagen back in the key nations.

Here lies a problem. It is in these largely uneventful stages where the groundwork is laid for what is possible at the next hinge moment. A lack of broader movement attention to in-between negotiations in Doha, Warsaw and Lima in 2012, 2013, and 2014, respectively, could prove devastating for realizing the change that we need in 2015 in Paris, when the next mitigation framework will be agreed to.

Success in Paris will be largely tied to civil society's ability to shift domestic interests during this interim period (particularly the United States and China). Civil society must be steps ahead of the negotiations working to transform state preferences away from fossil fuels and toward ambitious climate change action, so that lead states shift their positions in time to act collectively in the UNFCCC. Thus, there is a delay factor at work here. The People's Climate March in New York City in 2014 was an attempt to fill this gap.

Transnational mobilization will also be critical. Whether the UNFCCC will allow for more radical and broad-based participation is not clear. It is also not clear whether better-resourced groups in the negotiations will work to shift the political reality in the negotiations, rather than merely responding to it in a technocratic fashion.

Third, and finally, the UNFCCC is only one of many realms of governance that are relevant to climate change. Various other international governance frameworks are pertinent to action on the issue, including international trade regimes, financial institutions, scientific bodies, industry monitoring groups, and a host of other fragmented and layered governance systems. Thus far, civil society groups that work on climate change have primarily devoted their attention at the international level to the UN process while largely neglecting other relevant venues and bodies.

For example, the Trans Pacific Partnership trade pact being negotiated by the United States and countries including Chile, Japan, Mexico, Singapore, Peru, and six others will likely have major consequences for national approaches to climate change policy.[83] If approved, this will likely be the largest trade agreement in the world. Yet only a handful of environmental actors are currently engaging to shape or disrupt this process, which largely lacks transparency, while business actors are at the very heart of the negotiations.[84]

To be fair, the negotiations of this trade pact are mostly carried out behind closed doors, with very limited transparency and access for civil society. It makes sense why civil society has chosen to participate primarily in institutions like the UNFCCC that are more inclusive of their opinions, transparent, and easier to penetrate. But simply ignoring powerful and highly relevant governance processes due to access challenges is not a viable strategy for achieving climate justice.

In another example, in 2012 the United States filed a case at the World Trade Organization (WTO) to challenge India's use of subsidies and "buy local" rules in its domestic solar program. These new provisions of the Indian government are likely to curtail imports of thin-film solar cells from the United States. In its claim, the United States asserts that India's domestic content rules appear to have violated trade rules in the General Agreement on Tariffs and Trade (GATT), the Agreement on Trade-Related Investment Measures, and the Agreement on Subsidies and Countervailing Measures by giving more favorable treatment to domestic solar producers and products than to foreign ones.[85]

So here is a typical case of short-term trade interests in the name of free trade trumping emissions reduction priorities to address climate change. Yet civil society watchdog groups in this process are few, and they are not generating enough pressure on the GATT/WTO to meaningfully address related climate concerns. Unless climate movements orient themselves broadly to change the practices of international governance bodies such as the WTO that inhibit climate-friendly state actions, even realizing a bold and ambitious treaty in the UN may do little good. Civil society must broadly and strategically engage across a complex and fragmented global governance system, a daunting task.

### Moving the Unmovable

In the previous chapter, we argued that a new business class that views its interest in mitigating climate change is important, but not sufficient to address the climate crisis. The reason is that fossil fuel–based interests are too strong. In this chapter, we have turned our attention to transnational civil society. We have seen a largely technocratic, reformist, and in many cases, neoliberal approach by transnational civil society to climate change policy for more than two decades now, and it hasn't succeeded. Fossil fuel interests remain as powerful as ever, and carbon markets have been little more than a distraction from the real change that we need to see, which includes keeping the vast majority of remaining proven fossil fuel reserves underground.

This is perhaps not surprising. Scholars have found that transnational civil society has largely conformed to and helped to establish neoliberal approaches to policy in other international governance initiatives such as the Montreal Protocol[86] and the World Bank.[87] In other words, global environmental governance is both co-constituted and made legitimate by civil society as an expression of dominant systems of power (what global governance scholars in the tradition of sociologist Michel Foucault refer to as a process of "governmentality").[88] In doing so, mainstream civil society has internalized much of the dominant historic logic of neoliberal governance, including a commitment to market-based solutions, the imperative to sustain economic growth above all else, and a narrow view of the particular types of scientific knowledge and expertise that are deemed relevant to the policy-making process.

However, as we will discuss in chapter 8, civil society is not uniform, but rather a terrain of struggle between competing identities, politics, ideological constructs and strategies of engagement. But it remains deeply unequal. From a neo-Gramscian strategic perspective on power relations, mainstream civil society, with its commitment to working within dominant state and private sector power relations, governance models, modes of production and ideological structures, has barely scratched the surface of building a counterhegemonic movement capable of shifting the balance of forces on climate change. While we would likely be in a far worse position without some of the reformist interventions of big green organizations over the years, the next decade demands that we think, act, and fund far bolder efforts of social and political change.

We can no longer approach social, political and economic systems, and the hegemonic ideas that stand behind them, as unmovable. Indeed, even if the challenge we face is immense and unprecedented, social systems are far more malleable than ecological limits, which is what we are rapidly butting up against as we endeavor to remain realistic and reasonable in our incremental policy proposals. In order to boldly shift international policy at key hinge moments, we have argued that politics needs to shift in key domestic contexts, and this necessitates large investment in building broad-based movements that bring together unlikely actors around an alternative vision of how we can organize our societies.

We have also highlighted the fragmented and layered global governance systems that civil society must navigate in its efforts to curb climate change. While the UNFCCC remains an essential governance body, and what we see as our best hope for coordinating international action on this issue, we need a more strategic approach toward ensuring that this climate regime isn't simply upstaged by conflicting governance bodies that have greater authority, and offer less civil society access, in the international system. These are difficult puzzles, to which we will return in chapters 9 and 10. In the next chapter we examine the efforts of transnational advocacy networks representing particularly vulnerable peoples to find influence and realize rights in the UN climate change regime.

# 8

## Contesting Climate Injustice

### Movements from the Margins

It was December 8, 2009, during the first week of international climate change negotiations in Copenhagen. A large room was filled, mostly with white men in dark suits, who had gathered to participate in a question-and-answer session with the executive board of the Clean Development Mechanism (CDM). This board oversees the methodology for granting approval for funding of projects deemed to reduce greenhouse gas emissions.

Standing toward the back of the room, a very small woman from India wearing a vibrant yellow and blue sari spoke loudly into the microphone in the language Marathi. A woman next to her translated her message:

My name Baby Mohite and I'm from Pune. I'm a waste picker. And the work that I do is picking out waste, recyclables. ... How can you [the CDM board] possibly approve methodology [for waste incineration projects] that would make people like me lose my livelihood? And it's not just me, but thousands of people who make their living from waste. How could you possibly approve such a methodology? What would I do? Where would my children go? How would they go to school? You are taking away my livelihood.[1]

Mohite was in Copenhagen as part of a newly formed transnational advocacy network of waste pickers and allies. She was concerned that money flowing through the CDM, which was intended to reduce climate-changing emissions, was actually being used to fund waste disposal projects that would put her and her fellow informal recyclers out of business.

This chapter argues that the scholarship on transnational advocacy networks (TANs), such as the network of waste pickers and allies, has not articulated the diversity and range of rights struggles that take place in international regimes, particularly those representing marginal or vulnerable groups. In chapter 7, we explored the impact of transnational civil society on the issue of climate change broadly. In this chapter, we are particularly concerned with the possibilities for highly vulnerable and affected peoples to achieve gains in international climate change policy. A neo-Gramscian strategic view of power relations does not simply dismiss such actors as powerless, but rather directs our analysis to the possibilities for justice-oriented campaigns to find leverage in global governance processes. Studies of environmental justice have oriented our focus to those actors who are marginalized in society: if the needs of the least powerful are not met, then environmental problems will simply be shifted around to places where the poor and minorities live and work.[2]

We explore as case studies the interventions of three distinct networks in the UN climate change regime: gender equality advocates, Indigenous peoples, and waste pickers. Despite several differences among these networks, they have all sought to gain rights in the UN in order to address forms of marginalization and inequality experienced by their group, including disproportionate vulnerability to the impacts of a changing climate and climate change responses with adverse impacts—what the networks commonly refer to as "climate injustice." We chose these cases because of their importance and visibility in the climate regime, but also because of their differences, which allow a rich set of insights into the successes and limitations they tend to experience.

To assess the effectiveness of these efforts, we introduce the concept of *regime rights* and develop a framework for their analysis. Drawing on the case studies and relevant scholarship, we identify and discuss four main types of related struggles: for recognition, representation, capabilities, and extended rights. We argue that deciphering distinct forms of rights struggles in international regimes is critical for deepening understanding of what types of rights gains are more or less likely for networks representing marginalized peoples in the climate change regime, and beyond, to achieve.

In the next section, we conceptualize regime rights and develop the four-part typology. Then we describe the main successes and failures of the three networks in gaining distinct forms of these regime rights. In table 8.2, we categorize and assess the relative number of gains and the extent to which gains are consistent with network demands. In this way, we offer an approximation of network gains in each rights category ranging from weak to high. In the final two sections, we draw insights from these cases to discuss which types of rights interventions are likely to have impact and under what conditions. Understanding different types of rights struggles inside international regimes can greatly inform efforts of those seeking to improve the process, outcome, and impact of these treaties by allowing new voices to be heard, and, it is hoped, to be responded to.

## Regime Rights

The literature on international politics, and climate change politics in particular, has devoted scant attention to understanding the types of transnational advocacy network rights struggles in international regimes.[3] Even less attention has been directed to understanding the rights struggles of networks representing particularly marginalized and vulnerable actors.[4] Such a focus is needed to understand what types of rights demands have been successful, the processes that enabled such gains, and the relative impact of these gains. Addressing these demands may be the measure of a truly inclusive, sustainable, and effective regime.

We define regime rights as privileges for particular groups that enable certain behaviors and outcomes, and constrain others, within and extending from a given international treaty or other agreement. This is consistent with Iris Young's conception of rights, which emphasizes their relational nature. She argues, "Rights are relationships, not things; they are institutionally defined rules specifying what people can do in relation to one another. Rights refer to doing more than having, to social relationships that enable or constrain action."[5]

Regime rights struggles shape the opportunity structures of international regimes, which themselves represent "sets of implicit or explicit principles, norms, rules, and decision-making procedures around which actors' expectations converge in a given area in international relations."[6] As we discussed in chapter 2, scholarship from a neo-Gramscian perspective has emphasized that international regimes, and their relevant principles, norms, rules, and decision-making procedures, are highly contested terrains that serve to privilege certain actors and interests in the international or transnational sphere.[7] From this perspective, we can understand regime rights struggles as coordinated and competing efforts by actor groups to shape relevant regime structures in order to preserve or challenge the reproduction of established privileges. Moreover, regime rights as a set of group privileges may be strived for on social and normative grounds related to particular group identities, but they also may be advocated on technical and objective grounds, based on interpretation and reform of existing scientific or institutional grounds.

**Table 8.1**
Regime Rights Typology

| Regime rights type | Function | Relevant concepts in the literature |
|---|---|---|
| Recognition | Diffuse, codify, and institutionalize norms that link entitlements or protections to the group's identity. | Norm institutionalization (Finnemore and Sikkink 1998) |
| Representation | Enhance the institutional inclusion, participation, and representation of particular groups in regime politics. | Procedural gains (Betsill and Correl 2008) Global citizenship (Muetzelfeldt and Smith 2002) |
| Capability | Enhance the capability to act on representation rights, overcoming various forms of inequality. | Citizenship (Somers 2008) Capabilities (Sen 2001; Nussbaum 2001) |
| Extended rights | Establish international mechanisms that uphold rights at the local, state, and regional levels. | Boomerang effect (Keck and Sikkink 1998) |

The three cases reveal that regime rights struggles have taken varied forms in the UN Framework Convention on Climate Change (UNFCCC), have had disparate rates of success, and have likely varied in impact (as documented in table 8.2). Drawing on the cases and engaging with the relevant literature, we identify and discuss four main types of regime rights interventions (table 8.1).

First, gender equality and Indigenous peoples' networks have devoted significant attention to achieving *recognition*. These struggles have sought to diffuse, codify, and institutionalize certain norms, often established as rights in different international regimes, into the text of the UNFCCC and its subsidiary bodies. These norms link certain entitlements or protections to the identity of each group with reference to unique status, vulnerability, or capabilities. All three networks have also worked to leverage organizing in the climate regime to increase the recognition and visibility of the relevant groups, their rights, and positions in the broader public sphere, including the media, foundations, and other international organizations.

Network struggles for rights recognition correspond to the process of norm institutionalization articulated by Finnemore and Sikkink. Institutionalization, they assert, is important for broad-based norm adoption (defined as a standard of appropriate behavior for actors with a given identity), by clarifying the norm and what constitutes a violation, and for spelling out procedures by which norm leaders coordinate disapproval and sanctions for norm breaking.[8]

Gender equality and Indigenous peoples' networks in the climate regime have also worked extensively for greater *representation*: concrete gains in the formal inclusion, participation, and representation of particular groups in regime politics. This includes voting rights and a number of ways that groups formalize their participation in the regime, overcome "disenfranchisement,"[9] and erode the monopoly of state sovereignty in these arenas.[10] In the literature, this has been described as enhancing "civil participation,"[11] "institutional procedures,"[12] "procedural gains,"[13] and "global citizenship" in governance regimes—with the contention that civil society thrives at the global level through its interaction with strong, facilitating institutions of global governance.[14]

However, even when there has been modest progress in enhancing representation (such as gender balance in state delegations in the UNFCCC), such representation does not ensure empowerment or agency. Rather, a lack of capacity and various forms of inequality can limit the ability of actors to act on such rights. *Capability* thus refers to struggles that extend beyond institutionalizing a given right, to ensuring that groups have the ability to access and uphold a right in practice. This position holds that formal democracy may endow citizens with formal rights, but pervasive inequality within society limits the capability of citizens to act on these rights and creates conditions for uneven access to rights.[15]

As applied to the context of an international regime like the UNFCCC, formal institutional rights that are granted to participants, such as the right to voice an opinion in plenary sessions, the right to amend a proposal, or the right to vote on an outcome, may mean little in terms of influence if the capacity or necessary capital of individuals or groups is limited to act on these rights in practice. As Peter Newell argued, "Merely constructing more 'spaces' for civil society groups within international institutions does not address the inequalities within civil society that will continue to mean participation is unevenly distributed by region, issue, as well as other key social cleavages such as gender, race and class."[16] Thus, capability includes enhancing the capacity of groups to act through measures such as funding allocation and capacity-building meetings, social mobilization to challenge unequal practices, and leveraging institutionalized rights for concrete material gains.

Struggles for *extended rights* refer to efforts in the regime to realize mechanisms that uphold rights related to activities at the local, state, or regional level. This points to how networks are able to achieve distributive gains that enhance justice outcomes beyond the regime. An example of this is what Keck and Sikkink refer to as the "boomerang effect," where transnational advocacy networks converge on an international arena to put pressure back on states or localities in order to overcome repression and blockage.[17] These efforts are focused on the regime's influence on particular rights as they are governed and carried out beyond the internal regime's own practices. As the cases reveal, gaining extended rights may be viewed as particularly important when the regime itself is exacerbating an existing rights conflict or creating an entirely new one.

Extended rights interventions take two main forms, which are often interlinked. First, normative interventions rely on moral arguments and symbolic actions to contest the legitimacy of a given course of action. Second, technical interventions focus on achieving extended rights based primarily on reinterpreting or shifting scientific, "objective," and technical criteria or practice. Such rights struggles have been addressed in the literature as democratizing science movements that contest, reframe, and coproduce scientific knowledge around contested issues,[18] including in international regimes such as the Montreal Protocol.[19] These movements work to challenge and also leverage the process of scientization, which is "the transformation of political conflict ... into a debate among scientific experts, ostensibly separate from the social context in which it unfolds."[20]

We now turn to descriptions of these three struggles for rights, including those for gender equality, Indigenous peoples' rights, and the rights of waste pickers (informal sector recyclers) who make their living from waste.

### Gender Equality

There is no mention of gender concerns in the founding 1992 UN Framework Convention on Climate Change or in the Kyoto Protocol. However, organizing for issues of gender equality extends back to the initial years of international climate politics. Mobilizations have responded to two main overarching concerns of marginalization. First, because of the types of roles they are allotted in the household division of labor, women are affected disproportionately by climate change as compared to men.[21] They also contribute to climate responses in quite different ways than men do.[22] For example, women's daily chores in low-income countries often include the collection of firewood and water for cooking. Both are often made sharply more difficult in cases of climate change. Due to low socioeconomic status and lack of freedom of movement in public compared to men, women die in far greater numbers than men during and immediately following disasters.[23] Women also contribute to climate responses in different ways. For example, women and men perceive different risks related to climate change as important and attribute different meaning to the same risks.[24] Thus, women are not merely disproportionate victims; they offer unique perspectives and capabilities for offering solutions for building resilient communities.[25] The same could be said about developing solutions for reducing emissions of greenhouse gases. Compared to the other two cases discussed, there is great variation of vulnerability among women due to a broad range in class dynamics.

Second, to ensure more equitable and effective responses, gender equality is needed through greater representation and capability within the UNFCCC, a largely male-dominated space, and in policies that extend beyond this regime to respond to various forms of inequality on the ground. In doing so, advocates have often highlighted how norms of women's rights and gender equality are codified in various international institutions.[26]

It wasn't until 2001 in Marrakesh that gender concerns were integrated into UNFCCC texts. A decision was adopted that called for increased nominations of women to the UNFCCC and Kyoto Protocol bodies and tasked the secretariat with determining the gender composition of these bodies and bringing these results to the attention of parties.[27] However, the decision provided no mechanisms to ensure change in this area. It also did nothing to address the role of gender inequality in all aspects of climate policy, including mitigation, adaptation finance, technology use, and capacity building.

After sustained organizing by several organizations since 2002, the Women's Caucus was founded in Bali in 2007, providing gender equality advocates an official daily meeting in the negotiations.[28] This caucus then gained its own official constituency in the UNFCCC, the Women and Gender Constituency, which is a collaboration of networks including Women in Europe for a Common Future, Gender Climate Change, and Women's Environment and Development Organization (WEDO). These groups, along with other smaller groups, have engaged in activities such as issuing press statements, holding strategy meetings, lobbying state delegates, organizing side events, participating in protests, and advocating for the inclusion of gender-specific text in the negotiations.

One extended-rights engagement of gender equality groups (along with other constituents such as Indigenous peoples' groups) has been to seek to prohibit the addition of nuclear energy from eligibility in the Clean Development Mechanism (CDM). For example, Gender Climate Change presented the technical argument in a statement in 2007 that nuclear energy is not a clean energy source and that it has disproportionate impacts on women and children due to increased sensitivity to radiation and differing gender roles.[29] Other advocacy groups have also opposed its inclusion, and thus far, nuclear energy has not been included as an eligible technology.

Gender equality groups have devoted more concentrated energy to gaining other forms of rights. In the face of what had been seen as largely fragmented and watered-down referencing to gender and women's issues in the UNFCCC texts, the Women and Gender Constituency lobbied in Copenhagen in 2009 for inclusion of a shared vision preamble to the convention with the "full integration of gender perspectives."[30] Despite this call for recognition and the considerable mobilization of civil society groups behind it, the Copenhagen Accord contained not a single reference to women or gender.

The following year, a few gender-specific provisions were added in the Cancun Agreements, including the language that "gender equality and the effective participation of women and indigenous peoples are important for effective action on all aspects of climate change." In addition, gender sensitivity was mentioned in relation to adaptation, forest conservation, technology transfer, and capacity-building activities. The agreements also reiterated the 2001 call for gender balance and noted that gender is associated with vulnerability to direct and indirect climate change impacts and "enjoyment of human rights."

In the Durban Platform in 2011, the agreement reiterated the commitment to these issues and called for gender balance in relation to the new Green Climate Fund (GCF) and other select institutional bodies. The GCF has been a key priority for civil society groups since its development as a funding institution is designed to "promote the paradigm shift towards low-emission and climate-resilient development pathways." Gender equality groups such as WEDO have made various demands ranging from equitable representation in fund governance, extended rights such as a redress mechanism to address grievances by groups on the ground, and greater capability such as the development of effective mechanisms to systematically incorporate the input of women as stakeholders.[31]

Despite these efforts, the GCF has yet to adopt many gender equality provisions. There are only seven women among the forty-eight members and alternate members on the board—only 15 percent women.[32] Notably, gender advocates point to success in achieving commitments in the fund's mission statement to gender provisions such as to "encourage the involvement of relevant stakeholders, including vulnerable groups and addressing gender aspects." However, these commitments remain vague, and it is not yet clear how it will influence actual funding practices. And critically, as we described in chapter 5, the GCF remains a fund with limited funding.[33]

Beyond the GCF, some progress was made on issues of gender equality in the negotiations in Doha in 2012. Christiana Figueres, executive secretary of the UNFCCC, referred to a decision on gender there as the "Doha miracle." And as Mary Robinson, former Irish president and advocate of human rights, explained, "What we've now got is a decision that will bring gender rights into the bodies of the COP and the decisions of the COP, and it's quite obvious that this is going to make a huge difference because women are so central to making progress on climate change."[34] However, the text doesn't reveal much that is really new, except for the provision of a workshop and making women and climate change a standing agenda item. Initial language for "gender equality" was replaced at the last minute with the weaker concept of "gender balance."[35]

Yvette Abrahams with the organization Gender Climate Change described this as a "tick-box" approach, indicating that increasing the number of women in the UN process won't necessarily lead to gender equality. She explained: "What we have [here] is a gender decision that's not necessarily a feminist decision. … You can't just have representation in a male world."[36] And as articulated by the civil society newsletter *ECO*, the gender provision "was introduced under 'Any Other Business.' That means that the needs and concerns of half the world's population were not given a place of their own in the central agenda of the COP [Conference of Parties]."[37]

The following year in Warsaw in 2013, the first ever in-session workshop on gender was held. This led to a draft decision that reiterated a commitment to strengthening gender balance and gender-sensitive climate policy and provided proposals by parties on ways forward. Then in Lima in 2014, the last minute changes to the Lima Accord included language for regular opportunities for effective engagement of women, among other actors as part of a series of technical expertise meetings. Notably absent still is agreement on mechanisms to achieve the goals of gender equality in climate policy and adequate resources to facilitate movement in this direction.

Overall, gender equality groups have made some progress in pushing the regime to recognize the right to gender equality and the disproportionate vulnerability of women and modestly enhancing women's representation in the climate regime, including gaining an in-session workshop and agenda item on gender. But the promise in their formal gains remains largely unfulfilled.

### Indigenous Rights

As early as 1999, widespread engagement of Indigenous peoples' groups and networks emerged in relation to several perceived threats and opportunities in the UNFCCC. As part of the Quito Declaration, representatives of more than two dozen Indigenous organizations and local communities agreed that the UNFCCC and Kyoto Protocol "have been negotiated without the participation of the Indigenous Peoples and Organizations and do not take into account our rights."[38]

Indigenous peoples' network engagements have responded to three main concerns. First, due to being among the most marginalized, vulnerable, and impoverished peoples and bearing disproportionate impacts of climate change, they often have minimal access to resources to cope with the changes.[39] Second, Indigenous peoples' groups argue that the impacts of climate change mitigation strategies such as dam construction, monocrop plantation, and agrofuels have resulted in displacement of Indigenous peoples from their territories, as well as harm to the ecosystems on which their culture and livelihoods depend. Third, Indigenous peoples' networks have often argued that the Earth is a living being with rights that should be recognized in the climate regime. Moreover, they argue that they have protected the Earth from commodification and overexploitation for centuries, and their expertise and ecological services should be acknowledged, protected, and rewarded.[40]

In 2000, the International Indigenous Peoples Forum on Climate Change (IIPFCC) was established, a joint Indigenous peoples' caucus to participate in the negotiations. Despite various rights interventions,[41] no relevant progress was made in the negotiations for years. Then in 2007, with the leadership of Papua New Guinea and a new state network, the Coalition of Rainforest Nations, a market mechanism, Reducing Emissions from Deforestation and Forest Degradation (REDD+),[42] assumed a central role in the regime and was adopted as part of the Bali Action Plan. As part of this framework, text was included to address the needs of local and Indigenous communities when action is taken to reduce emissions from deforestation.[43]

Many representatives of civil society and states viewed the development toward a market mechanism for protecting forests as a positive development. However, just three months after the signing of the UN Declaration for the Rights of Indigenous Peoples (UNDRIP), Indigenous peoples' organizations immediately recognized REDD+ as a potential threat to their rights and territories that would result in land stealing and forced evictions.[44]

As REDD+ was developed, the Indigenous peoples' movement became increasingly mobilized and also fractured. There are many Indigenous peoples' networks active in the contemporary climate regime from regions around the world. Some groups and networks have focused their efforts to try to ensure Indigenous peoples' rights are protected in the design and implementation of REDD+.[45] They have fulfilled bureaucratic roles in the UNFCCC such as serving on the contact group on REDD+;[46] collaborated with agencies such as the Inter-American Development Bank to develop social safeguards;[47] conducted extensive media work, coordinated events, and organized protests to draw attention to their demands; and lobbied on the need for safeguards in the Green Climate Fund and UNFCCC texts.[48]

Others have mobilized in coalitions completely against REDD+ and the UNFCCC regime, standing adamantly opposed to any framework that commodifies forests and natural resources in the name of climate change mitigation.[49] They have focused on mobilizing protests and demonstrations, generating media attention, issuing statements, and challenging forest carbon market policies at the state level.[50] Several groups have taken a mixed approach, at times standing completely against REDD+ and other times working for adequate Indigenous peoples' safeguards within the mechanism.

Indigenous peoples' networks working for reform within the UNFCCC have mobilized extensively for an international framework that would ensure that their rights are recognized and extended to the national and local levels. This was a particular concern in relation to a history of exploitation of Indigenous peoples' communities and their lands and also a reaction to emerging cases of carbon traders, known as "carbon cowboys," who coerce and manipulate Indigenous peoples' groups to sign over their land rights.[51]

Indigenous peoples' organizations also argued that many national and local governments would not uphold the rights of their communities unless there was an international framework with enforcement and incentives capabilities. Some groups also focused on ensuring that adequate co-benefits are available to Indigenous peoples' communities that did seek to participate in REDD+ activities. As Victoria Tauli Corpuz, executive director of the network Tebtebba and cochair of the contact group on REDD+ of the UNFCCC[52], explained:

> The biggest threat is that if our rights are not recognized, then we have the potential of being evicted, displaced, suffering more from the impacts not only of climate change but also the solutions like if they are going to build more dams, if they are going to build more biofuels plantations, we are at very great risk from the solutions. ... For communities it's a matter of life and death, it's the collective survival of communities that is now at stake.[53]

The goal of these efforts was for the climate regime to recognize existing international frameworks such as the UN Declaration of the Rights of Indigenous Peoples and the right to free and prior informed consent to the development of their territories.[54] While they were originally successful in including language in negotiation texts in support of UNDRIP and Indigenous peoples' rights in 2008, this language was later removed due to opposition from states including New Zealand, the United States, Canada, and Australia, the major nonsignatories of UNDRIP.[55]

The following year in Copenhagen, despite formalizing the role of REDD+ into the newly configured climate regime, there was no mention in the Copenhagen Accord of Indigenous peoples or their rights. However, building on momentum that Indigenous peoples' networks had established in the draft negotiation texts,[56] the negotiations in Cancun the following year represented various breakthroughs. The agreements included explicit reference to UNDRIP, called for a process to ensure social safeguards and compliance in relation to REDD+, and included recognition of Indigenous peoples' rights in various subsidiary bodies such as those related to adaptation and technology transfer. However, the phrasing of commitments to safeguards and upholding UNDRIP was weakly worded as "should be supported and promoted," and there is no explicit mention of the right to free and prior informed consent.

Then in Durban in 2011, due to developing country opposition,[57] safeguards in REDD+ were interpreted to merely "encourage" countries to provide qualitative data on how safeguards were implemented. Notably absent was any form of compliance mechanism, baseline safeguards to be upheld, and a requirement for collection of data before, during, and after projects to measure and report on impacts. After relatively quiet negotiations for Indigenous peoples networks in 2012 in Doha, new language on safeguards was included as part of the Warsaw Framework for REDD+ agreed in 2013. Specifically, the framework says that countries "should provide the most recent summary" of how safeguards have been addressed and respected before receiving results-based payments for REDD+. This encouraged the development of what are known as safeguard information systems, and called for an information hub to be coordinated under the Convention to keep track of country reporting on safeguards. However, there were no guidelines created on the specific categories and types of information that these reports should capture.

In Lima in 2014, networks focused on advancing three main issues: more robust guidelines for safeguard reporting systems, the inclusion of consideration of non-carbon benefits of forests (meaning the other benefits that forests provide beyond storing carbon), and a proposal from Bolivia for what it called the Joint Mitigation and Adaptation Mechanism, which was to advance a non-market approach to REDD+. Despite these efforts, no conclusion or decision was reached on these issues in Lima.

While the UNFCCC has done little to develop safeguards, external implementing agencies, such as the World Bank's Forest Carbon Partnership Facility and UN-REDD's Forest Investment Program have taken some measures of their own. This has included developing criteria and safeguards related to Indigenous peoples' rights for countries to uphold before being eligible for REDD+ finance. However, implementing agencies have received numerous complaints from Indigenous peoples' and civil society organizations concerning violations of their own standards related to participation, transparency, consultation, and consent.[58]

The World Bank has also come under scrutiny for its unwillingness to adopt a standard of consent required by UN programs, as compared to its weaker standard of "consultation."[59] Additionally, many Indigenous peoples' rights advocates have pointed to the fact that the attention that networks have brought to the issue of safeguards has encouraged numerous national-level processes to develop stakeholder engagement, legal frameworks, and reporting and accountability systems. While some highlight the inadequacy of these national processes, others argue that they are an important step toward enhancing Indigenous peoples' rights.[60]

Indigenous peoples' organizations and allies have also made extended rights interventions to exclude dams and biofuel plantations as eligible projects in the CDM; to prohibit the addition of nuclear energy in CDM

eligibility; to amend the definition of forests in the forest emissions accounting framework in order to exclude monocrop tree plantations; and
to ensure that REDD+ does not allow for the conversion of forest ecosystems into plantations.[61] Notably, language was included in the Cancun
Agreements that REDD+ ensure the conservation of biological diversity
and not be used for the conversion of natural forests. In addition, biofuel
plantations on peat lands were excluded from eligibility in the CDM in
2010.

Another set of network interventions has focused on expanding representation rights in the regime. In this regard, some have called for sovereign decision-making power, equivalent to that of states. As Tauli Corpuz
explains:

> We are not just stakeholders, we are rights holders. We deserve a seat at the table,
> but until now that has not been provided, we are just Observers, yeah, just like
> anyone else. But we are not the same as NGOs, we have clear rights, we have
> clear constituencies, we have our systems, we have our governance systems, so it's
> not the same, that's why we deserve a seat at the table. We own natural resources
> and they are in our territories and these are the ones that are being used or being
> conserved outside our own control.[62]

Indigenous peoples' networks have not made progress on this demand.
They have also recently called for less far-reaching demands for increased
representation and inclusion in the climate regime, with limited success.[63]
Finally, despite advocating for voting rights on the board of the Green
Climate Fund, Indigenous peoples' organizations did not even gain status
as official observers. In response to this exclusion, IIPFCC has called for
less-far-reaching representation rights such as a civil society and Indigenous peoples' advisory board and capability-enhancing funding to support Indigenous peoples' participation.[64] They have also made extended
rights demands for direct access to funding and social and environmental
safeguards that link to an independent, accessible, and effective compliance mechanism. While the Fund's governing framework does have sections on both safeguards and accountability, the IIPFCC has stressed that
this framework must uphold international UN standards toward Indigenous peoples' communities. Thus far, the board has done little to respond
to Indigenous peoples' demands.

Overall, most of Indigenous peoples' network rights gains have been
in the form of recognition, an area where these networks have also devoted significant organizing attention. Limited progress has been made
on enhancing representation, capability, or extended rights. Notably,
both implementing agencies and national-level bodies have developed

safeguards for working in Indigenous peoples' communities, forms of extended-rights, although these safeguards have often come under attack for their deficiencies.

## The Right to Livelihood

Waste pickers have a shorter history engaging in this regime than the other two cases discussed. In Copenhagen in 2009, their first major engagement, the Global Alliance of Waste Pickers and Allies (GAWA), was formed. This happened when two international NGOs, the Global Alliance for Incinerator Alternatives (GAIA)[65] and Women in Informal Employment: Globalizing and Organizing, joined up with waste picker organizations in Latin America and India.

They have since organized delegations of waste pickers from three continents to climate negotiations in Cancun and Durban, with smaller groups also attending the intersessional negotiations in Bonn, Germany, and Tianjin, China. Organizers from GAIA and numerous waste pickers have also attended bimonthly CDM board meetings since 2011 to make relevant interventions. GAWA now has participating organizations from eighteen countries and a governing board representing networks from Latin America, Asia, and Africa.

Climate change negotiations have been a key avenue for transnational waste picker movement building. Unlike Indigenous peoples' groups, which are fragmented in their demands and strategies, waste pickers and allies remained a relatively cohesive, albeit small, group in the negotiations, despite some disagreements about strategies and goals. At times, the process of building a transnational movement has been elevated above advocacy efforts in the regime as a priority.

Strategies that GAWA (sometimes working solely as GAIA) has pursued within the climate negotiations fall into three main categories. First, the network has broadly sought to increase recognition of how climate change policy is undermining waste picker livelihood. This has taken shape as demonstrations at the negotiations in Copenhagen, Cancun, and Durban, where the waste pickers have sorted out recyclables from the trash in view of passing delegates. Waste pickers and advocates have also regularly held press conferences, participated in educational side events, lobbied delegates, released briefing reports and case studies, coordinated a display booth, and blogged about their issues.

One of the most prominent gains through these activities has been extensive international press coverage. This has increased recognition of the value of waste picking, the need for dignified work conditions, and the

protection of waste pickers' basic rights. During the negotiations in Copenhagen, waste pickers garnered coverage in at least forty media outlets.
Such international coverage is unprecedented for a group that is often
treated as invisible. For example, a story featured in the news agency
Agence France Presse begins, "Ignored, marginalised or despised in many
countries, waste pickers from Asia, Latin America and Africa have come
together in Copenhagen to lobby for recognition as unsung heroes in the
fight against climate change."[66]

Second, GAWA has sought to influence the Green Climate Fund to
secure a funding stream directly to waste picker organizations, thus enhancing extended rights. They have expressed that this is critical to ensure
that recycling and waste picker livelihood are not replaced by mecha-
nized disposal technologies. They have been particularly vocal against the
proposal for a funding modality to directly finance the private sector.[67]
Like the other groups demanding a fund that is favorable to marginalized
groups, they have yet to realize gains in this area.

Third, GAWA's most concentrated work has focused on gaining extended rights for waste pickers through reform of the CDM's waste
methodology. They have argued that most CDM municipal waste management projects are problematic for three main reasons:

- They help perpetuate waste management strategies that prevent truly
  sustainable and more cost-effective options.
- They usually threaten the livelihoods of waste pickers, some of the
  poorest people in developing countries' cities.
- They overestimate the GHG [greenhouse gas] reductions that can be
  attributed to these CDM projects.[68]

GAWA has challenged CDM waste management protocols and practices on several grounds. They have challenged a small-scale plastics recycling methodology as biased toward private large-scale recycling firms
and inaccessible to waste pickers due to onerous restrictions, high entry
costs, and limited benefits for small firms.[69] Another effort has challenged
subsidies to incineration technologies on grounds that it undermines the
Stockholm Convention on Persistent Organic Pollutants, a UN treaty.[70]
The network has also advocated to improve transparency and civil society intervention procedures at the CDM, an effort to enhance their capability rights in the regime.[71] These engagements did not result in notable
gains.

However, GAWA found some success in its main intervention: the
network worked to revise the waste-to-energy methodology to account

for preexisting waste picker recycling in baseline greenhouse gas calculations. This can have a substantial impact on analyses of which approach is more beneficial to the climate. Waste pickers made impassioned statements at CDM executive board meetings in Copenhagen and Cancun regarding the injustice of putting waste pickers out of work through the financing of carbon-intensive waste technologies. This was met by responses from the board that clarified that it is not within their authority to provide social safeguards.

However, in 2012, the CDM board said that revisions to the waste-to-energy standards would be made and tasked its methodological panel to propose an alternative methodology.[72] Subsequently, this panel recommended that the board revise the methodology to require that eligible projects don't reduce recycling.

GAIA then intervened with a statement calling on the board to approve the suggested revisions.[73] They mobilized an e-mail action alert that resulted in each board member receiving 300 e-mails from people around the world encouraging them to stop carbon credits for incinerators and landfills. Finally, just two days before the board meeting, GAIA released a sign-on letter to the European Commission urging it to end the purchase of CDM carbon credits from incinerators and landfills. The letter argued that the EU is financing projects through the purchase of CDM credits that would be illegal in the EU due to its various waste directives. The letter was signed by ten members from the European Parliament from various political parties and civil society organizations from twenty-three countries. In a news story about the letter, one parliamentarian said, "The CDM must stop issuing credits to counterproductive waste projects. Otherwise, parliament will be forced to cut off support."[74] As the largest market for CDM credits, weakening EU support to CDM waste projects posed a threat to the CDM's viability in this area.

In July 2012, the board approved the new recommended methodology. While this was viewed as a large victory, it is not clear that this condition will be respected and enforced due to the absence of a formalized procedure to ensure accountability. In addition to technical arguments, GAIA staff member Neil Tangri emphasized the political and normative war that they have raised at the CDM in advancing their demands:

What stirs them to action is precisely that [political pressure]. You can show them a project that can be absolutely devastating in terms of emissions ... and it will take them five years to get around to maybe doing something. But when we got out in front of the television cameras and we're saying they're taking away livelihoods of poor people, they paid a lot more attention. There is a disconnect between what they are allowed to take account of formally, and what actually motivates CDM panel members to do something.[75]

Overall, waste picker network engagement resulted in increasing the recognition of waste pickers as a constituency with rights in the media, with foundations such as the Clinton Initiative, and with institutions such as the World Bank.[76] Additionally, they have had moderate success in their extended rights interventions at the CDM, with which they have complemented engagement on scientific grounds with various strategies to apply political and normative pressure (see table 8.2).

## Assessing Gains

We have explored three case studies of networks working for climate justice in the UNFCCC process. Given what we have discussed, what types of rights gains have been achieved by the networks overall? In the cases discussed, most rights gains have been in the form of recognition. Indigenous peoples' and gender equality networks have devoted considerable time and resources to gaining recognition. Recognizing certain rights in regime texts has been a relatively simple way for the Conference of Parties to legitimize the regime activities in response to network advocacy efforts without necessarily changing actual practices.

While seemingly a weak gain, such recognition may be important for groups as a springboard for future rights gains internationally. It also may prove important in national contexts where groups can leverage codified international norms to put pressure on governments and corporations (what Keck and Sikkink call the boomerang effect). However, we still have limited understanding of the conditions under which such recognition is instrumental as compared to irrelevant, and thus it is not clear when this has been a worthy use of time and resources by the networks engaged.

In turn, the networks have made few inroads to achieving core representation and capability demands. Increasing representation of women has been taken far more seriously by the regime than improving Indigenous peoples' representation, despite considerable attention to the issue by both networks. This is not surprising, given that women already hold elite positions of authority in society and within the regime, including the position of secretariat, while Indigenous peoples and waste pickers do not. Also, increasing women's representation does not conflict with the principle of national sovereignty, whereas enhancing Indigenous peoples' representation presents a challenge to this bedrock concept of the UN. However, even with modest improvements in gender representation on state delegations, it is not clear that this has translated into greater capability for decision making by women in the regime. Nor is it evident that this has led to more gender-sensitive policies as they extend from the regime.

The networks have been only moderately successful in achieving relevant extended rights. Developing state delegates have often opposed extended social and environmental protections in the name of preserving state sovereignty. However, several international development agencies, which are often beyond the direct influence of developing state leaders, have enacted their own protocols and programs to incorporate extended rights related to Indigenous peoples' safeguards. In addition, numerous domestic-level processes have emerged with the expressed purpose of ensuring Indigenous peoples' rights related to REDD+ activities. While the adequacy of these measures has been widely critiqued by Indigenous organizations, this suggests that policy diffusion has likely been an outcome of Indigenous peoples' engagements in the climate regime.

Finally, the waste picker network GAWA intervened to reform the CDM to secure extended rights on scientific and objective grounds and also by challenging practices on normative and moral grounds. The CDM subsequently changed its methodology to be in line with network demands. However, the long-term impact of this gain is unclear. The absence of a formalized procedure to ensure accountability related to the new methodology remains a concern. In addition, with the CDM facing financial problems (due to a host of issues), the network's victory may be most relevant for setting precedent for other institutions to value informal recycling services.

We now discuss in the chapter conclusion what this analysis tells us in terms of the possibilities for advancing climate justice in international regime politics.

### Stuck at the Margins

The purpose of this chapter has been to elucidate the ways that networks have engaged in the international climate regime to contest varied forms of marginalization and inequality related to climate injustice. The cases explored include networks with very different organizational constructs, histories, class dynamics, strategic approaches, resources, political opportunity structures, and issue concerns. Despite these differences, the analysis in this chapter suggests that the pockets of climate justice network

**Table 8.2**
Regime Rights Gains in the UN Climate Regime

| Rights type | Extent that network demands have been met |
| --- | --- |
| Recognition | *Moderate-high*<br>• Indigenous peoples and gender equality rights frameworks from other international regimes have been recognized.[a] However, the recognition of the UN Declaration for the Rights of Indigenous peoples does nothing to change the status of nonsignatories, including the United States, Canada, New Zealand, and Australia.<br>• Language has been included related to recognition of Indigenous peoples' safeguards[b]; disproportionate vulnerability of Indigenous peoples, women, and children to climate change[c]; gender sensitivity[d]; the value of traditional forms of knowledge[e]; and the right to gender equality and full and effective participation of women and Indigenous peoples.[f] However, recognition language has often been watered down with phrases such as "should consider" and "it is important."<br>• The principle of free and prior informed consent has not been explicitly recognized for Indigenous peoples.<br>• Waste pickers have gained recognition as a constituency in the media, with institutions like the World Bank, and with foundations. |
| Representation | *Weak-moderate*<br>• Gender and climate change was added as a standing item on the agenda of sessions, and an in-session workshop on gender balance was established.[g]<br>• Indigenous peoples and women and gender equality caucuses and constituencies were established.<br>• There has been a modest increase in the number of female state delegates and country team heads. However, after two decades of negotiations, women make up only 32 percent of delegates and 19 percent of country team heads.[h] Only 15 percent of Green Climate Fund board members are women, and none are Indigenous.<br>• Despite extensive organizing, Indigenous peoples' groups have gained almost no representation rights in the regime. |
| Capabilities | *Weak*<br>• Third-party actors (such as World Bank and individual states) have organized and funded meetings for Indigenous peoples' organizations and relevant agencies.<br>• No trust funds or capacity-building mechanisms have been established specifically for Indigenous peoples or gender equality constituents in the climate regime.<br>• Responses to improve waste picker participation in the CDM, such as nonstate intervention practices, have been weak or nonexistent. |

Table 8.2 (continued)

| Rights type | Extent that network demands have been met |
|---|---|
| Extended rights | *Moderate* <br> • The Green Climate Fund governing framework has sections on safeguards and accountability; however, it is not clear how these will be interpreted and enforced. <br> • Implementing agencies for REDD+ have developed safeguards requirements for national and local governments; however, there are major shortcomings. <br> • REDD+ safeguards were established in the climate regime; however, they are weakly worded and without robust requirements. <br> • The Cancun Agreements say that REDD+ should ensure "this decision is not used for the conversion of natural forests."[i] However, this text is weakly worded and without accountability measures. <br> • The CDM executive board revised its baseline methodology to require that eligible projects don't reduce recycling.[j] However, there are currently no criteria to validate that CDM waste projects meet this requirement, and several other demands of GAWA at the CDM were not met. <br> • Biofuels from peat lands were excluded from CDM eligibility, and nuclear power was not granted eligibility. |

a. FCCC/CP/2010/7/Add.1.E; FCCC/CP/2010/7/Add.1.Appendix1/2.footnote 1; FCCC/CP/2012/8/Add.3.Decision23.preamble

b. FCCC/CP/2010/7/Add.1.Appendix1/2; Also see FCCC/CP/2010/7/Add.1.72

c. FCCC/CP/2010/7/Add.1; also see FCCC/CP/2012/8/Add.1.18.7aiii

d. See FCCC/CP/2011/9/Add.1.I.3; FCCC/CP/2011Decision3/CP.18.7b; FCCC/CP/2011/9/Add.1.B.3; FCCC/CP/2010/7/Add.1.E).

e. FCCC/CP/2010/7/Add.1.Appendix1/2.c; Also see FCCC/CP/2010/7/Add.1.12

f. See FCCC/CP/2012/8/Add.3.Decision23.preamble; FCCC/CP/2011/9/Add.1XII.71; FCCC/CP/2011/9/Add.1.V.31; FCCC/CP/2010/7/Add.1.7; FCCC/CP/2011/9/Add.1.I.3; FCCC/CP/2010/7/Add.1.Appendix1/2.D; FCCC/CP/2011/9/Add.1XII.71; FCCC/CP/2011/9/Add.1.V.31

g. FCCC/CP/2012/8/Add.3.Decision23.10

h. Women's Environment and Development Organization 2012.

i. FCCC/CP/2010/7/Add.1.Appendix1/2.e

j. Approved baseline and monitoring methodology AM0025 version 14.0.

influence have been small and have rarely (if at all), in Gramscian terms, touched core relations of power and inequality.

However, the networks investigated have made some important gains toward greater climate justice, which may very well have notable direct impacts on the lives of the marginalized. This chapter suggests that future research should provide greater insight into the specific conditions under which networks representing vulnerable actors are more or less likely to realize distinct forms of regime rights, particularly those that force change beyond measures at the margins.

The impending reality of a warming world necessitates strategies for building new coalitions in the present that seriously address the welfare of those most at risk of dispossession and disproportionate harm. A neo-Gramscian strategic approach to analyzing power relations directs attention to the ways in which those who are most disproportionately affected by the current political economic order engage to contest inequality and work for social and political change. The presence of these groups in policymaking processes offers an important check to the deep contradictions and inadequacies of the current development model. However, if this is not linked to a broader counterhegemonic mobilization and vision, such networks will likely continue to exist at the margins, achieving only small and incremental shifts in policy.

This analysis also points to the importance of understanding the intersections and gaps between governance processes at different levels. We have seen that international processes for rights safeguards may well have implications for domestic institutions and the ability of actors in domestic contexts to leverage these institutions for climate justice. But we still have limited understanding of the conditions under which these institutional intersections can be usefully leveraged by relevant social change coalitions.

Escalation in severity and frequency of disasters affecting marginalized peoples is a likely consequence of climate change in the coming years. Such hardship could inspire more militant and broad-based movements to build the pressure for stronger and more meaningful international and local rights for vulnerable groups. Such a context might also inspire growing movements that challenge neoliberal and market-centered models of development, which often neglect and even exacerbate current forms of inequality.

But it is perhaps just as likely that such a context of crisis would undermine organizing capacities of those on the margin while simultaneously encouraging those with power to focus on their own survival. For example, a warming world could shift the attention of powerful actors and more technocratic environmental NGOs to a narrow focus on geo-engineering strategies and protecting their own coastlines, while neglecting the plight of the most vulnerable. The next chapter plays out some of these scenarios to consider the implications of major decision points coming in the near future.

# 9

## Power in a Future World

### Scenarios Dark, Light, and Gray

Now it's time for the fun stuff: some unvarnished speculation about what's to come. In this chapter, we seek to do something we've not seen done in quite this way: exploring possible futures not built on computer modeling but informed by what all the foregoing suggests are likely or more hopeful outcomes in the next decade or two of climate politics. Futures research that develops storylines about possible outlooks often differ from what we do here in that our effort focuses on political drivers and equity implications rather than trajectories of emissions. We locked ourselves in a room, built a framework for the types of futures we could envision, and pushed ourselves to play out the likely outcomes.

There is much that we know is needed to successfully address climate change. Fundamentally, we need to switch to renewable energy sources and off the fossil fuels that are dumping gigatons of stored carbon into the atmosphere. Consumption levels and waste need to drop sharply, and that drop must begin quickly. We hear sometimes that there are lots of "cobenefits" to moving to a more "climate-sustainable" society. For example, shifting to sustainable communities can bring the potential for good green jobs, reduced commuting times, and less time and money spent caring for all of our material possessions: psychological studies have shown that these are often good for overall human happiness. Rather than taxing production or other goods, we should be taxing the "bads," like pollution and waste. As *New York Times* columnist Thomas Friedman put it, a carbon tax would be "win-win-win-win win" for the United States. It would reduce harmful emissions, generate funds for the deficit, "weaken petro dictators, strengthen the dollar, drive clean-tech innovation and still leave some money to lower corporate and income taxes."[1] But Friedman also noted that it is entirely off the table, while self-defeating proposals are on it.

This much is very clear: green futures don't happen automatically. Having the basics of development such as health care, housing, education, and basic energy is important, and in a perfect world one doesn't need to burn fossil fuels to have them. But getting from here to there will not happen without addressing power. In our understanding, power is the key variable in driving different scenarios: it can be a progressive or regressive force. As we discussed in chapter 6, there will be resistance by the massive private fossil fuel companies and vast state-owned petroleum bureaucracies in nations where oil extraction is controlled by the state. Those who are accustomed to cheap fossil fuels, and especially those profiting from their revenues, will likely resist the kind of change we need to wring the carbon from our economy. With this understanding of power in mind, the overall picture is that unless there is a strong challenge to these interests, this will lead us all to very unsustainable futures.

We begin the chapter with the dark and depressing scenarios, then we move on to the gray ones, then some quite brighter possible futures. We end up by reviewing the scenarios which we think are more likely and what we can learn from them, especially about how to get to the better ones.

We are in the uncomfortable position of writing scenarios that could well be removed from the realm of possibility by the time this book is published. For example, a key player blocking past international climate action has been the United States, including, to the surprise of many, President Barack Obama. After his largely unanticipated disappointing effort in Copenhagen in 2009, many hope that he can still help to address this tough global commons issue.

Indications in his second term provide some encouragement that he will follow in the footsteps of other leaders like Canadian Prime Minister Jean Chrétien on the issue of climate change. In 2004–2005, despite the fact that he was retiring and there was strong internal resistance, Chrétien pushed an increasingly oil-dependent nation to take domestic action on climate change, ratified Kyoto, and pushed his citizens to each strive for cutting their emissions by a ton of carbon—from 6 to 5 tons. All of this was undone by his successor, Stephen Harper, and so we are reminded that political opportunities are critical, but popular perceptions of the short-term national interest (driven by who is most effective at controlling the public discourse) often get the last word.[2]

A major unknown based on chapter 3's description of the hegemonic transition with the decline in US power is whether key nations will become more isolationist and statist, or more globalist with a reorientation of the national government's role. Globalist visions can emerge even while small groups of people control the direction of society, as transnational corporate interests and a transnational capitalist class drive a version of globalization that serves their ends by freeing themselves from regulatory states.[3] As Argentine-Brazilian political scientist Eduardo Viola points out, elites within nations can be either environmentally conscious or unconcerned about the global atmospheric commons, and these elites can be interested in globalizing their economies or putting up national protective tariffs.[4] During economic expansion periods, there are more generous attitudes toward imports and foreign investment, but in tougher times, major powers tend to protect their national interests.

However, the US-China partnership in 2014, where these two superpowers jointly announced future targets on climate change, suggests otherwise.[5] Though not adequate in scientific terms, this bilateral cooperation represents an important political breakthrough. Whether the United States can adhere to its targets through political transitions of the presidency and Congress is another question. Our hunch is that in the foreseeable future of insecurity and uncertain transition of global hegemonic powers (see chapter 2), nationalist strains will be more likely to dominate. But much depends on how civil society responds in this context to pressure these key states in one direction or another. In the medium or longer term, perhaps we might eventually reach a time when a globalist orientation in the United States becomes the norm.

Whether China will step up to take aggressive action nationally in the short-term period to slow its steep emissions rise and boost its leadership role in the UN Framework Convention on Climate Change (UNFCCC) negotiations presents another important question. There have been major smog inversions in Beijing and increasing pressure on the government to reduce its booming consumption of coal based electricity and vehicular smog.[6] In spite of vast investments in solar and wind energy,[7] China's carbon dioxide emissions have skyrocketed since about 2004: the country passed the United States in about 2007 in total emissions and in 2013 was estimated at having nearly double US emissions.[8]

At the same time, China has surged into world leadership in solar and wind power, and coal desulfurization. Is China's new target to end its upward trajectory in emissions by 2030 indicative of a meaningful shift? Will the several-year gap until there is an international framework in 2020 mean more skyrocketing emissions in the meantime? We do not see as inevitable the rise of strong civil society and democratic institutions in China (gaining economic freedoms appears to have priority over political freedoms), but the strongly centralized government there does allow some very aggressive planning and progressive action that is not often possible elsewhere. Concern over China's urban air pollution may drive a new level of political activism. The question is whether China's development model will be sustainable politically, economically, and environmentally.

One factor that might present a bigger question mark globally if we were doing this kind of work in the 1970s is population; today that is something less of an issue. Population growth rates are dropping quickly in most places as economic growth rises, but behaviors in some key countries remain uncertain. The UN Population Programme's medium-variant models project that global populations will rise from the 2013 population of 7.2 billion to level off at 10.7 billion people around 2100.[9] This is a lot to sustain, and it could be too much, depending on how we all consume the Earth's resources. Birth rates are dropping more quickly than predicted in some places, and beyond the death toll from the horrible AIDS epidemic in Africa, the economic advancement of women, and economic opportunities, a few more factors might drive them down further: endocrine disruptors, same-sex marriage, and renewed efforts to meet birth control needs worldwide. But much of the population trajectory remains highly uncertain.[10]

Carbon emissions, like other environmental impacts, are the result of three factors: more people ($P$ for population), the amount of stuff each consumes ($A$ for affluence), and how efficiently they consume it ($T$ for technology). The level of affluence in the future may be more important than the birth rate. The recipe for reducing emissions using this formula, Impact = $P \times A \times T$ must be on all fronts: education about reproduction and strong population control efforts (reducing the growth of $P$), very strong efforts to reduce consumption sharply in the wealthy countries and among elites in all nations ($A$), and factor 5 or 10 improvements in efficiency ($T$).[11] There will need to be strong collective action to make any or all of these three happen at a sufficient rate to prevent 1.5 or 2 degrees Celsius of warming and to secure those improvements in the face of a larger population and more money in the economy.

To understand our six scenarios, we developed a typology of climate futures. Collectively, humanity might move toward an adequate solution by reducing emissions sharply and in time with a pathway required by our best understanding of climate science to stay within our global carbon budget.[12] But whether or not we take adequate precautions to avoid

**Table 9.1**
Two Dimensions of Potential Climate Futures

|                 | Weak democracy/equity | Strong democracy/equity |
|-----------------|-----------------------|-------------------------|
| Strong adequacy | Exclusive action      | Climate justice         |
| Weak adequacy   | Exclusive inaction    | Democratic dysfunction  |

dangerous climate change, we can move forward in a way that is strong in terms of addressing social equity and democratic participation, or exclusionary and exclusive. These are of course continua combining many dimensions themselves, but for simplification, they produce four possible sets of scenarios: exclusive adequate action, where climate change protection is adequate but with limited attention to democratic process; climate justice, where democratic action combines with strong climate protection; exclusive inaction (the status quo), where climate change protection is inadequate and with limited democratic process; and democratic dysfunction, where justice and democracy are dominant modes of action, but they are uninterested or ineffective in addressing climate change (table 9.1). For each scenario we look at:

- who and what's pushing them to happen
- obstacles to their happening
- likelihood in our assessment
- what would drive it forward
- what it might look like
- outcomes—what it means for climate change and social impacts like inequality, international stability, and so on

A summary table on the six scenarios and the conditions is provided in table 9.2. We begin with the continuation of the status quo, the scenario of exclusive inaction. This is based on the dominance of an elite group of states that are unwilling to advance global action on climate change.

## Scenario 1: Exclusive Inaction

We'll get right to the really bad stuff. We envision as very possible a world where all negotiations break down and major nations grab at resources such as fossil fuels and water. This scenario we originally called Mad Max (or more accurately the sequel, Road Warrior) to reflect the dystopian future visualized in that movie. In that postapocalypse film, bands of renegades pillaged settlements that had any resources such as water and gasoline. Common sense would suggest that constraints on natural resources will inexorably lead to increased internal and international conflict. A growing cottage industry is springing up projecting how climate change will drive disasters and vulnerability leading to increased conflict, terrorism, and millions of transborder refugees. While wealthy nations may be able to obtain the fuels and materials they need, some poor nations will be virtually powerless in this system.

In this future, totalitarian regimes could become far more prominent. This scenario could also conceivably come to be within the current capitalist and (some would say nominally) democratic system. We are already seeing a new wave of land grabs to secure access to agricultural land and water and investment in extracting oil and coal and other precious resources.[13] The current wave is dominated by China and other BRIC nations (Brazil, Russia, India, and China), especially Brazil and Russia, but there is not too much difference from past colonial cycles dominated by the British, North Americans, or Japanese seeking access to energy and minerals. "Extreme energy" such as hydraulic fracturing is shifting geopolitical dynamics, and there will be strong geographical imperatives for businesses and nations to secure access to these resources across the planet.[14]

If we use fossil fuels at the current pace, we are headed to more scarcity and probably more conflicts, including between communities and capitalist interests and the governments that seek to make their businesses feasible and profitable. Land conflicts appear to be escalating, and other profound struggles are emerging from climate solutions, such as reducing emissions from deforestation and land degradation (REDD), creating or exacerbating conflicts over land, and competition between the production of biofuels and food (or for standing forests and the peoples who inhabit them, as discussed in chapter 8). There are also conflicts that are resulting from a changing climate, such as the dash for Arctic and Antarctic resources and transportation routes, and the growing flow of refugees from areas hit by drought and other climate change impacts. We can foresee conflicts over sea lanes and potential minerals at the poles, along with conflicts over renewable resources in places like river valleys for hydropower and forests for soybeans grown and palm oil.

We get to these awful outcomes in this scenario if we posit a continuation of the current situation: realist projections of short-term national interests prevail, and there are no binding or effective regimes to address the core of the climate issue. We could arrive at this negative outcome by two routes: the bottom-up route (such as that introduced in the Copenhagen Accord), which leads only to inadequate emissions reductions, or top-down structures with weak targets or weak compliance regimes (such as weak or nonexistent penalties when targets are not met). This might be the result of either faulty market approaches, as we've seen with several cases so far, or the result of weak national targets, as in the Kyoto Protocol.

This is our business-as-usual scenario, and breaking free of it will not be easy. National priorities tend to win out in the key countries (especially China and the United States), where no adequately strong social movements have been able to push governments to look beyond the perceived short-term interests of the politicians and business elites that influence them. These politicians do not wish to risk appearing radical in the sense of endangering economic growth and the jobs it is expected to create.[15]

There have been very few voices in the twenty-some years of climate negotiations that directly challenge growth-oriented market-based approaches. The growth imperative of capitalism is so hegemonic that it is almost never questioned. Bolivia's interventions in Cancun were among a few such statements over the past five years. At the conference there, lead negotiator Pablo Solón passionately defended the rights of Mother Earth, as was stated in the 2010 World People's Conference on Climate Change and the Rights of Mother Earth in Cochabamba, Bolivia.

With business as usual, we have seen that very little attention is paid to challenging the central structures of development. The machinery of science is focused on production, not environmental impacts.[16] That is, billions more in government and private research funding go to developing more extreme forms of fossil fuel extraction rather than, for example, developing cellulose-based biofuels.[17] Corporate efforts at greening may make marginal differences in climate impacts overall (especially in the face of rising costs, shortages, or great uncertainty), when massive amounts of carbon continue to be put into the atmosphere.

Our business-as-usual scenario of exclusive inaction may lead to 6 degrees Celsius of average warming by the end of the century. This is predicted by the mainstream International Energy Agency; and as cited earlier, this could mean higher increases in temperatures in Africa and South Asia, where it is difficult to see how life could go on in any fashion like the present.[18] Some theories of social change posit that things have to get worse before civil society will be sufficiently agitated to get active and fight for a global public good like preventing climate change. Or people

**Table 9.2**
Description of Characteristics of Climate Futures

| | Fair sharing of atmospheric space? | Broad participation in decisions by society? | International cooperation and coordination? | Strong civil society? | Social change focus? (versus primarily technological) | Public goods focus? (versus primarily private goods) |
|---|---|---|---|---|---|---|
| Scenario 1: Exclusive inaction (Copenhagen+) | No | No | Nonbinding | No | No | No |
| Scenario 2: Exclusive action (MEF, grandfathering) | No | No | Yes (but limited to the powerful) | No | No | No |
| Scenario 3: Technofixes: Geoengineering and CCS | No | Likely not | Likely no | No | No | No |
| Scenario 4: Global drive for cheap renewables | No | No | Potentially yes | No | No | No (could include subsidies) |
| Scenario 5: Going local: Fragmented; no adequate global governance | Likely no | Yes | No | Yes | Yes | Yes |
| Scenario 6: Strong national action with international treaty | Yes | Yes | Yes | Yes | Yes | Yes |

could be so busy trying to survive the disruption and large-scale inequality that they could be unable to organize. Under such conditions there would need to be massive transfers of funding for developing countries to adapt to the increased droughts, flooding, and heat waves. But in such a divided world, it is difficult to imagine there being substantially more funding for adaptation. All in all, this is a worst-case scenario, but it is also our expectation of world society's direction if business continues as usual.

### Scenario 2: Exclusive Action

Here we project ahead from what we see as lack of progress in the formal UN climate negotiations, leading to the potential delegitimization of the UN as a space for collective action, and its incapacitization to address the issue. Some argue that the UN is inefficient or unreasonably cumbersome, and the structural worldviews of the aggrieved parties has become too pronounced and bitter to find shared focal points on which to build a compromise agreement on climate change.[19] Those aware that international cooperation of some sort is needed are increasingly turning to "minilateral" or "plurilateral" forums to solve parts of the issue. This minilateralism could be of quite different types, and some forums already exist: the Asia-Pacific Partnership, sectoral groupings such as ones on black carbon or mini–nuclear generators, the G7, G8,[20] G8+5,[21] G20,[22] and the Major Economies Forum on Climate and Energy (MEF).

We'll focus here on the MEF. After "unsigning" the Kyoto Protocol immediately on assuming office in 2001, George W. Bush began developing a series of minilateral approaches, including, in 2007, the Major Economies Meeting on Energy Security and Climate Change. A White House press release from September 2007 argued that the meeting was "intended to reinforce and accelerate discussions under the UN Framework Convention on Climate Change and contribute to a global agreement under the Convention by 2009."[23] However a Heritage Foundation editorial from that time argued that the meeting reflected a completely different approach: voluntary targets that would foresee the pledge-and-review system adopted in Copenhagen.[24]

Upon arriving in the White House, President Barack Obama renamed the group the Major Economies Forum on Energy and Climate. This new MEF was officially launched in March 2009 "to facilitate a candid dialogue among major developed and developing economies [and] help generate the political leadership necessary to achieve a successful outcome at the annual U.N. climate negotiations."[25] The group, whose members include Australia, Brazil, Canada, China, the EU-27, India, Indonesia, Japan, Korea, Mexico, Russia, South Africa, and the United States, has met twenty times since then. If you add them all up, over four-fifths of all contributions to fossil fuel greenhouse gas emissions in the world are represented (81.3 percent in 2010). An ambitious agreement among these countries would cover five times more of global emissions than the current commitment period of the Kyoto Protocol, which covers only 15 percent of the global total.

In April 2013, together with Italian geographer Marco Grasso, Roberts put forward a proposal in a briefing paper for the Brookings Institution for how a deal among the MEF countries could break the impasse in the climate negotiations.[26] In this scenario, MEF countries would make a deal behind their closed doors and then go public with it, at the UNFCCC negotiations in 2015 in Paris, or some future negotiations. The authors proposed that early deal making might be limited to the seventeen countries in the MEF, accounting for 81.3 percent of global emissions, including the EU, United States and China, India and Brazil. They argued that the deal could then be brought back into the UNFCCC as soon as possible to quickly expand its reach, especially to the smaller wealthy countries. The briefing also called for allowing Least Developed Countries to continue to develop without emissions reduction obligations.

Is such an approach feasible, and if so, what might be its repercussions? While there was a time that it seemed unlikely that the biggest polluters could come to any agreement on reducing emissions, the November 2014 joint announcement by the US and China points to the possibility of future bilateral collaborations of this nature. Only if the United States and China unite on this issue, our thinking goes, can they impose a new framework approach on UNFCCC.

This minilateralism/MEF scenario with adequate action appears less likely to us than scenario 1. It could happen inside the UNFCCC or outside it. It could be binding and top-down as in the Brookings MEF proposal, but if the MEF is made up of only the biggest polluters, we consider it extremely likely these states would hog the emissions rights. This is essentially what the Brookings approach does: it allocates of 81.3 percent of emissions rights to the nations that have emitted 81.3 percent of the carbon over the past two decades.

A key question is what might be the response of the smaller nations that were not invited to the MEF meetings. Having not been consulted, the Alliance of Small Island States (AOSIS), the LDCs, or the Alliance of Bolivarian States (ALBA) might see an undemocratic and inadequate regime and choose to fight to block or reject any pact. Procedural justice is extremely important for building and maintaining international cooperation. Bringing these three negotiating groups together amounts to 98 countries, so any treaty without them lacks legitimacy and would never pass UNFCCC procedure. We see four outcomes relating to the 152 countries not in the MEF in this scenario, and especially regarding whether the LDCs will be extended atmospheric space and green technologies to develop, if and when their economies do start growing quickly. These we list from least to most likely.

First, the worst outcome would be an arrangement where the MEF gives itself all the remaining atmospheric space. We view this as very unlikely, since it would be dismissed as unjust for the remaining 152 countries, which would be left with no fuels to burn. If they are without allocations within the atmospheric space still available to stay under 2 degrees Celsius, they will immediately understand that they are being offered a scientifically inadequate treaty.

Second, and only slightly less draconian, is that the MEF emitters decide to allocate some atmospheric space to the poorest countries, given their need for growth to address basic social development needs. The big powers have only rarely shown this kind of generosity in the past and may not see a reason to in the future. However, climate change is a somewhat different issue, and China is emerging as a new hegemon but still has important elements remaining of its global South (G77) identity. It also has a culture that is more communitarian, at least internally.

Third, we see as somewhat more likely the scenario where the major polluters offer a strategic concession such as green technology transfer and aid to soften the blow of leaving little atmospheric space for the poorest countries to develop. This reflects behavior we saw in Copenhagen, in which major promises of finance were made, thus allowing the legitimization of a regime that LDCs and AOSIS nations viewed as inadequate and inequitable (as discussed in chapter 4). China and the United States are major aid providers and development investors, so we expect great hesitation to cross them among low-income nations reliant on that aid and investment. However, we see a fourth scenario as more likely: green technology transfer and development assistance promises are made but not fully met. Mitigation promises themselves are made but only partially and inadequately fulfilled, and the situation reaches a crisis.

This brings us back to our typology proposed at the start of this chapter (table 9.1), contrasting exclusive inaction—scenario 1's Mad Max future—and the exclusive action of this scenario's minilateralism. There is a split between different types of environmentalists in the North and in some other MEF countries. First are those who are focused on the sustainability part—that global greenhouse gas levels are dangerously high and rising so fast that any coalition of the willing that can make a difference should be created. A second group cares profoundly about reducing emissions in a way that is attentive to the need for development rights, equity, and justice.[27] Our observation is that the more powerful of these two groups in Washington, DC, in driving Congress and the administration are the bioenvironmentalists, who focus on efficacy, and not much on justice, as social greens do.[28] An important question is how strong climate justice groups and their discourse are in BASIC countries and in other key states in the MEF. A related scenario arises when failure to adequately mitigate emissions is met with the need and willingness to deploy what we call desperate technofixes: engineering the climate.

### Scenario 3: Desperate Technofixes

It is almost an inevitable outcome of scenario 1, where fossil fuel industry interests and our consumptive lifestyle habits are not questioned soon enough, and desperate measures are needed as overall temperatures soar or climate disasters reach tipping points.[29] Inaction is likely whether it is exclusive and the result of powerful nations blocking mandatory action or inclusive and democratic as the result of messy international processes and poor enforcement.

As discussed in chapter 5, the need and willingness to resort to desperate technofixes like geoengineering the climate fall into two big categories: solar radiation management (SRM) and carbon dioxide removal.[30] Blocking some of the sun's rays is the more controversial: it could take the form of sending jets over the high arctic to release trails of aerosols into the high atmosphere to block some of the sun's power or launching trillions of tiny mirrors into space.[31] Among the most controversial kinds of carbon absorption would be dumping massive amounts of iron onto the ocean shelves in order to to fertilize algae to gulp down large amounts of carbon dioxide into the deep oceans. Greater attention to date has been spent on building huge pollution collectors on power plants for carbon capture and storage (CCS), to pump it underground into deep formations.

We lump these two different sets of solutions together because both are technofixes that don't question or challenge powerful interests, especially fossil fuel companies. They are technological adaptations that pose high risks and face great uncertainties as to whether they'll work, especially at the scale needed to address climate change. Both have potentially destabilizing implications if they are attempted, and especially if they fail. For CCS, for example, the release of stored carbon is possible, and it could potentially be deadly if the carbon dioxide settles in low-lying areas where people, especially children, might pass through and suffocate. Concerns have been raised that solar radiation management might interrupt the monsoon rains in India, reduce solar energy for photosynthesis (and agriculture), and potentially destabilize ecological systems.[32] It also fails to address the long-term problem of rising carbon dioxide concentrations in the atmosphere and ocean acidification, which endangers key phytoplankton coral and shellfish species at the bottom of food webs.

Our most likely scenario is that things start getting bad: the climate has destabilized, food systems are failing in an increasing number of locations, and then some individual or nation unilaterally begins dumping iron in the ocean or starts spraying aerosols in the arctic to block the sun. This scenario could unfold in nontransparent and even covert actions, possibly done between allied nations, and testing could already be underway. Uncoordinated and ungoverned efforts are worrisome, but since there are grave concerns about even allowing scientific research in this area, we see coordinated global action on geoengineering as unlikely in the short or medium terms. In either case, geoengineering could have unforeseen consequences, causing desertification or extreme rainfall, freezing extremes in the arctic, or disruption of equatorial climates.[33] The global climate system is so complex that attempting to tinker and manage it with interventions is almost certainly an example of great hubris. Unfortunately we see this scenario as increasingly likely, given the direction of the world economy, political developments we have described and technological developments already underway.

This scenario could unfold in many ways as the result of several of the other scenarios. CCS has received billions in stimulus money and state subsidies in the United States and elsewhere.[34] If the global shift from coal to natural gas continues at the current pace, we expect pressure on the natural gas industry to also sequester much of its carbon dioxide, in addition to preventing methane leakage throughout the life cycle of production.

However, our assessment is that CCS will ultimately prove too expensive, too risky, and not effective enough to have a major negative impact on emissions. In particular, we expect much carbon dioxide will be captured at the site of combustion and used for "enhanced oil recovery" and prospecting for geological formations, practices that will actually increase emissions. The oil extraction industry is taking this approach as a way to make CCS affordable, but it means spending more energy to extract more fossil energy to power the process of capturing the carbon. Estimates for decades have been that capturing carbon, transporting it to a stable geological formation, and injecting it underground will likely take substantial extra energy and expense. So in putting forth this effort, much more coal or other fossil fuel must be mined and drilled, which means steeply increasing energy, financial, and social costs to exploration. This is the worst kind of treadmill: mining more to sequester more.

If CCS exceeds our dour expectations and overcomes its energy and cost downsides, then we still are locked into our old fossil fuel–based infrastructure to deliver our basic needs and power the economy. Doubling down on CCS in the United States, for example, means reinvesting in a generation of energy provision that could have been retired at the end of its useful life (coal plants are forty-two years old on average).

Natural gas has recently boomed, largely because the infrastructure exists to deliver it to consumers. Unfortunately, when life cycle "fugitive" methane emissions are taken into account, it is not clear that natural gas production is preferable to coal production in terms of its greenhouse gas impact.[35] This is particularly true in the short term, when methane is much more potent as a greenhouse gas than carbon dioxide.[36] Solar and wind renewables need extensive infrastructure, including a very different setup for storage of intermittent energy and the construction of transmission and transportation infrastructure for its unique needs. In short, it matters which path one is on for what outcomes are available. Switching from high-carbon to low-carbon pathways of development is critical but likely to be difficult and expensive. Making the switch will be more difficult after another generation of investments in the carbon-based fuel sources that CCS will allow and require.

If CCS continues along its current trajectory, which has involved continual delays in reaching commercial viability, then the whole project represents a massive boondoggle: billions of dollars have been invested in a technology that never will pay off. This is a huge opportunity cost, since those billions could have gone to major breakthroughs in renewables, energy efficiency and approaches to reduce demand and consumption. The promises of CCS have also served to distract regulators from reshaping markets away from fossil fuels. However, the continued investment in CCS is also very likely, given the projected growth of coal use globally and the planned construction of twelve hundred new coal plants that are in the project planning pipeline today.[37]

One potentially positive development that might raise the possibility of actual negative emissions is the production of second- or third-generation cellulosic biofuels (which capture carbon dioxide from the atmosphere), coupled with the sequestration of the carbon from their combustion. This could include sugar cane ethanol in places like Brazil, where there is a strong net positive energy production from cultivation and processing (as opposed to biofuels from, for example, corn and soy oils). Other options include algae and other emerging cellulosic biofuels.

This bioenvironmentally positive scenario of biofuels linked to CCS raises three issues. First, land use is critical. If it becomes too profitable to grow biofuel crops, they will displace forests or food crops, or both, as has already been the experience. Beyond this ecological cost, however, there are potentially vast social costs in displacing communities and driving up food prices, which can be destabilizing. There is also the need for energy and pollution accounting over the full life cycle, including transportation of crops, fertilizer production and their greenhouse gas emissions, cultivation, and processing energy use.

CCS is currently limited to stationary facilities that produce electricity unless some way emerges to capture carbon in vehicles. All in all, there is the sense that this is unlikely to be a winning proposition at the scale needed, or that if it does have an overall net positive balance, there will be losers as well as winners. In this way, biofuels coupled with CCS resembles other CCS and geoengineering solutions: riddled with injustice, inequity, and inefficacy issues. Perhaps it could be effective in just the right places under just the right political, social, and environmental conditions. We worry then that this is not a solution that can be scaled up well.

Many influential commentators, such as key experts from the Environmental Defense Fund, the Natural Resources Defense Council and the Breakthrough Institute, have proposed that CCS is almost inevitable as a transition strategy toward a low-carbon economy.[38] We see the adoption of technologies like geoengineering and CCS as likely because more of the limited public dollars are being spent on responding to fossil fuel issues to not threaten that industry, while relatively little is being spent on replacement of fossil fuels. However our observation is that CCS technology is perennially described as "ten years away" from implementation.

Those who see solar radiation management geoengineering as useful or even okay to consider see it as better than letting the warming climate drive a series of worsening disasters.[39] Some see geoengineering as far less expensive than mitigation and adaptation.[40] And some argue that mitigation targets are now too ambitious to be met without it, such as those proposed in the 2013 draft of the next Intergovernmental Panel on Climate Change (IPCC) report. However some experts argue that geoengineering lacks an exit option (once started, it cannot be stopped without even greater risks), and there are concerns about its fairness, justice, and distributional effects. Geoengineering may be used to bring rain to local and regional climates, but local benefits could mean harm to other localities.[41] And finally, geoengineering may hamper mitigation efforts by keeping the costs of fossil fuels low enough that renewables cannot compete.

CCS itself raises some new geopolitical issues. For example, the injection of carbon into deep underground reserves could spread risks across borders. This raises another issue of governing geoengineering: it could be governed globally if the UN Nuclear Non-Proliferation Treaty's provisions signed in 1959 on nonmilitary uses of the atmosphere are respected. However, it appears that nations are not waiting for international efforts to develop geoengineering technologies and approaches, and so it seems by default that some geoengineering will occur without international controls. It is entirely possible that some geoengineering efforts will go entirely unregulated.[42]

Among the geoengineering skeptics are two groups.[43] One group seeks to stop all engineering, and the other sees geoengineering as risky but worth pursuing nevertheless and seeks to regulate it globally. The first group's efforts focus on making discussion of the issue unacceptable politically, comparing research on geoengineering to genetically modified organisms, cloning, or tactical nuclear offenses. Some geopolitical realists increasingly see CCS and geoengineering as inevitable.

Overall, we believe that geoengineering and CCS, if played out fully, do not offer a realistic solution in terms of science, given the limits of our understanding of ecosystems geological formations where the burned fuel needs to be pumped underground. It may be true that we have surpassed our ability to prevent dangerous climate change by staying below 2 degrees Celsius or even catastrophic warming of over 5 or 6 degrees Celsius (acknowledging that no one can say which adjective and probability goes with which temperature rise or whether any single number explains local experiences). Geoengineering is likely but potentially nightmarish. A very different approach would be a global effort to produce renewable energy cheaply, to which we turn next.

## Scenario 4: Riding Renewables

A dream held by a small group of visionaries since the 1970s is that zero carbon energy sources could be developed that would make fossil fuels unnecessary. We see this scenario as a potentially required outcome of scenarios 2 and 6, our very different cases where strong state and interstate action challenges the powerful fossil interests (and those who think we cannot live without fossil fuels). A major shift in how public resources, private sector, and university research efforts are currently spent is also a prerequisite of this scenario. Renewables, although potentially centralizable, appear to require a major restructuring of the energy grid to take onboard power from microproducers such as household or commercial solar, wind, or small hydro, and store their intermittent power for times of peak load. A major investment will be needed to achieve this restructuring, as well as the policy and legal conditions to allow it (e.g., feed-in tariffs, siting in the face of Not In My Back Yard movements, trade restrictions, and environmental permitting). This will be more difficult in some locations and nations than others, depending on national will, institutional barriers, and incentives.

Depending on which country one is observing, renewables industries are currently relatively weak and small in terms of their political power. Major obstacles exist for these industries to push forward new technologies and achieve the infrastructure and legal restructurings needed in the face of huge fossil fuel–based lobbies and already committed public coffers. For example, five times the amount of public subsidies are still being given annually to fossil fuels as compared to renewables.[44] Total investments are hundreds of times higher for fossil energy, leading to an uphill struggle for renewables in transforming the energy landscape of major nations. Finally, fossil fuel extractive companies have spent billions finding and researching major hydrocarbon reserves and will not want to lose these investments. These are already incorporated into their companies' market values.[45] The point here is that fossil subsidies are major and entrenched.

All of these obstacles are on the side of energy supply, but there are huge gains to be made on demand-side reduction. The efficiency debate is decades old but still fairly revolutionary: some very easy but many longer-term and more difficult investments could be made to reduce energy needs drastically without sacrificing quality of life. These reductions could be driven by the installation of solar or wind power and the realization by users of how far they are from being carbon neutral and just how much waste is in the system. Factor 4 reductions in energy use are quite feasible, and Amory Lovins and other efficiency mavens have talked about factor 10 improvements, mostly achieved through better design. The "negawatts" produced could boldly improve humanity's chances of preventing dangerous levels of warming, but only if aggressively adopted along with some way to avoid the "rebound effect" or "Jevon's paradox" of increased consumption, which frequently comes with improved efficiency—as people drive efficient cars more or have more money to spend after realizing savings.[46]

One important point here is that reduced consumption and profound behavioral change, arguably a prerequisite of any sober assessment of the dire predicament in which we find ourselves, is rarely discussed in the climate policy debate. The occasional scolding column by George Monbiot in the *Guardian* takes on those who understand climate change but continue to jet around the world for work or vacation, but a wider discussion rarely occurs. Small groups in a few countries have formed CRAGs—carbon reduction action groups—that support each other in setting steep carbon use reduction goals and trading internally to meet them. The absence of discourse on radical reductions in consumption is partly explained by Kari Norgaard's research on the collective denial of the reality of climate change;[47] what is more difficult is imagining and creating the conditions for reversing this avoidance.

There are a number of other key concerns on the issue of a bold push for renewables. Cheap large-scale energy storage is needed and potentially could be developed for smoothing the very uneven production of renewables such as wind and solar power with shifting weather. Nuclear energy is lower carbon (plant construction and maintenance of course have substantial climate impacts), but seems unlikely to have a rapid renaissance due to fears of contamination and exceedingly high costs. The Fukushima disaster in Japan led that nation and Germany to turn off numerous nuclear power plants, and Hurricane Sandy led to greater fears of sea-level plants facing flooding in future climate disasters. In Germany, the turn from nuclear power has encouraged adoption of solar, since strong tax credits and feed-in tariffs allow solar producers to sell their energy back to utilities. Nevertheless, in the short term, Germany is also increasing its use of coal.

In addition to wind and solar, we think other renewables have huge potential, such as wave power, tidal power, microhydro, geothermal, and anaerobic digestion of food and green waste for biogas (particularly in developing countries like India). Each raises concerns and has limitations, but together they could make a large dent in human fossil fuel consumption.[48] We need to factor in any carbon budget for the future that much of the transition to renewables requires energy for the manufacture and installation of technology, and seemingly this would have to come from fossil fuels. This includes mining lithium for batteries and silicon for solar panels, which is an important and very little discussed concern.

An important question is how to drive a renewable revolution. One can support regulatory regimes that require utilities to incorporate energy from nonutility producers (net metering in households) or the use of cogeneration plants to capture and use waste heat from electricity generation. US states and other jurisdictions are incorporating renewable energy portfolio standards to force utilities, local governments, or the state more broadly to source an agreed percentage of zero-carbon fuels. Clean energy standards are similar but far less strict, in that they also often include some combination of nonrenewable sources, such as CCS, natural gas, waste incinerators, and nuclear power. Other regulatory routes to push this transition are possible, through regulations such as the Clean Air Act, which is currently being used by the Obama administration to make electricity production from coal infeasible or more costly in the United States.

There are also carbon taxes or cap-and-trade programs, both of which, if aggressively implemented, would encourage the switch to lower or zero-carbon technologies. Direct subsidies, such as production tax credits, could reduce production or installation costs for renewables. Incentives such as feed-in tariffs offer encouragement for renewables project developers by providing a guaranteed pricing system. Research and development strategies also may drive the transition to a low-carbon economy by drastically decreasing the costs of renewables.

Finally, decreasing the lifecycle emissions associated with product consumption also needs to be a priority. Redirecting public subsidies away from greenhouse gas intensive waste disposal technologies such as landfills and incinerators (including "waste-to-energy" technologies), and toward waste reduction, reuse, recycling, and composting programs could also make a significant dent in greenhouse gas emissions, while providing large employment benefits.[49] Cities such as Oakland and San Francisco in California have taken the lead on such initiatives, raising the bar on what is deemed to be possible in this area.[50]

In this renewables scenario, we need to consider who and what are pushing such a transition to take place. The coalition that might do so has been described as an amalgamation of "Baptists and the bootleggers,"[51] because it includes social movements attacking coal and other fossil fuels on moral grounds (the Baptists), and firms that stand to make a good deal of money on the transition to renewables (the bootleggers). Increasingly, international aid agencies such as the World Bank, the Asian Development Bank, Inter-American Development Bank, export agencies, and of course UN agencies are pushing for investments in renewables. Some of their impact will be in making these investments a safer bet.

Yet there is often resistance by managers in key positions to new technologies. They fear what they don't know, such as was rumored to be the case of Detroit auto plant managers who long resisted switches to diesel, hybrid, or other radically new engine types. This applies also to federal, state, and local governments, corporations, and utilities and influences how they assess risk. We have seen in our state of Rhode Island how laborious it has been for a large-scale wind project, Deepwater Wind, to get permitted. The need to regulate development also has made it difficult for development banks to lend money to decentralized renewable energy systems and makes them harder to manage. Large financial institutions, whether the World Bank, regional banks, or even neighborhood commercial banks, often prefer to lend money in larger lumps to avoid the transaction costs of small-scale lending.

A second barrier to efficiency is the simple point that utilities generally can make more money only by selling more energy. This is true as well of waste haulers, which are paid by the ton or cubic yard they take to the landfill or incinerator. Conventional energy is grossly underpriced in many locations (compared to its full social cost), which makes it very difficult for renewables to gain a foothold. There are also cost hurdles for poor people and communities, for example, the capital neccessary to insulate one's house or to install compact fluorescent lamps versus using cheap incandescents. We have huge sunk costs in our existing electrical grid and are stuck to some extent with the path dependency that comes with that. There are structural barriers, such as renters who are unable to insulate and open windows, and who can't plant shade trees in their yards. As a results, whole neighborhoods suffer from the heat. Locational instability, such as transience in jobs and residence, driven in part by economic restructuring, leads to stark inequality between communities, with some able to make the leap to renewables and others being driven into greater fossil fuel dependency and bearing more of the costs to maintain the grid.

What would drive a renewables revolution forward, and what might it look like? Social movements need to be strong and impatient, and there needs to be the right market dynamics for technology development to make renewables cost competitive. A few major technology break-throughs, including on storage of intermittent energy like wind and solar, would help, and some major economies taking bold steps could also put some wind in the sails of this effort. China is making a great leap forward with its government-sponsored renewables, and it appears that the country is following through on its five-year plan in this regard. However, it is also investing in some very dirty energy sources, and China is not alone in this regard. Cheap renewables will be a help, but not enough by themselves, to drive the revolution.

What is the likelihood of an all-out push for low or zero-carbon energy systems in the short term? Our assessment is that due to the factors we have noted, the progress will likely be spotty and partial. In some developing countries, fossil fuels infrastructure and electric grids are distant, so renewables such as wind and solar can be adopted more quickly. The strong shift in the major multilateral development banks such as the World Bank under its new president, Jim Yong Kim, and the new funding mandate from the Inter-American Development Bank to lend more for climate change (and less to greenhouse gas intensive fuels such as coal) provide strong signals to developing countries in search of money. Other lenders exist, weakening that signal. So we see renewables advancing but not displacing much of the fossil energy sources.[52] And their adoption is not likely to be timely enough to displace fossil fuels as needed to avoid dangerous climate change. However, that depends strongly on the other scenarios. If scenarios 2 or 6 take off, we will see far greater take-off of renewables.

What might a renewable revolution look like? We believe it would be quite different in different places, and depending on whether we head down a more corporate path to renewables and efficiency or take a community small-scale approach.[53] Visually we could see differences across a landscape: Do we see panels and miniwindmills across housing and farms, or just big utility-scale wind or solar farms on major corporate buildings? There are appeals in both: small-scale green power would likely create more jobs, and potentially more satisfying work with a ladder that extends lower into poorer communities providing a series of important social benefits.[54] However, given all the obstacles we have laid out here, the uncomfortable truth may be that the transition to renewables might be more likely if corporations see potential profit in this transition and are able to get the regulatory and managerial landscape changed. A key question is whether the two can coexist: Would small-scale renewables represent a deterrent to the corporate investors? Would a corporate dominance in renewables make small-scale investing unviable? Projecting outcomes of a renewables revolution depends on the social, political, and economic context in which it takes place. Social inequality, international stability, and ultimately levels of carbon in the atmosphere will in turn depend on the social context of an erstwhile green power transition.

**Scenario 5: Going Local**

A view of a number of scholars and activists is that the whole UN-led effort to address climate change is so completely set up to benefit corporations rather than environmental or social needs that an entirely new approach is needed.[55] The argument is that it is best to build change from the grassroots, including connecting community-led and sustainable counterhegemonic development models around the world, through translocal movement building and a radical reclaiming of the global commons.[56] This perspective emphasizes the importance of grassroots organizing and of small-scale and local democratic decision making. It encourages communities facing similar issues of inequality and injustice around the world as a result of corporate-led global capitalism and neoliberal privatization to work together to build alternative development models. This approach often puts limited emphasis on or directly rejects the importance of realizing state, national, or international climate change policy.[57] Examples of solutions that are pursued include building zero-waste communities, local sustainable agriculture, transition towns committed to moving off fossil fuels, worker cooperatives, sharing economies, local financial currencies, community-based adaptation, and community efforts to protect common pool resources such as water, among many others. One value of this approach is that studies show that people who see viable solutions to global warming are more likely to admit it is a real and a human-caused problem.[58] These approaches also offer tangible, inspiring, and positive examples that challenge the premise that there is no alternative to free market capitalism and its ever-increasing consumption.

The translocal approach explicitly or implicitly suggests a radical departure from what we spent the first six chapters of this book discussing, but it is consistent with some of the local and national efforts we explored in chapters 7 and 8, outside of the UNFCCC. Who and what are pushing in the direction of local or translocal solutions? The actors are diverse, from anarchists fighting corporate and government control to local governments in the international group ICLEI—Local Governments for Sustainability (see chapter 7), to international networks and social movements focused on connecting and supporting community models that challenge powerful fossil fuel, corporate, and other private interests. Many of these actors were present from around the world at the 2010 Cochabamba People's Conference, calling for more grassroots and non-corporate approaches to addressing the issue of climate change. Social movements such as those described in chapter 8 are making important gains, from the expanding peasant-led movement Via Campesina and Indigenous movements resisting REDD-type market-based systems, to waste pickers fighting against the Clean Development Mechanism projects that damage their livelihoods.

What are some obstacles to this approach of adequately and equitably responding to the climate crisis? We would list a rather formidable set of actors. First are those at the top: corporate power focused on market-based and reformist approaches to climate change, and financial capital increasingly seeing future revenues in climate insurance products, carbon market derivatives, and so on. Second are the influential private foundations, even the most progressive of which have concentrated on reformist politics and market solutions, in particular.[59] In most parts of the world, the social movements that have focused on climate change have been relatively weak and fragmented, and our observation is that foundations have tended to support the most moderate and free market friendly of these.

The World Bank and other international financial institutions have favored rather centralized investment and solutions, and there is little reason to expect a sharp turn away from these preferences without a much stronger movement that pushes them to do so. Other development banks exist further outside the influence of civil society and often have less consideration for social safeguards than the World Bank. Another chronic problem is that politicians often believe that they are going to be judged by the rate of economic growth that is achieved while they are in office. Efforts to calculate a genuine progress indicator by the NGO Redefining Progress and the government of Bhutan's use of a national happiness index are both excellent efforts to shift the focus of planning by expanding beyond narrow indicators like gross domestic product. These approaches seek to assess progress in more holistic ways, including accounting for environmental impacts and other negative outcomes.

In short, the hegemony of neoliberal approaches and the unquestioned importance of economic growth and the assumption that it will solve social problems makes radical community-led alternatives difficult to implement on a large scale, but not impossible. There is widespread fear of anarchist approaches, which are often misunderstood or misrepresented.

Many social movement organizations in the global North and South have been built on the expectation that the state is needed to redress inequality. Certainly with more disasters tied to climate change, and the stark evidence they reveal about social inequality and vulnerability, there is an opportunity for social movements to make the connection with the need for a more just global system. With the push for extreme energy and resources and the drilling and digging they require, we are seeing more protests at sites of extraction, transport, and disposal. Examples include the movement Idle No More, antidrilling protests in Colombia, and the growing number of activists against hydraulic fracturing (fracking) in the United States.

We commend this perspective's attention to the need for radical re-structuring of the system that created climate change in the first place. We also agree about the need for change to come from local embodiments of what's at stake in the climate change issue and from people building alternative communities. As Robert Putnam has argued, climate nego-tiators cannot agree to anything that is not in their national interest.[60] Currently, national interests in many key major economies are seen as requiring the protection of a country's sovereignty against restrictions from foreign influence on their core economic and political decision mak-ing. The obvious response is that national interests will not change by themselves; they have to be redefined in a much broader sense, and this will occur only through progressive social movements. Finally, we think that because they are so wedded to the incomes and the cheap economic growth they've delivered, states are unlikely to act first when it comes to curbing fossil fuel use.

However, we believe that such a decentralized approach, without being accompanied by efforts to realize centralized and top-down legal frameworks to address climate change, seems unrealistic for addressing the problem in two respects. First, because it is by definition piecemeal, such approaches are unlikely to arrive at an adequate solution in ecologi-cal terms, keeping global average temperatures below catastrophic levels. For this reason, we would unfortunately place this scenario in the "demo-cratic dysfunction" cell of table 9.1. Second, because it relies so strongly on decentralized forces, these approaches may not lead to an equitable or enduring outcome. States still hold the monopoly on force in a given piece of territory. As a result, we might see strong upheaval at times, which some groups might channel for certain gains, but we fear they will be less likely to result in positive overall and enduring change without also engaging the state and international institutions. On inequality, this model may be much better than alternatives in the short term, but it could be risky for international stability without broader institutional coordination.

We agree with geographer James McCarthy that it is puzzling that so many contemporary radical responses to corporate-led global capitalism are dismissive of the role of the state (and also international institutions) as a necessary part of effective resistance.[61] We do not see the community or the local as inherently more democratic or a necessarily more effective manager of common pool resources than the state. Though we need more strong community-based models of equity and sustainability, we believe that broader institutional frameworks are also desperately needed despite their current flaws and state of corporate capture. While there have been immense shortcomings of both state and UN processes to date, we are concerned about losing an international framework and institution for coordinating nations that establishes international standards and oversight. That is what we address in our final scenario.

### Scenario 6: Global Climate Justice

Predictably, we have saved the scenario we prefer for last. And certainly it has some utopian elements, but we wish to lay out what might be required to get there. The major pieces of this scenario are a global agreement based on a fair sharing of the remaining capacity of the atmosphere and oceans to safely absorb greenhouse gases. In this sense, this solution is high on both table 9.1 dimensions: adequacy and democracy/equity. To protect humanity, the ability of the UN system to police emissions is strengthened; to do so, nations agree to fair levels of effort in reducing emissions and financing adaptation, both based on their responsibility for the problem and their capability to act. There would need to be substantial corporate involvement in creating the new low-carbon economy, but their role must be in following and providing support, not in capturing and distorting it for their own exclusive benefit. Especially in the poorest cities of the world, community-derived and -driven solutions need to lead the way. Multilevel climate justice is needed—from fair UN processes and outcomes to national decarbonization strategies all the way to local processes and outcomes.

As we've pointed out in the other scenarios, getting to an adequate and equitable climate treaty will require a rapid and dramatic transformation in how key powerful state leaders define their individual, national, and collective interests. Some nations are already there, quite willing to sacrifice on these levels in the shorter term for a more secure future for humanity. We believe more will follow later, especially if the powerful ones such as the United States and China move convincingly in this direction. For this to happen, clearly the political calculation needs to be different than it is today in the key major economies, and this requires strong, strategic, and unco-optable social movements.[62]

A series of things will need to happen at the local, national, and international levels to bring us to the vision of scenario 6. At the local level, we would need to see the reorganization of some social systems away from high-carbon pathways, and these changes would have to be affordable, especially in developing countries. For this reason, some of these local efforts will need to be driven by what happens at the national and international levels, including through effective technology transfer and finance from the global North to the South (as has been long agreed in the UNFCCC, chapter 5). However, there also have to be strong social movements pushing on their local governments to spend these funds and deploy these technologies well, and on actors throughout global commodity chains to responsibly bring products to market. We explored in chapter 7 the effectiveness of some of these, such as the Sierra Club's Beyond Coal campaign and more decentralized and less funded networks that are working on the same issues.

We would add a strong equity part to this scenario, which we believe is needed for an enduring solution. Equity needs to be addressed both internationally in the formula by which nations share the burden of emissions reductions and nationally, where domestic political consensus can be built around a fair approach. We believe the international framework is needed to avoid local abuses of less powerful social groups and ecological systems. The provision of innovative international financing, such as an airline levy, financial transaction tax, or from other sources, will be needed and will have to avoid flowing through national treasuries where entrenched interests often take precedence.

At the national level, we need strong and much more broadly based social movements that emerge to define their common interests around mitigating climate change. This barely exists now. Climate change needs to become a bread-and-butter issue, seen as having near-term and immediate impact on people's lives. In the United States, very few movements gain this level of salience when addressing a long-term problem; the exception might be the rise of the national deficit as a top five issue, and we believe quite a lot can be learned from that effort of the conservative Right.

Several additional shifts are needed at the national level, and we'll focus on the United States on this. Superstorm Sandy in 2012 and other freak weather opens the possibility of evidence-based communications strategies, framing the rising number and intensity of climate-related disasters in progressive, not regressive, ways. Responding to climate change needs to be redefined as a jobs-positive issue. There needs to be a strategic debunking of climate denialist movements in key places like the United States, Poland, Australia, Canada, and Russia. Legal cases may help to change the calculus within corporations about whether to deny the inevitable rise of climate change. This would especially be true given reputational risks of firms that may be hiding their involvement in climate-denial organizations.[63] And based on our discussion of renewable energy, we need a multipronged strategy to foster renewables development and prevent the exploration of fossil fuels. The People's Climate March in September 2014 showed some promise in these regards. Over 400,000 marched in the streets of New York City for climate justice and against fossil fuels two days before 130 heads of state met at the UN headquarters. Finally, much of the success of the former issues will depend on our ability to sever the influence of corporations over political processes. Campaign finance reform is one place where that could make an important difference.

At the international level, we need a more democratic approach. Civil society at the national level must first drive states to come to UN negotiations with clear mandates to aggressively address the issue and demand that other nations do the same. CAN-International, an effective network of environmental organizations serving as a watchdog in international climate politics, has often been exclusive and focused on technical parts of the solution. While no doubt there are coordination challenges, a shift to more diverse and inclusive civil society networks can help states and the UNFCCC develop and carry out good solutions and push for progressive norms on all the issues of Indigenous rights, gender equality, and so on at the global level.

The likelihood that this scenario will come to pass in our assessment is still fairly low, but it could emerge as more likely with more disasters like Sandy and the shifting political calculus of young generation activism so apparent in New York in September 2014. It will be critical to create a new class of workers in green jobs whose interests are in keeping climate action moving forward and willing to mobilize to protect government programs and laws that favor this new economy over the old.[64] One substantial risk is the slicing off of different groups of supporters from

**Table 9.3**
Our Evaluation of the Scenarios' Ability to Address Climate Change

|  | Democratic in process? | Adequate? (time, ambition, safety) | Equitable in outcome? | Efficient? |
|---|---|---|---|---|
| Scenario 1: *Exclusive inaction* (Copenhagen Plus) | No | No | No | No |
| Scenario 2: *Exclusive action* (grandfathering by central governance) | No | Yes | No | Yes |
| Scenario 3: *Technofixes* (Hail Mary extreme technology) | No | Likely no | No | Yes |
| Scenario 4: *Global drive for cheap renewables* | Maybe | Maybe | Depends | Uncertain |
| Scenario 5: *Going local*: local and translocal without adequate global governance | Yes, probably | Likely no | Likely no | No |
| Scenario 6: *Global climate justice*: Strong national action with international treaty | Yes | Yes | Yes | Maybe |

collective solutions to climate change by individualistic approaches such as the provision of private insurance against impacts.

Funding for the green transition could certainly come from redirecting harmful subsidies, for example, $1.7 trillion a year in global military spending[65] and $700 billion a year in fossil fuel subsidies.[66] Capturing even a substantial fraction of these revenues will not happen without significant resistance from the entrenched interests in maintaining them, however, even as humanity races to the edge of the climate cliff.

**Wagering on Warming Worlds**

We have laid out six possible future scenarios, focused on two key dimensions: whether they are based on democratic process and lead to equitable solutions, and whether they lead to solutions that are adequate in terms of keeping global concentrations of greenhouse gases to safe levels. We described two broad sets of likely scenarios based on weak global equity and limited democracy (the left side of table 9.1), varying in whether action is adequate for addressing climate change. The business-as-usual scenario, unfortunately, is the first: exclusive inaction based on the Copenhagen model of voluntary national actions, which do not add up to adequate protection of the atmosphere. This likely scenario fails on all four dimensions of our evaluation on its ability to meet the requirements of an enduring and effective climate scenario: democratic in process, adequate and timely in emissions reductions, equitable in outcomes, and efficient in process and action (table 9.3).

The current conjuncture during our writing of this chapter in early 2015 is one where trust in the UN system to deliver an adequate climate treaty is at one of the lowest points we can remember in twenty years of observing it. For this reason we see scenario 2, exclusive action through minilateral groups such as the Major Economies Forum, the G8, or even the G2 (US and China), as more likely than our ambitious scenario 6. As tables 9.2 and 9.3 lay out, scenario 2's exclusive action fails in most terms of justice but may be more obtainable and could lead to a more adequate solution from an overall mitigation perspective. Beginning the path toward an adequate and equitable solution through the MEF, the G8+5, the G20 or the G2 would require bringing a deal from these groups back into the UN system, at which point it may face strong resistance from excluded actors.

Scenarios 3 and 4 are technological paths (extreme technofixes like geoengineering in scenario 3 and a renewables revolution in scenario 4) that could unfold in ways that are more or less adequate and equitable, more democratic or less democratic in process, and more or less efficient. Our language in evaluating them in table 9.3 therefore is more guarded. Given the polarized positions for and against them, it is fairly difficult to imagine a democratic process for governing geoengineering and carbon capture and storage. Their mobilization is likely to cause unequal impacts on different populations, and one can imagine poor and marginalized peoples being unable to deflect the siting of facilities or buffer themselves from the impacts of whatever technologies are mobilized. Renewables can be deployed in ways that create wide social benefits including new employment and development opportunities, or they can be held in a few hands with inequitable impacts.

Our final two scenarios are both strong in terms of democratic process and equitable outcomes (the right side of table 9.1), but they differ in their adequacy for responding to climate change. In imagining pathways that would lead to these two outcomes, we envisioned one way as a radically democratic approach favored by some leading thinkers in the climate justice field, who prioritize boldly local or translocal action. This scenario 5 we see as unlikely to lead to an adequate outcome that meets our criterion of keeping relatively safe and stable global mean temperatures. To do so requires swift and widespread action, and we see the hyperlocal or translocal approach (without also having a strong international component) as likely to be spotty and potentially halting in pace.

Our final scenario is one that many climate activists have been advocating for over the past two decades: strong national action with an ambitious and binding international treaty, enforced by a UN with teeth, that is, real power to enforce an agreement on who would do what about climate change and by when.[67] Critically, this involves aggressively challenging fossil fuel interests, not just cozying up to them for a compromise position. Such an outcome may not be likely, but we still believe it is critical. It may well require a radical reform of the United Nations itself, so that the process is more tractable, while retaining the trust of the vast majority of the world's nations. This will be no easy feat. We turn now to our final chapter, where we consider paths forward, given all we have churned over so far.

# 10

## Linking Movements for Justice

### Action among the Ruins?

In October 2012, a major storm that would become known as Superstorm Sandy was conceived in the Caribbean. It whipped through Jamaica, leaving 70 percent of the population without electricity; ravaged Haiti, killing more than fifty people and exacerbating a cholera epidemic in the country; and pummeled 15,000 homes in Cuba, killing eleven people. It gained force as it joined up with other converging weather systems, becoming what many called a "Frankenstorm"—a strange cocktail of largely unprecedented size and force. With epic winds and storm surges, it pounded the US Northeast coast, including New York City, just a week before the presidential election. In the United States, it killed over 140 people, left thousands homeless, and caused at least $70 billion in damage.

Sandy connected the fates of people in the developed and developing world in a kind of vulnerability interdependence, making clear that natural disasters could not be just "their" problem. The global center point of the financial world, Wall Street, was plunged into darkness, vulnerable to the hand of climate change just like everyone else.

Was this storm the result of climate change? Like a baseball player on steroids up to bat, that single home run might have been hit anyway, but we can't ignore the newly bulging biceps or the statistically improbable records being set each year. In a context of warmer seas, the increased moisture in the atmosphere, and the atmospheric blocking effect that has shifted the jet stream south, climate change leaves its imprint on many extreme weather events.

So how will a warming world that has impacts (albeit disproportionate) for poor and rich, marginalized and privileged, and South and North influence relationships of power and our ability to tackle this problem?

Do we simply need bigger and worse disasters in the core economies to shift the political calculus on climate change, as some suggest? Do we just need to hit rock bottom before we take action? Will disasters like Sandy make increasingly clear the inherent unsustainability of limitless global capitalism and its partner in crime, increasingly "extreme" fossil fuel extraction?

These are crucial questions. Our short answer is that disasters create new spaces for political action and discourse, but these spaces are subject to existing power relations. Sandy led to a growing discourse on climate change in the United States that some have attributed to Obama's embracing climate change efforts as a core priority in his second-term inauguration address. However, as activist-scholar Naomi Klein adeptly reveals in her book *The Shock Doctrine,* more often than not, disasters of any form are seized on by those with power to carry out reforms that would not be popularly embraced in periods of relative calm, democracy, and stability.[1] Antonio Gramsci knew this well. As he explains:

A crisis cannot give the attacking forces the ability to organize with lightning speed in time and space; still less can it endow them with fighting spirit. Similarly, the defenders are not demoralized, nor do they abandon their positions, even among the ruins, nor do they lose faith in their own strength or their own future.[2]

Following on our exercise in developing scenarios for a warming world in the previous chapter, we argue that disasters may well prove instrumental for revealing the contradictions of our political economic system. That hegemonic economic system's proponents repeatedly claim that it is limitless in its benefits, compatible with ecological realities if allowed to function unfettered, and the only real means of organizing modern society. However, revelation about the system's unsustainability will not occur—or lead to the radical change that we need—unless political forces broadly mobilize to effectively counter the competing narratives and structural privileges of those who still benefit from the existing system.

Most central, private and state-owned energy companies, slaves to quarterly earnings reports to shareholders, owners, or political elites have sunk far too much capital in continuing to extract every last bit of accessible fossil fuel from the ground for them to simply walk away or even to admit the error of their ways. Simple math reveals that these companies have already invested in extracting and burning far more fossil carbon than we can safely pump into the atmosphere and oceans.[3]

Many politicians, relying on campaign contributions (or other payments) from these same companies or the support of their employees to stay in office, will not cut ties with the fossil fuel industry, even in the face of convincing climate change science and escalating disasters on the ground. Some large environmental NGOs are also too dependent on mainstream foundation dollars, too committed to their existing world-view, and too attached to maintaining a seat at the table with the elite and the powerful to demand the types of structural changes that are needed. As a result, people need to be organized and mobilized long before major storms wreak havoc on symbolic and material centers of power like New York City, or the nation's underbelly in New Orleans, momentarily lifting the veil that orients how we see and understand the modern world.

Gramsci told us that it is on the terrain of "the incurable structural contradictions" of the existing order that the forces of opposition or-ganize.[4] How they go about doing so, we argue, will largely determine the fate of human civilization. We desperately need a new power politics in order to adequately address the climate crisis. It will not happen any other way.

As this book has revealed, we are beyond the point of win-win-win solutions; there will also be losers, and no transition will occur without a plan to minimize the influence of those who need to lose power, and offer workers in fossil fuel industries a viable and just transition. In the next three sections, we revisit power in the context of a warming world. The world order is changing, and that has reshaped how climate action can occur. This creates new opportunities and challenges inside and out-side the UN. Some climate justice activists have seen clearly for at least a decade what many of the rest of us are now only coming to accept: that technofixes and state and NGO hobnobbing alone will not stop climate change; only broad-based local and transnational social movements that intensely challenge the interests of the powerful can do that. However, being right does little good if it doesn't lead to the change that we need. To this end, the global climate movement and allies will have to take a more strategic approach moving forward and despite its massive deficien-cies to date, now is not the time to abandon the UN process.

This chapter is divided into four sections. In the next section, we revisit our argument that the world order has undergone profound changes in the contemporary period that fundamentally alter the terrain on which movements for climate justice engage. This creates new opportunities and challenges both inside the UN climate regime and in domestic contexts where climate change responses come to life. In the two sections that follow, we evaluate why responses to climate change in this context have been so inequitable and inadequate thus far. First, we argue that given the evidence from the previous five years, states alone are not likely to come to an effective agreement without being pushed to do so from a powerful outside force. Second, we turn our attention to nonstate actors, including the private sector and civil society. We argue that a focus on incremental market-based reform has proven insufficient for challenging the enduring power of fossil fuel interests and will continue to be inadequate without a bolder, more inclusive, and aggressive response. We end the chapter and the book with a discussion of how the climate movement has been divided and offer some initial ideas on how those divides might be overcome, paving the way for a new historic bloc to emerge capable of rising to the challenge of preventing widespread ecological catastrophe and advancing global climate justice.

## Out of Order

In the period since the pivotal UN negotiations in Copenhagen in 2009, the global political context has shifted in important ways since the signing of the Kyoto Protocol a dozen years before. Building on the scholarship of neo-Gramscian scholar Robert Cox, we identify major shifts in the contemporary period in four main areas: global political economy, geopolitics, ecological conditions, and transnational civil society. In each area, important tensions are largely structuring the limits and possibilities for action on climate change moving forward.

First, this context includes a wobbly and wounded neoliberal doctrine in the aftermath of the 2008 Great Recession. The Washington consensus and commitment to free markets is no longer seen as having all of the answers. There have also been some important trends toward a more heterodox and multipolar form of political economic organization. We can no longer look only to financial centers of New York and London to understand this decentered and complex new global economy; rather, emerging economies such as China, India, and Brazil are increasingly fulfilling core development functions in the global economy.

This context has led to discursive openings for state and nonstate actors to call for new forms of regulatory and social action. For example, the state of Bolivia, networks of Indigenous peoples, and various civil society organizations have raised bold critiques of the neoliberal development model and its deleterious impacts on global ecologies. However, we've also seen responses by several key states of pursuing austerity measures to limiting state intervention toward environmental goals. Thus, it's not at all clear which way the pendulum will swing in our increasingly heterodox global economy: toward new socially and ecologically embedded models of economic organization or hyperneoliberal forms of governance focused on enhancing investor rights above all else.

Second, and related, the Great Recession has arguably intensified what was already a shift in geopolitical power relations in the new multipolar order of the United States, China, and other newly emerging economies. This includes the hegemonic decline of the United States, the rise of China and an increasingly multipolar interstate order, and the fragmentation of the global South's identity along various lines, with new interstate class dynamics threatening longstanding ideals of developing world solidarity. Perhaps most important from a climate perspective is the growing insecurity of the United States in the face of its economic and political decline in relation to China.[5] As a result, the United States has been unwilling to make costly emissions reductions that may put it at an economic disadvantage; China, and its coalition partners of India, Brazil, and South Africa, have yet to demonstrate the capability and interest to lead constructively.

As for the countries in the global South most harmed by climate change and least responsible for causing the problem, although there is the beginning of shared identity rooted in disproportionate vulnerability to ecological harm and some notable forms of resistance, they have yet to demonstrate a full willingness to challenge the new major world polluters (especially China, Brazil, and South Africa).[6] This is perhaps not surprising given their dependence on old allies from the South in terms of financial resources and ties of political power (seeing the value of negotiating in a larger bloc such as the G77), but it has meant that international politics is often still negotiated in outdated North–South terms.

Third, we are witnessing the collapse of critical ecological support systems and a rush for the development of unconventional or extreme fossil fuel technologies to extend our ability to access reserves previously considered out of reach. This is having impacts on geopolitical power relations, ranging from resource wars, to political struggles over the limits to geoengineering solutions, to contestation over the rights to remaining ecological or emissions space, to rebuilding efforts after unprecedented large-scale disasters, to the apparently inevitable disappearance of whole countries under rising seas. In this context, we have seen previous leaders on climate policy such as Canada reverse their positions while pursuing aggressive development of unconventional fossil fuel resources. Some other nations seem to be retreating from bolder negotiating positions due to renewed dependence on natural resource extraction as

China invests heavily in extraction and infrastructure in South America and Africa.[7]

The overall trend in fossil fuel energy investment has remained constant despite rapid growth in renewable energy and carbon markets, with investors continuing to invest heavily in fossil fuel development.[8] We know that big industry (and its Wall Street partners) has already sunk capital to unearth amounts of fossil energy that far exceed what the Earth can sustain without warming beyond thresholds of no return. And melting ice sheets in the Arctic and elsewhere are making long-inaccessible fossil resources available for those countries that are aggressive and well positioned to take advantage of them, introducing new tensions to the region. Indeed, human interaction with global ecological systems is playing a largely overlooked role in shaping international political processes.

Fourth, transnational civil society is increasingly not confined to the borders of the territorial state. New forms of civil society organization, in the form of international NGOs, transnational advocacy networks, translocal movements, and globalized social movements and labor parties, offer the potential to challenge fossil fuel interests and realize new forms of rights for the most vulnerable. This occurs through various forms of engagement, from idea generation and expertise to disruptive politics.

However, the normative and celebratory view of civil society as rights-bearing citizens who serve to counterbalance the market and state is rarely substituted for what exists in reality in terms of power imbalances, inequality, corruption, co-optation, and exclusionary practices. While there are examples of growing movements representing the interests of vulnerable populations, in some cases, NGOs and advocacy networks are actually the promoters of key tenets of the neoliberal project rather than its main force of resistance. But despite its many challenges, transnational civil society presents the most viable means to spur a new politics of global solidarity on the issue of climate change. Such an effort will need to brew a consciousness and political force that spans our fragmented world order, compelling us to view mitigating climate change as worthwhile in terms of social and economic justice and indeed necessary to our collective global survival.

### Like a Pillow

Within this shifting world order, there have been notable efforts by states, market actors, and civil society to push for a more effective and equitable response to climate change, both within and outside of the UN process. These efforts have come up far short of what is needed. What explains these shortcomings to date?

To refresh: At the international level, since the negotiations in Copenhagen in 2009, the UN climate regime has taken a notably neoliberal turn. A voluntary pledge-and-review emissions reductions framework, newly dubbed in the UNFCCC negotiations as intended nationally determined contributions, and a focus on leveraging financial markets for climate gains, has replaced the legally binding regulatory framework established by the Kyoto Protocol. A new emissions reductions framework is supposed to be agreed on in Paris in 2015, but short of a major political shift, it is unlikely that this will provide sufficiently robust binding greenhouse gas emission limits on the largest polluting countries or establish pledges sufficient for maintaining reasonably safe global average temperature rise. Overall, these new trends do not suggest a departure from a commitment to unbridled economic growth, the power of relying on the market to solve problems, and ever-more extreme forms of fossil fuel development. Development continues as if ecology doesn't matter.

However, the inadequacy of this approach has not gone uncontested by the states that are most vulnerable to a changing climate and least responsible for causing the problem. We have seen the emergence of challenges to this inaction by new and emboldened negotiating groups in the developing world, such as the Independent Association of Latin America and Caribbean States, the Association of Small Island States, and the Least Developed Countries group. These groups are increasingly aligning with the EU, which recently committed to a 40 percent reduction of their emissions by 2030 compared to the 1990 level. Low-income states were presented with a choice to decide between a new and inadequate mitigation framework or no international mitigation framework at all. Compulsory, structural, and institutional forms of coercion ensured that there were high costs and limited benefit for low-income states to withhold their consent. The influence and unity of low-income states had been weakened by the fact that their strongest allies in the South—the BASIC coalition of Brazil, South Africa, India, and China—forged a new pact of freedom from binding constraints with the United States in Copenhagen. Their eventual consent in the negotiations in Cancun, Durban, and Warsaw reflected the reality that for weak actors, "bad rules that are universally acknowledged are better than no rules."[9]

However, the consent of low-income states was also contingent on the provision of strategic concessions framed as rightful forms of governance.

When low-income states consented in Cancun, their acquiescence was justified on grounds of legitimacy; it wasn't simply a bribe that was being offered and accepted, but was presented as a "rightful" program for international financial support of adaptation. In the end, the concession of climate finance, and adaptation finance in particular—despite its ambiguous terms—was embraced by weak and strong states alike as a core area of progress in the climate change negotiations. Wealthy states made a promise of $30 billion in fast-start financing during the 2010–2012 period and $100 billion in climate financing a year by 2020. In our view, most of those who are most responsible for climate change and capable of supporting adaptation actions have fallen far short of their obligations.[10] Despite clear language in the Convention that wealthy countries agreed they had a responsibility to provide adequate funding to developing countries to adapt to climate change, there remains an ever-widening chasm between funds that are needed and what has been promised and delivered.[11]

The lesson here is that weak states have the ability to strategically exert influence on some second-tier issues in the negotiations like adaptation finance; but structural change on the difficult core issue of adequate emissions reductions will not likely come through their resistance and dissent. Their consent reflects Robert Cox's assessment of multilateralism in which "hegemony is like a pillow: it absorbs blows and sooner or later the would-be assailant will find it comfortable to rest upon."[12] State delegates like Ian Fry of Tuvalu who make bold statements about the injustice of the regime can help to catalyze momentum, but holding ground on such positions will require far stronger movements standing behind them. One further issue of concern is that competition among developing countries over scarce financial resources to mitigate and adapt to climate change threatens to be a wedge that further fractures solidarity among negotiating blocs in the global South.

Importantly, policy decisions are ultimately made in national contexts, and thus domestic politics remains central to action. There are promising signs here, such as China's joint announcement with the United States on reducing future emissions; however, there are also worrying trends, such as its massive investments in some new coal-fired power plants. Moreover, the fragmented global governance system is hardly currently capable of effectively and equitably managing issues such as climate-induced migration, climate-related security issues, disappearance of states under rising sea levels, fragmented intergovernmental structures for disaster management, geopolitical conflicts over the thawing Arctic, and the role of insurance companies and private actors in climate change adaptation. And there exist gaping governance, political, and social challenges related to large-scale technological attempts to engineer the climate.

In sum, given the particular structural conditions of the contemporary international order, states alone show very little promise of arriving at a sustainable, effective, and equitable international climate treaty or, more broadly, international governance of climate-related issues. States experiencing the impacts of climate change most severely are also the ones that have the least power in the negotiations and are particularly vulnerable to co-optation in the current historical context.

Given the downward trend of the US economy and its difficult institutions for decision making on international treaties, the global hegemon shows little promise of emerging from its position in the negotiations as a brake on enduring ambition without being pushed hard to do so, despite modest progress on greenhouse gas regulatory policy at home and its 2014 policy coordination with China. As an emerging global leader with an economy that largely thrived during the Great Recession, China offers perhaps the most promise of being in a position to shift the negotiations in a more promising direction. Overall, a bolder catalyst beyond state coalitions will be needed. Thus, we now shift our attention to nonstate actors.

**False Solutions**

Many scholars argue that a new class of business interests that defines climate change mitigation in their financial interest is our best hope as the catalyst for the change necessary to address climate change. Indeed, as climate science has become increasingly difficult to dispute, fossil fuel associations have become less visible in the international negotiations. However, while there have been fragmentation and diversification in the approach of different business actors in international climate politics, we do not see significant evidence that the obstructionist forces of fossil fuel lobbies have waned in power. Fossil fuel industries still compete on a highly unequal playing field, continually subsidized by the very governments that negotiate international climate treaties. The shift of many in the industry to a carbon market approach, once admonished by some of the more obstructionist fossil fuel companies, has not proven a threat to fossil fuel interests, and the biggest actors have continued to reap record profits with no sign of slowing down.

The continued dominance of fossil fuel industries in international climate politics has been made possible by three processes. First, what we call the *carbon trading diversion* has involved an overwhelming focus by state, business, and civil society actors in the UN negotiations on market-based strategies to reduce carbon emissions, with limited attention to directly challenging fossil fuel industries. This has enabled a debate focused on sustainable projects, while leaving powerful interests largely untouched. Second, *tacit power* points to the fact that the structural advantage and political capital of powerful and established industries such as fossil fuel companies make it challenging to uproot them in domestic contexts. As a result of their powerful lobbies, state decision makers often define their policy options narrowly. Third, the *business boomerang* points to how fossil fuel lobbyists engage at the UNFCCC, as a hub for decision makers around the world, to build far-reaching social and political capital needed to influence domestic policy, which is instrumental in shaping the international climate positions of key states.

As a result, we argue that the change that is needed will not be designed by business coalitions seeking new markets. The existing investments made by fossil fuel interests are simply too high. This is not to say that green business is not needed. Indeed, any coalition that challenges fossil fuel interests must have viable technologies and well-organized industries to push forward an alternative vision, policy approach, and infrastructure. Technological breakthroughs such as lower-cost batteries for storing intermittent renewable energy are essential. But a solution to climate change will not emerge through a market logic alone.

This turns our attention to the role of civil society. The literature on civil society in international climate change politics has not fully accounted for the causes of the failure of civil society to adequately influence emissions reductions action in the contemporary period. We highlight three main deficits. First, despite diversification of civil society actors involved in the UN negotiations, resources and links to power still rest overwhelmingly in the hands of professionalized NGOs that are taking a more reformist approach. Civil society has internalized, and in some cases helped to deepen, the dominant historic logic of neoliberal governance, including a commitment to market-based solutions, the imperative to sustain economic growth as the primary goal, and a narrow view of the particular types of scientific knowledge and expertise that are deemed relevant to the policy-making process. This has meant that civil society's leverage has been largely confined to issues with which they can persuade powerful actors that it is in their interests to pursue a particular policy option. They have largely failed to establish the capability to undermine powerful interests or push states to move beyond their nationally defined economic self-interest.

Second, and related, civil society has failed to take a coordinated and viable strategy for building strength to realize influence at hinge moments in the negotiations. There has been little coordination between mobilization in national contexts in order to ensure that key states such as the United States and China are prepared in advance to take action when new international treaties are being negotiated. And third, civil society has primarily devoted its attention at the international level to the UN climate processes, while largely neglecting other highly relevant international governance frameworks, including international trade regimes, financial institutions, and scientific bodies.

Groups such as Indigenous peoples, the urban poor, informal workers, women, and rural farmers all stand to be disproportionately harmed by a changing climate and also by the collateral impacts of what they call "false solutions." Those are responses to climate change that serve to reproduce or further entrench existing relationships of environmental inequality, placing a disproportionate burden on those that are marginalized in the global social, political, and economic hierarchy.

Through case studies of interventions in the negotiations of waste pickers, gender equality, and Indigenous peoples' networks, we found that most rights gains have been in the form of recognition in the UNFCCC texts of basic rights such as the importance of gender balance or respecting Indigenous knowledge systems. Such measures have been a relatively easy way for the Conference of Parties to legitimize the regime activities in response to network advocacy demands, without necessarily changing actual practice. The cases also show some potential for gains for marginalized groups when networks can simultaneously pose a threat to disrupt the viability of a particular regime body, delegitimize its actions through shaming techniques, and provide an objective rationale for the institution to change its course. This is a tall order, particularly when it often means challenging the interests of materially strong industry or state actors.

Overall, the analysis suggests that while there is some potential for the strategic agency of nonstate actors to outmaneuver their better-resourced adversaries in regime politics, these pockets are small and have rarely touched the core relations of power or caused substantial emissions reductions. While there has been much diversification and fragmentation of both business coalitions and civil society in international climate politics, the interests of fossil fuel actors continue to reign supreme. The most environmentally and socially vulnerable civil society actors, such as In-digenous peoples and waste pickers, while not without influence, have found only limited ability to influence the regime in their favor on the more substantive issues. Big green NGOs, while often having strong ties to both powerful state and business actors, have been unable or unwilling to push for more radical forms of change. As a result of the largely fragmented and unequal condition of climate change civil society, those on the inside, while more flush with resources, have limited leverage to put actual pressure on states that would cause the transformational change needed. Rather, they serve as mainly purveyors of ideas without real power.[13]

## A New Historic Bloc

The main argument of this book is that climate change is a problem that is deeply enmeshed in global power relationships specific to the contemporary world order. In this context, conditions of global inequality have intensified through economic globalization: despite bold promises of trickle-down prosperity and some notable gains for the poor, boosts in global economic growth in the past three decades have done far more to expand the profits for the extremely wealthy than to ameliorate the suffering of the world's poorest people, and least of all to reduce the gap between them. This inequality is already being exacerbated by climate change: those at the bottom are suffering worst, eroding the development gains of nations and whole regions, despite their having virtually no role in causing the problem. The solutions offered thus far have not meaningfully challenged the power structures that sustain this inequality and the interests that continue to benefit from it.

In chapter 9, we offered a series of highly depressing scenarios and a few that were more optimistic. With our six possible future scenarios, we focused on two key dimensions. On the one hand, do they envision a democratic process that will lead to equitable solutions? On the other hand, will they lead to solutions that are adequate in terms of keeping global concentrations of greenhouse gases to sustainable levels? Unfortunately, the business-as-usual scenario fails on both accounts.

One scenario rose to the top for us as by far the most preferable moving forward: global climate justice. This is the one that many climate activists have been working for during the past two decades: strong national action with an ambitious and binding international treaty, enforced by a United Nations with the power to enforce an agreement on who would do what about climate change and by when. Critically, this involves aggressively challenging fossil fuel interests, not just cozying up to them for a compromise position. Such an outcome may not be likely, but we still believe it is critical. It may well require a radical reform of the United Nations itself, while retaining the trust of the vast majority of the world's nations. This will be no easy feat.

If warming continues unabated, the commonsense discourse that the global market, if harnessed properly, can solve society's problems and lift all boats will ring increasingly hollow and may be threatened by competing narratives. A societal backlash against the shortcomings of the so-called free market (what political economist Karl Polanyi called a "double movement") may create space for a return for more state-centered Keynesian approaches to governing and re-embedding the market, or new forms of social organization altogether.[14] Alternatively, we fear that the withering of hegemonic social control may be replaced by forms of fascism reminiscent of Spain under Franco or Italy in the time of Mussolini and Uganda under Idi Amin: states that rely increasingly on military and coercive force rather than the power of legitimate and democratic rule. Modern forms of fascism would likely include a heightened role of corporations collaborating with the state to maintain favorable markets by whatever means necessary.

A warming world also means that the ecology of the planet will inevitably become less predictable, and this may be heightened by climate engineering experiments that have unforeseen consequences, as well as yet more disproportionate impacts on the poor and marginal. There is the risk that if society reaches certain tipping points, powerful actors double down on the extreme forms of energy extraction, making the case that limiting fossil fuel emissions as a mitigation strategy is now futile. Alternatively, escalating climatic instability could mean a broader positive response by civil society to challenge the status quo of fossil fuel development and the gospel of economic growth as measured by GDP as the only indicator that matters.

Constructively addressing the problem of runaway global climate change will require the emergence of what Gramsci calls a new historic bloc: a broad and powerful coalition of civil society, state, and market actors with the capability to advance a radically different development vision and pathway. Writing before the climate crisis was at the current level, Giovanni Arrighi and Beverly Silver also saw a crucial role for a transnationally linked civil society in keeping these transitions from descending into violent interstate conflicts.[15] Any such effort to constructively address the climate crisis, we argue, will have to contend with the major shifts and tensions in the contemporary world order that we have outlined.

Importantly, we believe that due to the entrenched positions of both states and powerful business actors, the main catalyst and vision for such a historic bloc will have to come from civil society. That effort must look profoundly different from the dominant approach that we have seen and been part of during the past two decades.

We next identify what we see as three main shifts that transnational civil society will have to undergo in order to advance international climate justice in this context.

## Uniting the Unusual

A historic bloc capable of challenging and uprooting powerful interests in our entrenched and unsustainable economic model can emerge, we contend, only through important strategic shifts in transnational civil society. Here we point to three main changes that are needed together to achieve international climate justice; we call them *linking movements.* First, we need movements that adeptly link policy advocacy efforts with powerful grassroots movement building, counterhegemonic discourse, organizing, and protest. Existing efforts of civil society to address climate change, as well as the majority of funding resources, have focused on incremental policy reform. This approach has presumed that the needed change will come as a product of having good ideas, bringing the right powerful people together in the room, inspiring entrepreneurialism, or as a concession to strengthen government or corporate legitimacy.

However, this is not how a problem as deep, complicated, and interest laden as climate change will ultimately be addressed. There is too much at stake for fossil fuel companies, and the millions of actors that define their own interests in our fossil fuel economy, to address climate policy as an insider compromise. As evidenced by collapsing carbon markets, reformist insider politics that work to address climate change primarily through compromise can potentially do more harm than good if they lock in place an inadequate framework that limits policy options moving forward. This is not to say that we don't need ambitious policies at the federal level in polluting states; we do. But the right policies will come only through efforts that link policy advocacy with movement politics. This includes the reframing of the major ideas by which we organize society (especially challenging the logic of economic growth as our main indicator of progress), broad-based protest, grassroots electoral politics, and establishing the means to disrupt harmful economic activity.

Importantly, as evidenced by the more than 400,000 people who participated in the People's Climate March in New York City in September 2014, climate change is no longer understood in the United States as merely an environmental issue. Rather, we are witnessing the beginning of perhaps a broader climate movement that defines climate change in terms of social and economic injustice. This didn't happen by accident. The People's Climate March represented one of the first effective collaborations between environmental groups and environmental justice organizations. In particular, the organizing committee for the march included the newly formed Climate Justice Alliance working closely with larger and better-resourced organizations such as the Sierra Club, Avaaz, and 350.org. As part of this, some traditional environmental groups such as the Sierra Club are redirecting resources that were previously devoted to Washington, DC, advocacy work, to supporting local grassroots struggles through organizing, legal, and communications strategies, including disrupting the siting of coal and natural gas infrastructure projects. At the same time, the organization has not totally abandoned federal regulatory efforts, but rather is using local activism to strengthen its political muscle and vision in Washington.

We argue that other organizations, and importantly large foundations, should follow this lead and invest heavily in long-neglected movement-building efforts—not just in the United States but also in other major polluting countries worldwide. This is not easy work: major differences in worldviews; a history of neglect of the environmental issues facing poor people, communities of color, and Indigenous peoples; time constraints; and the complexity of balancing locally defined interests with national and international political realities stand as important obstacles to success. Movement building is messy work. Nonetheless, if the past decade has taught us anything, it is that old school reformer environmental efforts will not be enough. We need environmental organizations and leaders willing to relinquish their seat at the table of the powerful elite in exchange for building broad-based political muscle from below.

Second, and related, movements are needed to bring together various constituents that are understood as having divergent interests—what Elizabeth Desombre has called "Baptist and bootlegger coalitions."[16] In terms of climate change, there is often a perceived conflict between employment opportunities and environmental aims. Laborers are often pitted against environmentalists and fail to find common ground. Actors that stand to benefit from the status quo often manipulate these dynamics. As a result, we need to build coalitions of workers, industry representatives, citizens, and political leaders that are equipped to effectively challenge the pervasive myth that there is no alternative to fossil fuels, unsustainable consumption and our current model of economic development.

Simultaneously, these coalitions must offer solutions that resonate with the ideological beliefs and perceived material interests of a broad cross-section of society. This is what Antonio Gramsci refers to as a counterhegemonic force capable of forming a new historic bloc. Moreover, while economic growth is the dominant frame of our existing global political order, it is not the only frame that can serve to unite the interests and capture the imagination of such a coalition. Rather, alternative visions are needed for how we can respond to the contradictions of an economic model that does not adequately meet the basic needs of many of the world's people, perpetuates inequality, and is incompatible with ecological limits.

There are few examples of such large-scale counterhegemonic movements in modern history. Scholars and activists typically point to the civil and gay rights movements or tobacco legislation as the models to turn to for addressing global climate change. However, we see abolition of slavery as a more accurate comparison, given the broad economic base and ideological system that was established around that industry and had to be overcome.[17] Indeed, fossil energy in some ways has taken the place of slave labor.

The main point here is that to have a chance at advancing the radically more environmentally sustainable agenda that is needed now, the climate movement must establish a broad-based coalition that cuts across class and identity divides to bring together unlikely market, civil society, labor, and state allies. Such a movement will likely have to offer a viable transition for those actors that currently define their interests within the existing petrodevelopment model. This is, of course, much easier said than done. But no one said that challenging the world's most powerful industry and questioning the central logic by which we organize society was going to be easy.

A third dimension of movement building that is needed is the ability to link social movement actions spatially across the globe. Social movements and social movement organizations have yet to fully respond to the global reorganization of production relations that has occurred since the 1970s—what geographer David Harvey called the shift to "flexible accumulation."[18] Namely, production of goods is now organized in flexible global supply chains, allowing corporations to source and produce goods in contexts where it is most efficient and affordable to do so, which is often where environmental and social regulations are minimal. Consumers of these goods are now farther in distance from the impacts of what they consume, with limited direct feedback to influence changes in behavior.[19]

This points to the fact that climate change is inherently a transnational problem in its causes and consequences. In this context, purely national forms of advocacy are no longer sufficient because the problems we face have taken on particularly global dimensions. Moreover, a fragmented and largely incomplete global governance architecture has emerged that has furthered the interests of a powerful investor class[20] while largely failing to address our most pressing issues of environmental and social inequality. In other words, problems have globalized, but resistance movements haven't quite caught up.

This globalized context offers social movements with new challenges in terms of accessing power and strategically coordinating actions across borders. However, it also offers a host of new possibilities as marginal actors in different countries simultaneously point to the same global economic and political processes as perpetuating systemic forms of inequality, including climate injustice. To respond, we need movements that establish new transnational identities of solidarity linking suffering in distant parts of the world; strategically coordinate actions to challenge unsustainable and unjust practices across distinct leverage points in global commodity chains; and effectively navigate and leverage fragmented global governance structures to re-embed economic processes in social and ecological systems.

In recent years, we have seen a few campaigns emerge that are demonstrating the power of activism across the commodity chain on issues such as global food justice,[21] the origin and disposal of waste,[22] and the transnational production and transport of fossil fuels.[23] Indeed, new communications technology has made transnational movement coordination more accessible than ever before. However, these movements continue to be the exception, not the norm, and they often function on shoestring budgets. Most civil society organizations and their funders continue to focus on spatially and thematically isolated conceptions of problems. Thus, stronger and smarter movements are needed that challenge the externalization of environmental and social costs of products from the extraction of raw materials, to production, transport, consumption, and disposal. Commodity chain activism must extend beyond national contexts to address the global reach of production networks across political and material geographies.

In sum, we argue that global climate justice necessitates a radical transformation from isolated, fragmented, and top-down civil society efforts that conform to dominant relations of power, to social movements that link grassroots activism to legislative efforts; unite unlikely, broad-based, and diverse counterhegemonic coalitions; and respond strategically to globalization's spatial reorganization of environmental problems.

To conclude, in February 2013 (after another difficult year at the Doha negotiations), a call for participants in an upcoming World Social Forum session in Tunisia went across NGO electronic lists. It argued that we need to "go beyond our usual strategies and see how we can win concrete victories on the ground by working together, across sectors, across movements, old and new, linking social struggles with environmental struggles bringing together trade unions, peasants, women, indigenous, migrants, faith communities, indignados, occupy movements, Idle No More and other climate and environmental activists." We would add even more groups and class fragments, enlightened bureaucrats, business transformers, and academics to this list of the linking movements needed.

We see such a broad, courageous, creative, and open-hearted linking of forces as our best, and perhaps our only, chance in the context of the greatest challenge we have faced together as humanity: living together in a warming world. As the slogan of the People's Climate March in New York and to twenty-six hundred other events in 162 countries in September 2014 put it, "To Change Everything, We Need Everyone."[24] The success of the march at linking movements was the result of its deliberate inclusiveness and nonjudgment about different reasons for being concerned about climate change. We need everyone.

# Notes

## Chapter 1

1. ABC News Online 2009.

2. Our book *Leaders from a Fragmented Continent* (Edwards and Roberts 2015) describes in some detail how Mexico led this effort.

3. Only Bolivia refused to accept the Cancun Agreements.

4. The Kyoto Protocol is not completely defunct; the Second Commitment Period of the Kyoto Protocol, agreed to in Doha in 2012, covers a handful of developed countries whose share of total world greenhouse gas emissions is less than 15 percent. The Durban Platform for Enhanced Action, established in December 2011, challenged the pledge and review system. Its mandate is to develop a protocol, another legal instrument, or an agreed outcome with legal force under the convention applicable to all parties, which is to be adopted by 2015 and to come into effect and be implemented from 2020.

5. The website Climate Interactive (2013) estimates that current pledges as of April 2013 will lead to a global average of 4.5 degrees Celsius of warming.

6. Intergovernmental Panel on Climate Change 2007b.

7. For example, over the period 1980 to 2011, Least Developed Countries experienced 51 percent of all deaths from climate-related disasters, while constituting only 12 percent of the world's population. Thus, people in these countries are four times more likely to die from climate-related deaths than those in other parts of the world (Ciplet, Roberts, Ousman, et al. 2013).

8. Garnaut 2008; Olivier, Janssens-Maenhout, Muntean, et al. 2014. Since 2008, there has been a slowing in this trend, apparently due to higher levels of carbon dioxide absorption in the oceans and an economic recession. Outlooks still predict troublesome emissions growth patterns in the coming decades (Intergovernmental Panel on Climate Change 2013, 2014).

9. *Bloomberg* 2012.

10. Gareau 2013.

11. Bodansky 2001; Gupta 2010; Okereke 2009; Ott, Sterk, and Watanabe 2008.

12. Intergovernmental Panel on Climate Change 2013; Hulme 2010; Lynas 2008.

13. Cook, Nuccitelli, Green et al. 2013.

14. Steinberger and Roberts 2010; Steinberger, Roberts, Peters et al. 2012; Lamb, Steinberger, Bows-Larkin et al. 2014; Jorgenson 2014.

15. Intergovernmental Panel on Climate Change 2007b.

16. On the latter see Brauch 2009; O'Brien and Barnett 2013.

17. UN Development Programme 2006; Roberts and Parks 2007; DARA 2012.

18. Ciplet, Roberts, and Khan 2013.

19. LDCs account for only .34 percent of cumulative global carbon dioxide emissions excluding land-use change and forestry from 1850–2011. Data compiled with World Resources Institute's *"Climate Analysis Indicators Tool"* Washington, DC (2015). Available at http://cait2.wri.org.

20. Den Elzen and Höhne 2008.

21. UN Environment Programme 2010.

22. Messner, Schellnhuber, Rahmstorf et al. 2010; Anderson and Bows 2011.

23. Naomi Klein's 2011 piece in the *Nation* agrees that these changes are monumental, requiring far more profound revisions in capitalism than most environmentalists acknowledge.

24. Baer, Athanasiou, and Kartha 2013.

25. Stern 2007.

26. EcoEquity 2009; Price Waterhouse Cooper 2014.

27. Note that we focus here on global climate justice, but acknowledge that we are still largely focused on agreement between states (international affairs). We believe that there are other national, transnational, and subnational means by which climate justice can and should be advanced. Indeed, in chapters 6 through 8 we move beyond the state and UN realm to look at the role of other actors in climate governance. However, we believe that achieving an international framework to address climate change in a just manner, between states, is a necessary condition for adequately and equitably responding to climate change. Three of the six dimensions of climate justice that we discuss focus on the mitigation side of climate justice, two on process, and one on adaptation. These six dimensions are some of the major elements, but there are dozens, even hundreds, of more detailed parts of the negotiations that one could focus upon. There is some overlap with Shue's (1992), Müller's (1999), and Paterson's (2001) elements of a just climate policy. Many elements could be added, especially on the governance, collection, and distribution of adaptation finance, technology transfer, and intellectual property rights, on what counts for carbon sinks and agriculture, reducing emissions from deforestation and forest degradation, and many elements of process.

28. As one anonymous reviewer of Robert's 2011 Global Environmental Change article usefully pointed out, there is much room to debate whether it is just to

grant such exclusive rights to national states (as opposed to, say, communities) and what such equal voice might mean (e.g., speaking rights or veto power).

29. United Nations 2008.

30. The peaking question raises several subissues. The IPCC reports have repeatedly put that tipping point at a temperature rise of 2.0 degrees Celsius above preindustrial levels, and newer data and meta-analyses indicate that 1.5 degrees Celsius or 1.8 degrees Celsius would be far safer (Richardson Steffen, and Liverman 2011). The 2007 IPCC report also summarizes scientific studies to say that emissions reductions will have to be 80 to 95 percent below 1990 levels by 2050 and at least 25 to 40 percent below that baseline by 2020 (Intergovernmental Panel on Climate Change 2007a). The IPCC targets are almost certainly too modest. A just solution that avoids the worst damage to coastal populations requires that global emissions peak and begin to decline by 2015 according to the IPCC, and there is a global consensus that the wealthy countries need to peak much sooner than the poor countries.

31. These groups include nearly all the members of the Climate Justice Now! coalition, plus many others. See http://www.climate-justice-now.org/category/climate-justice-movement/cjn-members.

32. In chapter 6, we argue that the focus of major environmental NGOs on carbon trading as a technical solution has come at a cost of not effectively challenging powerful interests to keep fossil fuels in the ground.

33. Ciplet, Roberts, and Khan 2013.

34. Sen 2010, 251.

35. For example, Rawls 1999.

36. Sen 2010, 106.

37. See Roberts and Parks 2007 for their earlier work on fairness claims in the UNFCCC negotiations.

38. See Bodansky 2001; Paterson 2001; Okereke 2009; Gupta 2010.

39. The "systems governance" literature has developed these points for over a decade; see Bierman et al. 2010; Dellas, Pattberg, and Betsill 2011.

40. See, for example, Sierra Club's website on the "coolest schools"; Sustainable Endowments Institute's Green Report; and the Princeton Review assessment of environmental efforts by schools.

41. The organization's website clarifies that "ICLEI originally stood for the `International Council for Local Environmental Initiatives,'" but in 2003 the organization dropped the full phrase and became ICLEI–Local Governments for Sustainability to reflect a broader focus on sustainability, not just environmental initiatives.

42. Signatories voluntarily commit to ten action points to advance local climate action, including "the reduction of emissions, adaptation to the impacts of climate change and fostering city-to-city cooperation." In just over a year, over 250 cities from 50 countries had signed on to the agreement.

43. Global Cities Covenant on Climate, "The Mexico City Pact."

44. Liverman and Boyd 2008.

45. Built on the model of cap and trade, the actions following from Assembly Bill 32 established a cap on emissions based on returning the state to 1990 levels by 2020: 427 million metric tonnes of carbon dioxide equivalents, which they calculated as 28 percent below the business-as-usual scenario of 600 million metric tonnes. Still, this does not meet the original US Kyoto pledge of 7 percent below the 1990 baseline. See Assembly Bill 32 at http://www.arb.ca.gov/cc/ab32/ab32.htm.

46. See RGGI 2013a, 2013b.

47. Berners-Lee and Clark 2014, 91.

## Chapter 2

1. This corresponds roughly with Clapp and Peter Dauvergne's (2005) typology of different approaches to environmental issues (Clapp and Dauvergne 2005).

2. O'Neill, Balsiger, and VanDeveer 2004.

3. Thompson 2006.

4. Harrison and Sundstrom 2010.

5. Roberts and Parks 2007; see also Okereke 2009.

6. See Keohane 1982.

7. See Biermann 2007; Andonova, Betsill and Bulkeley 2009; Bierman, Betsill, Gupta et al. 2010; Bierman, Pattberg, Van Asselt 2009.

8. Lovins and Cohen 2011, 4.

9. Newell and Paterson 2010.

10. Bond 2012; Klein 2014.

11. Levy and Egan 2003, 820.

12. Meckling 2011, 7, 9; Falkner 2000.

13. Ford 2003.

14. See Okereke, Bulkeley, and Schroeder 2009.

15. See Andonova, Betsill, and Bulkeley 2009.

16. A growing number of works take up the role of nonstate actors in transnational networks in international climate politics, including development banks and cities (Bulkeley and Newell 2009) and elite transnational advocacy coalitions (Meckling 2011).

17. Gramsci [2012] 1971. Neo-Gramscian scholars Levy and Egan (2003, 820) coined the term *strategic power*.

18. Cox 1987, 357; Murphy 1998; Robinson 1996; Levy and Newell 2005.

19. Gramsci [2012] 1971, 235.

20. Ibid., 178, 244.

21. Levy and Newell 2002, 48.

22. Ibid.

23. Levy and Egan 2003, 808. Also, see Gramsci [2012] 1971, 80: "The normal exercise of hegemony on the now classical terrain of a parliamentary regime is characterized by a combination of force and consent which form variable equilibria, without force ever prevailing too much over consent."

24. Hurd 1999.

25. There are numerous works that reveal the importance of discursive struggles and legitimacy in shaping environmental governance. See Litfin 1994 and Gareau 2013 on the Montreal Protocol; Hess 1992 on governance of chlorofluorocarbons; Goldman 2007 on World Bank water policy; and Bernstein 2001 on international environmental treaties broadly, to name just a few.

26. See Gareau 2013 on the Montreal Protocol, Goldman 2007 on World Bank water governance and Okereke, Bulkeley, and Schroeder on international climate change governance.

27. Sending and Neumann 2006; Gareau 2013; Goldman 2007; Okereke, Bulkeley, and Schroeder 2013. Also see Lipshutz 2005. For original conceptualization of governmentality, see Foucault [2003] 1976.

28. Sending and Newman 2006.

29. See Jasanoff 2013.

30. Reus-Smit 2007, 44; Franck 1990, 24; Hurd 1999.

31. Reus-Smit 2007, 158.

32. See Levy and Newell 2005, 51.

33. Biermann, Pattberg, Van Asselt et al, 2009.

34. Levy and Newell 2005; Levy and Egan 2003; Andrée 2005.

35. Cox 1992: 177.

36. Ibid.

37. What Robert Cox (1987) refers to as problem-solving theories.

38. Cox 1992.

39. Ibid.

40. Gramsci [2012] 1971, 161.

41. We are referring here to insights from work in the world system tradition by, for example, Arrighi 1994; Wallerstein 2011; Chase-Dunn 1998; and Bunker 1985.

42. See for example Rice 2007; Jorgenson and Clark 2009; Shandra, Leckband, McKinney et al. 2009.

43. Roberts 2001; Roberts and Parks 2007, 2009.

44. Levy and Egan 1998, 339.

45. Wheeler and Hammer 2010; Olivier, Janssens-Maenhout, Muntean, et al. 2014. It's important to note that emissions on a per capita basis remain far higher in the global North. An exception may be China's emissions, reported in 2014 by the EU-JRC to be nearly equal to the EU average.

46. Roberts and Parks 2007, chap. 2.

47. Najam 2005.

48. For an introduction to this line of argument on development, see Roberts and Hite 2007.

49. Prebisch 1950; Cardoso and Faletto 1979; see also Bunker 1985.

50. See Karl 1997.

51. However, inequality between nations has declined largely as a result of China's growth.

52. Boyer 2000; Newell and Paterson 2010.

53. Sklair 2009

54. Smil 2005.

55. Bernstein 2001; Gareau 2013.

56. Ibid.

57. Helleiner, Pagliari, and Zimmerman 2009.

58. *Los Angeles Times* 2008.

59. Pepinsky 2012.

60. Cahill 2011.

61. Bhattacharya and Dasgupta 2012.

62. UN Conference on Trade and Development 2012.

63. Economic Coalition for Latin America and the Caribbean 2011; Edwards and Roberts 2015.

64. Cahill 2011; Crouch and Colin 2011.

65. Önis and Güven 2011.

66. Pieterse 2011.

67. Wade 2011, 351.

68. Hurrell and Sengupta 2012, 465–466.

69. Arrighi's 1994 book described transitions in global hegemony from Genoa, the Dutch, British, and now American cycles. This is not to say that the current cycle will inevitably follow the pattern of the last three, but a growing number of indicators suggest it may. See Arrighi and Silver 2001; Arrighi, Silver, and Brewer 2003.

70. Arrighi 1994, 271.

71. Arrighi and Silver 2001, 270–271.

72. Ibid., 271.

73. Keohane 1980.

74. Arrighi 1994, 301.

75. Arrighi and Silver 2001, 322.

76. Shuhan and Marcoux 2010. It is likely that far more than this was given by the OPEC donors, but much is not officially reported.

77. Arrighi and Silver 2001.

78. See Roberts and Parks 2007.

79. Overholt 2010; Nye 2010.

80. Arrighi and Silver 2001, 278.

81. Ibid., 279.

82. Ibid.

83. Wade 2011; Nye 2010.

84. Wade 2011, 347.

85. Roberts 2011; Vihma, Mulugetta, and Karlsson-Vinkhuyzen 2011; Ciplet, Roberts, and Khan 2013.

86. See Deffeyes 2011; Leggett and Ball 2011; Patzek and Croft 2010.

87. Tucker 2013.

88. McKibben 2012.

89. Rainforest Action Network, Sierra Club and BankTrack 2014.

90. DiMuzio 2012, 365.

91. York 2012.

92. Füssel 2009.

93. Le Quéré, Raupach, Canadell et al. 2009; Garnaut 2008.

94. Ciplet, Roberts, and Khan 2013.

95. Paterson and Newell 2010, 10.

96. Keck and Sikkink 1998; Kaldor 2003; Evans 2008.

97. Kaldor 2003.

98. See Chandhoke 2005.

99. See Castells 2011.

100. Evans 2008.

101. See Habermas 2011.

102. Castells 2011.

103. Somers 2008.

104. Meckling 2011.

105. Clapp and Swanston 2009.

106. As Thomas Friedman (2006) might claim.

## Chapter 3

1. Michaelowa 2010; Winkler and Beaumont 2010.

2. For example, in the working negotiation texts, short-term emissions reductions went from "[25–40%]" reductions from unspecified "[ZZ levels]" to "[50]% by [2017[[2020]." Long-term reductions went from "[75–85 by 2050]" to "[more

than 100% by 2040]." By contrast, at COP3 in Kyoto in 1997, a deft handling of the final night negotiations by COP president Estrada led to the agreement of the Kyoto Protocol (Depledge 2005). Or, when in Cancun in 2010, a few countries raised objections to adoption of the text, the Mexican foreign minister, Patricia Espinosa, ruled them out, saying that consensus does not mean unanimity.

3. These numbers are from the texts revived at the June 2010 Bonn meetings, an attempt to restart the UN's negotiating after Copenhagen.

4. Democracy Now! "Voices from Small Island States" December 17 2009.

5. Najam 2005; Najam, Huq, and Sokona 2003.

6. There are many examples one might cite where some countries should have resisted the overall G77 position but did not. A little-known but still relevant one was back at COP4 in Buenos Aires in 1998, when the host country and Argentinian COP 4 president Maria Julia Alsogara proposed that developing countries take on voluntary commitments for reducing their emissions. Other G77 members rejected it, fearing that it might be a slippery slope to binding commitments for all of them in the future.

7. Najam 2005.

8. Vihma 2010. See also Bodansky 2009 and Christoff 2010.

9. Najam 2005; Roberts and Parks 2007. Our 2007 book, *A Climate of Injustice,* described and documented the enduring importance of their experience of exclusion to understand why trust is perpetually so low in climate change negotiations and why they break down repeatedly.

10. See the Editors' Introduction in Roberts and Hite (2007) for a review.

11. Wallerstein 1988; Chase-Dunn 1998.

12. This concept builds on the rural labor reserve literature, of which there was a substantial amount in the 1970s and 1980s, especially in Latin America and Africa. The term *subsidy from nature* was used by Hecht, Anderson, and May 1988.

13. The closeness of Saudi Arabia to the oil industry, including the US petroleum industry and even the US government, was manifest in their common positions (Depledge 2008). Newell (2000) claims that lobbying by the oil company interests is not needed since interests of domineering OPEC members are closely aligned with US interests from the outset. The US rejection of the Kyoto Protocol was welcomed by Saudi Arabia, which ratified it only at the last moment when it became clear that with Russia's ratification, the protocol was coming into force and the Saudis were eager not to be excluded from potential CDM project revenues from carbon capture and storage. By its detractors, Saudi Arabia's U-turn to becoming one of the diehard champions of the Kyoto Protocol is motivated only by its eagerness to prevent balanced progress on the Long-Term Cooperative Action and Kyoto Protocol tracks, which is consistently raised as a requirement of progress on either. The US/Saudi alliance appears to have reemerged recently as Saudi Arabia sided with the United States as the only two countries that raised objections to the work of the Transitional Committee designing the Green Climate Fund (Yamin 2011).

14. Personal communication with Roberts, 2005.

15. Kasperson and Kasperson 2001; Intergovernmental Panel on Climate Change 2007a; 2013, Roberts and Parks 2007.

16. While billions in assistance were promised in Copenhagen in 2009 and Cancun in 2010, the fact that there was no baseline of effort or of funding agreed (Stadelmann, Michaelowa, and Roberts 2011) and no compliance regime in adaptation shows that the voices of the most vulnerable continue to be ignored in the power politics of high emitters in the climate change negotiations (Khan and Roberts 2013).

17. Earth Negotiations Bulletin 1995.

18. Jackson 1993. However, vulnerability differs within these countries (see Adger and Vincent 2005).

19. Kasa, Gullberg, and Heggelund 2008.

20. Khan 2013.

21. Dubash and Morgan 2012.

22. Smith 2010.

23. Climate Interactive 2013.

24. Roberts 2011.

25. These funds would be placed under the proposed Copenhagen Green Climate Fund, with a high-level panel under the COP to study the implementation of financing provisions. Also mechanisms of technology transfer and forestry (REDD +) have been envisaged.

26. *Guardian* 2010a.

27. Lumumba Di-Aping COP Briefing. December 11, 2009.

28. China is not a strong supporter of the others' gaining permanent seats in the Security Council, though in late March 2011, China expressed support for India's bid for a seat. President Obama expressed his support also for India's seat while visiting India in 2010, and the United States also expressed appreciation for Brazil's wish in March 2011 for its Security Council aspirations.

29. Later in 2010, South Africa joined the group (now BRICS).

30. Yamin 2011, 5.

31. Hochstetler 2013 (personal communication).

32. China consumes nearly as much coal as the rest of the world combined (US Energy Information Administration 2012).

33. *New York Times* 2014a.

34. Steinberger et al. 2012.

35. Kasa et al. 2008. For example, since 2006, China and India have been members of the US-initiated Asia-Pacific Partnership for Clean Development and Climate, which established eight government and business task forces to promote and transfer clean technology. Another example is the Australia-China Partnership on Climate Change (2003) and the EU-China Partnership on Climate

Change (2005). The United States also signed a Memorandum of Understanding on Cooperation in Climate Science and Technology with China in 2009. China has signed similar agreements with nations across Latin America (Edwards and Roberts 2015).

36. Heggelund 2007; Kasa et al. 2008. A new "Like Minded Group" of nations including most of BASIC is taking a strong position against binding commitments for any developing countries in the foreseeable future.

37. *International Business Times* 2014; *Bloomberg News* 2014.

38. Chayes and Kim 1998.

39. Economy 1997.

40. Heggelund 2007.

41. Alessi 2012.

42. We take up these issues in Edwards and Roberts 2015.

43. *New York Times* 2014a.

44. World Bank 2014b.

45. *Guardian* 2014a.

46. *Guardian* 2013a. Ranking in per capita emission assessed using 2010 data.

47. India's National Action Plan on Climate Change adopted in 2008 contains measures for increasing energy efficiency, renewable energy, and even a domestic cap-and-trade program as incentive for emission reduction for nine energy-intensive sectors. It was the first country to establish a ministry of nonconventional energy sources (Mehra 2009).

48. Robinson 2010.

49. Roberts and Edwards 2015.

50. *Times of India* 2009.

51. When Ramesh deviated from the official script in a ministerial statement in Cancun and said that "all countries must take on legally binding commitments in an appropriate legal form," this was interpreted domestically as India's agreement to assume legally-binding commitment (*Financial Express* 2010).

52. Hultman et al. 2010, 2011.

53. International Energy Agency 2007.

54. *Guardian* 2009b.

55. Brazil also sided with the United States in pushing for inclusion of all six major greenhouse gases under the Kyoto Protocol process, contrary to the G77 and China position, which focused on carbon alone.

56. See Jacobsen 1997.

57. Edwards and Roberts (2015) discuss Brazil's position at length in *Leaders from a Fragmented Continent*.

58. Nikolas Kozloff (2012) argues that far from exhibiting Third World solidarity, China and India have conspired with the United States to thwart the

EU in reaching a meaningful deal. Then BASIC expanded its bloc into "BASIC Plus," inviting the group leaders of the G77, AOSIS, LDCs, OPEC, and other groups as observers, simply in order to preserve its tarnished image with its G77 partners.

59. Kathy Hochestetler, personal communication, October 11, 2013.

60. Arrighi and Silver 2001, 271.

## Chapter 4

1. The Kyoto Protocol is not completely defunct: the Second Commitment Period of the protocol, agreed to in Doha in 2012, will cover a handful of developed countries whose share of total world greenhouse gas emissions is less than 15 percent.

2. Current pledges are estimated to lead to a global average of 3.5 to 4.5 degrees Celsius of warming (Climate Interactive 2013; International Energy Agency 2014a).

3. Füssel 2009.

4. Roberts and Parks 2006.

5. Ciplet, Roberts, Ousman et al. 2013.

6. We focus on state actors because they have formal voting rights, and thus can formally offer or withhold their consent.

7. Exceptions include Larson 2003; Deitelhoff and Wallbott 2012; and Betzold 2010.

8. For an overview, see O'Neill, Balsiger, and VanDeveer 2004.

9. Lindblom 1965, 227.

10. Notable studies on low-income or small state participation in multilateral regimes find that these states mitigate their disadvantages in material resources in three main ways: through prioritization by focusing scarce resources on shaping particular issue areas while neglecting other issues; coalition building in order to strengthen their representation and facilitate information exchange (Björkdahl 2008); and developing ideational resources, persuasion-based strategies (Panke 2012), and translating moral leadership into discursive power (Deitelhoff and Wallbott 2012). Studies in this area also highlight shortcomings of small states in finding influence, including lack of financial and negotiating resources, inability to offer side payments for support, and inability to credibly threaten other states (Panke 2012).

11. Gramsci [2012] 1971, 161.

12. Forgacs 2000, 422–423.

13. Gramsci [2012] 1971, 161.

14. Mouffe 2013, 36.

15. Gramsci [2012] 1971, 235.

16. Finnemore and Sikkink 1998, 891.

17. What Gramsci [2012] 1971 termed a process of "trasformismo" (58).

18. Gramsci [2012] 1971, 161.

19. Ibid., 182.

20. Christian Aid 1999.

21. Martinez-Alier 2003.

22. Roberts and Parks 2009.

23. Note that other countries excluded from this designation, such as Guatemala, Colombia, Bolivia, and Pakistan, have opposed this category. See Ciplet, Roberts, and Khan 2013.

24. Decision CP.13 1e(i).

25. Alliance of Small Island States 2009; African Group proposal 2009; Bolivia 2009.

26. The LDC nations themselves have grown more forceful and organized, and there was over a decade of support for the group developing its demands from the European Capacity Building Initiative, directed by the Oxford Institute of Energy Policy. These consisted of workshops that gathered LDC representatives for one or two weeks to learn climate science and policy and develop strategic interventions in the negotiations.

27. Third World Network, December 10, 2009. Updates.

28. Among other issues such as respecting of international property rights regimes in the transfer of technology.

29. Third World Network, December 10, 2009. Updates.

30. Ibid. Updates.

31. *BBC News* 2009.

32. The Ad Hoc Working Group on Long-term Cooperative Action and the Ad Hoc Working Group under the Kyoto Protocol.

33. UN Framework Convention on Climate Change 2009.

34. Ciplet, Fields, Madden et al. 2012.

35. *Hindustan Times* 2009.

36. CAN-International Press Conference, Copenhagen, December 18, 2009. Available at http://cop15.meta-fusion.com/kongresse/cop15/templ/play.php?id _kongresssession=2750&theme=unfccc.

37. Third World Network, December 20, 2009.

38. YouTube 2009.

39. *Guardian* 2010a, 2010b; *New York Times* 2010.

40. Author conversations with LDC delegates December 2010, 2011.

41. *TerraViva* 2009.

42. *Ethiotube* 2009.

43. Author conversations with LDC delegates, December 2010, 2011.

44. Bodansky 2010.

45. The $5 billion was in funding from 2000 to 2009. *Washington Post* 2010.

46. *Guardian* 2010c.

47. *Democracy Now!* 2010.

48. *Guardian* 2010d.

49. Ibid. 2010e.

50. Ibid. 2014b.

51. Third World Network, "Attempts to Make the Copenhagen Accord a "Plurilateral Agreement," December 21, 2009.

52. Ibid., December 21, 2009.

53. USCAN 2010.

54. Ibid. 2010.

55. Skype interview by Timmons Roberts and Guy Edwards with Luis Alfonso de Alba, March 2013, and personal communications, April 2014.

56. Ibid.

57. Author conversations with LDC delegates in 2010 and 2011.

58. Our own research group, the Climate and Development Lab at Brown University, has issued a half-dozen policy briefings analyzing whether these promises were met and identifying specific problems with the language and nature of the promises. See Stadelmann, Roberts, and Huq 2010; Ciplet, Roberts, and Stadelmann 2011; Ciplet, Roberts, and Khan 2011; Ciplet, Fields, Madden et al. 2012; Ciplet, Roberts, and Ousman 2013.

59. *Reuters* 2010.

60. For example, *Economist* 2010.

61. Ciplet, Roberts, and Stadelmann 2011.

62. Donor Panel in UNFCCC, December 2011.

63. *Guardian* 2011b.

64. UN Framework Convention on Climate Change 2009.

65. United Nations Framework Convention on Climate Change 2011.

66. *Guardian* 2012a.

67. Originally introduced in the Cancun Adaptation Framework.

68. United Nations Framework Convention on Climate Change 2013/

69. For example, statements made in a Central American Integration System side event, November 23, 2013.

70. Decision 2/CP.19.

71. United Nations Framework Convention on Climate Change 2014a.

72. Observations by the authors in UNFCCC plenary December 13, 2014. Also see Federico Broccheri, "ADP Text: Who's in Favour, Who's Against," Twitter, December 13, 2014. Available at https://twitter.com/federicoclimate/status/543891049824145409.

73. Ibid.

74. See Adopt a Negotiator 2014.

75. Ibid; UN and Climate Change 2014a.

76. United Nations Framework Convention on Climate Change 2014.

77. United Nations Framework Convention on Climate Change 2015.

78. Kartha and Erickson. 2011.

79. Ciplet, Roberts, and Khan 2013.

80. Ibid.

81. Oxfam 2012.

82. *Guardian* 2010f.

83. Jansen 2009.

84. Bhattacharya and Dasgupta 2012.

85. Ibid.

86. Vihma, Mulugetta, and Karlsson-Vinkhuyzen 2011.

87. Ciplet, Roberts, and Khan 2013.

88. Gupta 2000; UNfairplay 2011.

89. Data compiled with World Resources Institute's "*Climate Analysis Indicators Tool*," Washington, DC (2015).

90. Khan 2013.

91. Shadlen 2004, 81.

92. Cox 1993.

93. Levy and Egan 1998, 339.

94. Edwards and Roberts n.d.

95. Third World Network, December 20, 2009.


## Chapter 5

1. Welz 2009.

2. *Hindustan Times* 2009.

3. Intergovernmental Panel on Climate Change 2007a (Glossary).

4. Kasperson and Kasperson 2001; UN Development Programme 2007; Roberts and Parks 2007.

5. Sen 2010.

6. Kuik et al. 2008, 328.

7. Schipper 2006.

8. Kates 2000, 1.

9. Schipper 2006, 85.

10. Ayers and Dodman 2010.

11. Pilifosova 2000.

12. Anderson 1998, 13; Burton 1996.

13. Schipper 2006; Ayers and Huq 2009.

14. Smit and Pilifosova 2003.

15. Anderson 1998, 13.

16. This argument still is used by contributor nations to the 2008–2012 fast-start finance as an explanation for why they are funding more mitigation activities in the shorter term.

17. Gupta 1997; Schipper 2006.

18. Shue 1999.

19. Author personal communication with EU negotiator, Buenos Aires, 2003.

20. CAN-International ECO 5 June 2010.

21. AOSIS proposed this during discussions under the Intergovernmental Negotiating Committee, which was formed for drafting and negotiating a convention in 1990–1991. This was the reason that an insurance mechanism was incorporated in Article 4.8 of the Convention.

22. Decision 5/CP7.

23. The Adaptation Fund was to be funded by a 2 percent levy on the sale of certified emission reductions through the Clean Development Mechanism. Later we describe the rise and fall of these sales.

24. The first large transnational event that explicitly used the term *climate justice* occurred in the Hague in 2000, corresponding to the UNFCCC negotiations there. It was organized by CorpWatch and called the "Climate Justice Summit Program" in November of that year. Earlier climate justice–themed protests occurred around the world in the early 1990s.

25. Khastagir 2002.

26. Ibid.

27. Ibid.

28. Ott 2003.

29. Simms 2001 is also an early statement of the ecological debt concept.

30. Okereke Mann, Osbahr, et al. 2007, 4.

31. Ibid.

32. In Nairobi the governance structure of the Adaptation Fund was also nearly finalized.

33. Decision CP.13 1e(i).

34. In the twelve-paragraph Copenhagen Accord, adaptation was explicitly discussed in four paragraphs, with paragraph 3 exclusively focusing on adaptation. However, this paragraph again links adaptation to the "adverse effects of climate change" with "impacts on response measures."

35. Decision CP. 16 2 (b).

36. The Adaptation Framework is a body of sixteen experts tasked with promoting the implementation of adaptation under the convention.

37. Intergovernmental Panel on Climate Change 2007a.

38. Agarwal and Narain 1991; Shue 1999; Roberts and Parks 2007; Caney 2010.

39. An important footnote in the 2007 IPCC Fourth Assessment Report suggested that emissions reductions by the wealthy nations would need to be on the order of 25 to 40 percent by 2020 and 80 to 95 percent by 2050, to allow global reductions of 50 percent by that date.

40. Ciplet Roberts, and Khan 2013.

41. Stadelmann, Michaelowa, and Roberts 2011.

42. Rubbelke 2011.

43. Although AOSIS had been insisting on this issue for discussion since the early 1990s, including their proposal of an international insurance pool, it is only after almost two decades that the UNFCCC has internalized this agenda item, starting in 2012.

44. Singer 2002; Traxler 2002; Gardinar 2004; Roberts and Parks 2007; Baer et al. 2008.

45. Dellink, den Elzen, and Eiking et al. 2009.

46. Grasso 2010, citing Jagers and Duus-Ottestrom 2008: 577.

47. Grasso 2010: 75.

48. Sen 2010: 251.

49. Article 4.1.e.

50. Article 4.1.f.

51. Article 2.3.

52. Article 3.1.

53. Article 11.3(d).

54. See Khan and Roberts 2013.

55. United Nations Environment Programme 2014, 33.

56. United Nations Environment Programme 2014, 26.

57. Ibid.

58. Italics added for emphasis.

59. See Stadelmann et al. 2013.

60. Stockholm International Peace and Research Institution 2014.

61. Buchner, Falconer, and Hervé-Mignucci et al. 2012.

62. Moore (2012) brings in two framings of adaptation that have important implications on how adaptation is being and is likely to be funded. These framings are "adaptation as development" and "adaptation as restitution" in terms of compensation. Since low adaptive capacity is linked with low levels of social development, and it is not feasible to distinguish between impacts of climate change

and those of climate variability, development agencies would prefer to rethink climate aid as simply "good development." For example, the US says in its 2009 submission to the Ad Hoc Working Group on Long Term Cooperative Action in the UNFCCC: "We have an interest in ensuring our funding is used effectively, and we are convinced that this can best be done if adaptation funding is integrated into broader development assistance." In turn, the framing of adaptation as restitution (in the form of compensation) is most actively supported by the AOSIS and the LDCs, but completely rejected by the Annex 1 countries.

63. Stadelmann, Roberts, and Huq 2010.

64. Ciplet, Roberts, Stadelmann et al. 2011.

65. Oxfam America 2012.

66. German Advisory Council on Global Change 2010.

67. Ciplet, Fields, and Madden et al. 2012.

68. Ciplet, Fields, Madden et al. 2012.

69. OECD Development Assistance Committee 2014, 8.

70. See Hicks, Parks and Roberts et al. 2008.

71. Decision 11/CP1.

72. Article 4.9.

73. Paragraph 1ci.

74. Paragraph 1cvi.

75. Paragraph 8.

76. CAN-International ECO 2 December 2010.

77. Najam 2005; Roberts and Parks 2007.

78. Clémençon 2006.

79. Global Environmental Facility 2011.

80. Adger 2006; Adger, Brooks, and Bentham et al. 2004.

81. Decision 1/CMP.4.

82. AFB/B.12/5.

83. Harmeling 2010; Yohe, Malone, and Brenkert et al. 2006; Klein 2010; DARA 2012.

84. Müller 2013 describes why this is important.

85. 2008 Accra Agenda for Action.

86. Decisions 5/CP.7, 6/CP.7, 7/CP.7 and 10/CP.7.

87. This is consistent with Article 11, which creates a financial mechanism for implementation of the Convention under the guidance of and accountable to the COP.

88. Global Environment Facility 2010.

89. Fankhauser and Martin 2010.

90. Article 11.2 of the Convention.

91. Ciplet, Fields, and Madden et al. 2012.

92. Ciplet, Roberts, Khan et al. 2011.

93. ActionAid 2010; Oxfam International 2010.

94. UN and Climate Change 2014b.

95. The White House 2014.

96. Personal communication with the Global Environmental Facility staff, 2012.

97. World Bank 2014a.

98. Krasner 1985.

99. Intergovernmental Panel on Climate Change 2013.

100. Intergovernmental Panel on Climate Change 2013.

101. Intergovernmental Panel on Climate Change 2012, 5.

102. International Energy Agency 2013; World Bank 2012.

103. US Federal Advisory Committee 2014.

104. World Bank 2012, Executive Summary. See also Lynas 2008.

105. World Bank 2012, Executive Summary, xiii, xviii.

106. Smith 2007.

107. Myers 2002, 609, 611.

108. Decision 1/CP.18, para 7vi.

109. Podesta and Ogden 2008.

110. Walter 2012.

111. Ibid.

112. Biermann and Boas 2010.

113. World Bank 2000, 1–2.

114. Kotin 2012.

115. Kraska 2011.

116. Shepherd 2009.

117. Launder and Thompson 2010.

118. Ibid; Gardiner, 2004, 2010; Shepherd 2009; Hamilton 2011, 2013.

119. Preston 2013.

120. Svoboda et al. 2011.

121. World Bank 2010, chap. 6; Parry 2009.

## Chapter 6

1. See *People and Planet* 2009.

2. Meckling 2011.

3. Lovins and Cohen 2011, 4.

4. Rule 7.2. Note that two-thirds of parties may override the participation of a given group.

5. The footnote to Rule 30.

6. Vormedal 2008, 50.

7. Ibid.

8. Levy and Egan 1998, 342; Lund 2013, 4.

9. Lund 2013, 9.

10. Ibid., 4; Princen, Finger, Clark et al. 1994, 35; Meckling 2011, 39.

11. The analogy is from the US experiment with the prohibition of alcohol in the 1920s and 1930s (DeSombre 2000). Bootleggers were those who illegally produced and/or smuggled alcohol, sometimes by strapping bottles to their legs.

12. Meckling 2011, 40; Lund 2013, 4.

13. Pulver 2005, 54.

14. Ibid.

15. Newell and Patterson 2010, 8.

16. Ibid, 9.

17. Lovins and Cohen 2011, 4.

18. For example, see Mol 1997 and response by York and Rosa 2003.

19. World Commission on Environment and Development 1987.

20. Meckling 2011.

21. Levy and Egan 1998; Princen, Finger, Clark et al. 1994.

22. Vormedal 2008.

23. Clapp 1998.

24. Karliner 1997.

25. Spaargaren and Mol 1992.

26. York and Rosa 2003.

27. Levy and Egan 1998, 2003; Newell and Paterson 1998; Leggett 1999; Levy and Newell 2002, 2005.

28. Levy and Egan 1998, 346.

29. Ibid.

30. Ibid, 348.

31. Now called the Environmental Defense Fund.

32. Pulver 2005, 53.

33. Meckling 2011.

34. Ibid.

35. Ibid., 28.

36. Vormedal 2008.

37. Dunlap and McCright 2012; Brulle 2014.

38. Personal communication with Roberts 2005.

39. *Guardian* 2014c.

40. *Guardian* 2012b.

41. Oil Change International 2012.

42. *Bloomberg New Energy Finance* 2014.

43. Ibid.

44. York 2012.

45. *Climate Progress* 2013

46. Center for American Progress 2014.

47. Stanford Report 2014.

48. Rainforest Action Network, Sierra Club, and BankTrack 2014.

49. International Energy Agency 2014b.

50. *Bloomberg* 2013.

51. As the World Bank's research partner, Ecofys explained in 2013: "Unlike in previous years, the report does not provide a quantitative, transaction-based analysis of the international carbon market as current market conditions invalidate any attempt and interest to undertake such analysis" (Ecofys 2013).

52. *Guardian* 2012c. From a high of $20 to less than $3 for each credit.

53. *Financial Times* 2013.

54. See Lohmann 2006; Michaelowa and Purohit 2007.

55. *Guardian* 2013b.

56. Cited in the *Guardian* 2013b

57. Schneider 2011.

58. See *Sixty Minutes* 2012.

59. Angus 2010; Lohmann 2006, 2008; Lohmann, Hällström, Österbergh et al. 2006.

60. United Nations Environment Programme 2010.

61. World Wildlife Fund claims that "Gold Standard certification is awarded to those that were certified as meeting social and environmental standards developed by the Secretariat, overseen by an independent Technical Advisory Committee and verified by UN accredited independent auditors. The certification process uniquely requires the involvement of local stakeholders and NGOs" (WWF n.d.).

62. Caney and Hepburn 2011.

63. There are numerous examples of Indigenous peoples being displaced by carbon traders. See the story on the "carbon cowboy," Daniel Nilsson, on *Sixty Minutes* (2012).

64. Sterk and Wittnben 2006, 276.

65. Pearson 2007.

66. Anderson 2012.

67. Meinshausen et al. 2009.

68. McKibben 2012.

69. For example, International Energy Agency 2012. The agency says, "No more than one-third of proven reserves of fossil fuels can be consumed prior to 2050 if the world is to achieve the 2°C goal, unless carbon capture and storage (CCS) technology is widely deployed" (3).

70. Heede 2014.

71. Berners-Lee and Clark 2014, 87

72. Berners-Lee and Clark 2014, 91

73. As a recent statement of over ninety civil society organizations states, "It is time to stop fixating on 'price' as a driver for change. We need to scrap the ETS and implement effective and fair climate policies by making the necessary transition away from fossil fuel dependency" (Carbon Trade Watch 2013).

74. For example, see *Economist* 2009.

75. Levy and Egan 1998.

76. *Guardian* 2013b.

77. Vormedal 2008, 47; Also see Newell and Paterson 1998; Betsill 2006: 192; Levy and Newell 2005.

78. Center for Public Integrity 2011.

79. Center for Public Integrity 2009.

80. Ibid.

81. Ibid.

82. Keck and Sikkink 1998.

83. Clabough 2013.

84. See *Frontline* 2012.

85. McKibben 2012.

## Chapter 7

1. See *Guardian* 2013d.

2. This oil has a high carbon footprint for two main reasons: it is very carbon intensive to extract given that it is found under forested areas and often far underground, and it is carbon intensive to refine due to being mixed with sand.

3. *Huffington Post* 2013. "Obama golfed with oil men as climate protestors descended on White House." Obama's golfing partners included Jim Crane and Milton Carroll, directors at Western Gas Holdings.

4. Weather Channel Interview 2013.

5. Roberts and Ciplet observed different parts of the protests and arrests. A video is at http://www.telegraph.co.uk/earth/copenhagen-climate-change-confe/6799264/Copenhagen-climate-summit-1000-anarchists-arrested.html.

6. *Guardian* 2010h.

7. *Guardian* 2010g.

8. *BBC News* 2003.

9. Earth Day Network n.d.

10. See Harrison and Sundstrom 2010; DeSombre 2000.

11. We acknowledge that there are also numerous other types of climate movements, ranging from faith communities, to transition towns, to city government efforts, that we can't deal with in full here due to lack of space.

12. Martinez-Alier 2003.

13. Martinez-Alier 2013.

14. Koopman 2012.

15. Roberts and Thanos 2003.

16. Lang and Xu 2013.

17. *Guardian* 2013c.

18. Hansen (2013) writes, "From 2007 to 2009, only 15 percent of environmental grant dollars were classified as benefiting marginalized communities, and only 11 percent were classified as advancing "social justice" strategies, a proxy for policy advocacy and community organizing that works toward structural change on behalf of those who are the least well off politically, economically and socially." Also see *Huffington Post* 2012a and Brulle 2000.

19. See Piven and Cloward 1979.

20. Hansen (2013) writes, "In 2009, environmental organizations with budgets of more than $5 million received half of all contributions and grants made in the sector, despite comprising just 2 percent of environmental public charities." Also see Brulle and Jenkins 2005.

21. Meckling 2011, 155.

22. See the Rising Tide spoof at http://understory.ran.org/2007/12/03/hoax -website-spoofs-us-greenwashing-gang.

23. For example, see the 2008 statements of the Environmental Justice Forum on Climate Change.

24. *Washington Post* 2013.

25. Meckling 2011.

26. *Guardian* 2012c.

27. Ibid. 2013b.

28. Third World Network 2013.

29. *Guardian* 2013b.

30. Rainforest Action Network 2013.

31. Friends of the Earth 2010.

32. CNN 2012.

33. *Stanford Report* 2014.

34. *Guardian* 2014d.

35. *New York Times* 2014b.

36. Arabella Advisors n.d.

37. *Huffington Post* 2012b.

38. For example, see the sign-on petition by labor leaders calling on labor to oppose the Keystone XL pipeline. Available athttp://www.labor4sustainability.org/articles/petition-against-keystone-xl-and-tar-sands-pipeline-please-sign-now.

39. Lund 2013.

40. Burgiel 2008; Betsill and Corell 2008; Humphreys 2004.

41. Lund 2013, 5, citing Burgiel 2008.

42. Gareau 2013.

43. Betsill and Corell 2001, 95.

44. Ibid.

45. Bernauer and Gampfer 2013, citing Princen, Finger, Clark et al. 1994; Yamin 2001.

46. *Worldwatch Institute* 2010.

47. Mitchell 2011; Piewitt, Rodekamp, and Steffek 2010.

48. Bernauer and Gampfer 2013 citing Keohane et al. 2009; Scholte 2002.

49. Bernauer, Böhmelt, and Koubi 2013.

50. Newell, Tussie, and Cox 2006; Dombrowski 2010.

51. Newell 2008, 127 citing Marceau and Pedersen 1999, 43.

52. Levy and Newell 2005; Corell and Betsill 2001; Pulver 2005.

53. Betsill and Corell 2008.

54. Humphreys 2004.

55. Finnemore and Sikkink 1998; Meckling 2011.

56. Lund 2013, 6, citing Gulbrandsen and Andresen 2004.

57. Newell 2005.

58. Unmüßig 2011

59. See Rising Tide 2000.

60. CorpWatch 2002.

61. Pulver 2005, 53.

62. Neumayer and Plümper 2007.

63. *Climate Justice Now!* 2007.

64. Ibid.

65. Heinrich Böll Foundation 2008.

66. Designed by marketing giant Havas Worldwide, with clients from Coca Cola to McDonald's. See *Havas Worldwide* 2009.

67. UN Framework Convention on Climate Change 2011.

68. See "Intergovernmental Organizations and the Climate Change Process," available at http://unfccc.int/parties_and_observers/igo/items/3720.php.

69. Fisher 2010.

70. Ibid.

71. Ibid.

72. Ibid.

73. Author observation.

74. CNN 2009.

75. *Daily Mail* 2009.

76. *Guardian* 2009b. Several of Ciplet's colleagues who were part of the waste picker network were cornered just behind him and handcuffed during the march on the cold Copenhagen streets, despite merely walking as part of the protest and chanting peacefully.

77. Fisher 2010, 16.

78. See the video at http://climatedevlab.wordpress.com/2011/12/09/occupy-cop17/#more-520.

79. Author observation of what was discussed during the protest.

80. In the end, the UNFCCC did not carry out this threat.

81. *Climate Connections* 2011.

82. See Dunlap and Mertig 2014

83. For example, see Climate Progress 2013; East Asian Forum 2012.

84. *Huffington Post* 2012c.

85. Ibid.

86. Gareau 2013.

87. Goldman 2007.

88. Sending and Neumann 2006; Gareau 2013; Goldman 2007; Okereke, Bulkeley, and Schroeder. Also see Lipshutz 2005. For original conceptualization of governmentality see Foucault [2003] 1976.

## Chapter 8

1. CDM 2009.

2. Bullard 2000; Roberts and Toffolon-Weiss 2001.

3. As discussed in Ciplet 2014.

4. As discussed in ibid.

5. Young 1990, 25. Also see Somers and Roberts 2008, 412–413. For discussion on various competing conceptions of rights, see Morris 2006.

6. Krasner 1982.

7. Robert Cox (1992) argues that "multilateralism can be examined from two main standpoints: one, as the institutionalization and regulation of established order; the other, as the locus of interactions for the transformation of existing order" (163).

8. Finnemore and Sikkink 1998, 900.

9. Fisher and Green 2004.

10. Held 1999.

11. Clark, Friedman, and Hochstetler 1998, 4.

12. Keck and Sikkink 1998, 5.

13. Betsill and Corell 2008.

14. Muetzelfeldt and Smith 2002, 55.

15. Somers 2008; Nussbaum 2001; Sen 2001. These authors develop distinct capability perspectives outside of the international regime context.

16. Newell 2008.

17. Keck and Sikkink 1998.

18. McCormick 2007; Jasanoff 2013.

19. Gareau 2013.

20. Kinchy 2012, 25.

21. Neumayer and Plümper 2007.

22. Terry 2009 citing Gustafsod 1998; Bee, Biermann, and Tschakert 2013.

23. Neumayer and Plümper 2007.

24. Gustafsod 1998.

25. Bee et al. 2013.

26. See the Convention on the Elimination of All Forms of Discrimination against Women (1979), the Beijing Platform for Action (1995), and the Millennium Development Goals (2000), the UN Entity for Gender Equality and Empowerment of Women, and the Declaration on the Elimination of Violence against Women (1993).

27. FCCC/CP/2001/13/add.4.

28. Alber 2009, 60.

29. Gender and Climate Change Network 2007.

30. Women and Gender Constituency 2009.

31. Women's Environment and Development Organization 2010.

32. Green Climate Fund 2013.

33. Ciplet, Roberts, and Khan 2013.

34. Cited in Reuters 2012.

35. See the draft UNFCCC text with "equality" at http://unfccc.int/resource/docs/2012/sbi/eng/l36.pdf.

36. Cited in Reuters2012.

37. *ECO* 2012.

38. Joint Statement of Indigenous Peoples Networks 2000.

39. Ford 2012.

40. Doolittle 2010.

41. Examples are the Hague Declaration (2000), the Bonn Statement (2001), the Marrakesh Statement (2001), Indigenous Peoples' Caucus Statement on Climate Change (2002), the Milan Declaration (2003), and the Declaration of COP 10 (2004). Joint Statements of Indigenous Peoples Networks 2000–2004.

42. The plus sign stands for "the role of conservation, sustainable management of forests and enhancement of forest carbon stocks."

43. Decision 2/CP.13.

44. Joint Statement of Indigenous Peoples Networks 2007.

45. Major networks include the Accra Caucus, the Indigenous Peoples' Caucus, and the REDD+ Safeguards Working Group.

46. For example, Victoria Tauli Corpuz with the network Tebtebba.

47. For example, the network COICA, a coordination body of nine national Amazonian Indigenous organizations.

48. International Indigenous Peoples Forum on Climate Change 2012a.

49. For example, the No REDD Alliance.

50. For example, No REDD+ 2012.

51. *Sixty Minutes* 2012.

52. Her position has since changed; she is now UN special rapporteur on the rights of Indigenous peoples.

53. David Ciplet interview with Victoria Tauli Corpuz, December 2011.

54. The declaration recognizes the right of Indigenous peoples to free, prior, and informed consent to development of their territories.

55. Joint Statement of Indigenous Peoples Networks 2008.

56. UN Framework Convention on Climate Change 2009, cited in Schroeder 2010, 327.

57. David Ciplet interview, December 2011.

58. Martone and Griffiths 2013.

59. World Bank and UN-REDD Programme 2011.

60. David Ciplet interviews, 2011–2014.

61. Indigenous Peoples Forum on Climate Change 2011.

62. David Ciplet interview with Victoria Tauli-Corpuz, December 2011.

63. International Indigenous Peoples Forum on Climate Change 2012b.

64. Ibid. 2012a.

65. David Ciplet worked for GAIA from 2006 to 2009, previous to this research.

66. Agency France Presse 2009.

67. Global Alliance for Incinerator Alternatives 2011.

68. Carbon Market Watch 2011.

69. Comment to AMS-III.AJ, CDM Small-Scale Methodology for Recycling Solid Waste.

70. CDM EB 55th meeting.

71. Ibid.

72. CDM EB 58th meeting.

73. CDM EB 59th meeting.

74. Parliament.com 2012.

75. David Ciplet interview with Neil Tangri, March 2013.

76. Ibid.

## Chapter 9

1. Friedman 2013, 1.

2. Harrison and Sundstrom 2010

3. Sklair 2009; Held and McGrew 2007; Held 2010.

4. Viola 1998.

5. *New York Times* 2014a.

6. Chris Buckley, 2015 "Documentary on Air Pollution Grips China." Sinosphere/ New York Times blog. March 1, http://sinosphere.blogs.nytimes.com/2015/03/01/documentary-on-air-pollution-in-china-grips-a-nation/?_r=0.

7. See, e.g., *Climate Wire* 2013.

8. Janssens-Maenhout Olivier, Marilena Muntean, et al. 2014, figure 2.2.

9. However, their low-variant model has populations peaking about 2050 and dropping through the rest of the century. The high variant puts populations near 15 billion by 2100. UN Department of Economic and Social Affairs, Population Division 2013.

10. UN Department of Economic and Social Affairs, Population Division 2013.

11. Erhlich and Holdren 1970; Commoner 1971. See also York, Rosa, and Dietz 2003, 351–365.

12. German Advisory Council on Climate Change 2009.

13. See, e.g., Moyo 2012; *Christian Science Monitor* 2011; Smaller and Mann 2009.

14. Harvey 1982; Walker and Storper 1991; Edwards and Roberts 2012.

15. Useful here is literature on the growth machine–a pro-growth coalition in localities and nations (Logan and Molotch 1987)—and on the need for politicians to create a climate for the accumulation of capital (Harvey 1987).

16. Schnaiberg 1990; Rudel 2013.

17. Here, an interesting mix of writers argue for investment in alternative technologies, such as McKibben (EcoWatch 2014) and Nordaus and Shellenberger 2013.

18. International Energy Agency. 2012; Intergovernmental Panel on Climate Change. 2013.

19. Roberts and Parks (2007) discuss the importance of focal points and the role of inequality in driving a lack of them.

20. This includes Canada, France, Germany, Italy, Japan, the United Kingdom, and the United States. Note that Russia is no longer an official member.

21. This refers to the G8 countries plus the five leading emerging economies including China, Brazil, India, Mexico, and South Africa.

22. This group includes twenty major economies including Argentina, Australia, Brazil, Canada, China, France, Germany, India, Indonesia, Italy, Japan, Mexico, Russia, Saudi Arabia, South Africa, South Korea, Turkey, the United Kingdom, the United States and the European Union (represented by the European Commission and European Central Bank).

23. White House 2007.

24. Lieberman and Schaefer 2007.

25. Major Economies Forum website, http://www.majoreconomiesforum.org.

26. Grasso and Roberts 2013, 2014,.

27. Athanasiou 2012.

28. This typology is from Clapp and Dauvergne 2005.

29. *Guardian* 2011a; Hamilton 2011.

30. Shepherd 2009.

31. *Guardian* 2010i; Hamilton 2011.

32. Hamilton 2011.

33. *ClimateWire* 2014.

34. Folger, Peter. 2014. The FutureGen Carbon Capture and Sequestration Project: A Brief History and Issues for Congress. Washington, DC: Congressional Research Service.

35. See debate between Howarth, Santoro, and Ingraffea 2011, 2012 and Cathles III, Brown, Taam et al. 2012.

36. Scientists' letter 2014.

37. Bradbury and Obeiter 2013.

38. See Environmental Defense Fund n.d.; Block n.d.

39. Hamilton 2011.

40. This section relies upon Royal Society 2009; ETC Group 2012; Parson and Keith 2013; Hamilton 2011a; 2011b; Vidal 2011.

41. Intergovernmental Panel on Climate Change 2014.

42. One example is the actions of the entrepreneur Russ George, who dumped 100 tons of iron dust into the ocean off western Canada. *New York Times* 2012.

43. Preston 2013; Hamilton 2011.

44. *Guardian* 2012d.

45. Greenpeace has listed fourteen major fossil projects that if completed would represent "game over" for the climate (Voorhar and Myllyvirta 2013). Bill McKibben's powerful "Do the Math" article (McKibben 2012) and tour has made a similar point, with a smaller list of projects.

46. York 2010.

47. Norgaard 2009.

48. See Pacala and Sokolow 2004.

49. Tellus Institute with Sound Resource Management 2011.

50. For example, you can read about Oakland's "zero waste" program at http://greencitiescalifornia.org/best-practices/waste-reduction/oakland_zero -waste.html.

51. See DeSombre 2000.

52. York 2010, 2012.

53. An illuminating case from corporate versus community-based approaches to recycling in this regard is in Weinberg, Pellow, and Schnaiberg 2000.

54. Ibid.; Jones 2009.

55. For example, Klein 2014; Bond 2012.

56. See Klein 2001, 82.

57. For example, Klein 2014 pays limited attention to the role of state or international bodies as a part of the solutions she proposes.

58. Norgaard 2009.

59. Brulle 2000; Brulle and Jenkins 2005.

60. Putnam 1988.

61. McCarthy 2005 19.

62. In this case, reflexive modernization as laid out by Ulrich Beck (1994) and Anthony Giddens (1994) actually functions.

63. Though they are currently able to hide their support for attack think tanks through shelters such as the Donor's Trust.

64. Jones 2009.

65. Stockholm International Peace and Research Institution 2014.

66. Oil Change International 2012.

67. For the phrase "UN with teeth," Roberts acknowledges lectures by Christopher Chase-Dunn from the late 1980s in the Sociology doctoral program at Johns Hopkins University.

68. Lawyers are grappling with the connotations of this last option. India insisted on hammering on this option only to keep space for not agreeing to more

concrete legally binding options. However, the trend in negotiations after Durban indicates that even this last option may not be achieved by 2015.

## Chapter 10

1. Klein 2007.

2. Gramsci 2012 (1971).

3. McKibben 2012; Anderson and Bows 2011.

4. Gramsci 2012 (1971), 178.

5. Roberts 2011.

6. We are reluctant to include India in this group, which remains extremely low in terms of per capita emissions.

7. Edwards and Roberts 2015.

8. York 2012.

9. Shadlen 2004.

10. Ciplet et al. 2012.

11. Certainly improving understanding of what is needed for adaptation to climate change is increasing the amounts expected, but scaling up to $100 billion a year by 2020 would entail far more funding than we have seen to date.

12. Cox 1993.

13. In conducting research and delivering policy briefings at the talks, we understand our own positions are usually of this nature.

14. Polanyi 1944.

15. Arrighi and Silver 2001

16. Desombre 2000.

17. See Hayes 2014.

18. Harvey 1987.

19. Princen, Maniates, and Conca 2002

20. Gill 2002.

21. For example, the transnational network on food justice, Via Campesina.

22. For example, the transnational network on waste issues, the Global Alliance for Incinerator Alternatives.

23. For example, the Indigenous rights network, Idle No More.

24. People's Climate March website 2014. Available at http://peoplesclimate .org/lineup. The march was divided into six main blocs that sought to give place to all types of people concerned about climate change: (1) Frontlines of Crisis, Forefront of Change (The people first and most harmed are leading the charge: Indigenous, Environmental Justice, and other Frontline Communities); (2) We Can Build the Future (Every generation's future is at stake, we can build a better one—Labor, Families, Students, Elders and More); (3) We Have

Solutions (A just transition is possible—Renewable Energy, Food and Water Justice, Environmental Organizations and More); (4) We Know Who is Responsible (Let's call out those who are holding back progress: Anti-Corporate Campaigns, Peace and Justice and More); (5) The Debate is Over (The facts are in! Taking action is a moral necessity: Scientists, Interfaith and More); and (6) To Change Everything, We Need Everyone (Here comes everybody! LGBTQ, NYC Boroughs, Community Groups, Neighborhoods, Cities, States, Countries and More).

# References

ABC News Online. 2009. "Future Not for Sale: Climate Deal Rejected." December 20.

ActionAid. 2010. "ActionAid Proposes a New Climate Finance Mechanism."

Adger, Neil W. 2006. Vulnerability. *Global Environmental Change* 16 (3): 268–281.

Adger, Neil W., Nick Brooks, Graham Bentham, Maureen Agnew, and Siri Eriksen. 2004. *New Indicators of Vulnerability and Adaptive Capacity*. vol. 122. Norwich: Tyndall Centre for Climate Change Research.

Adger, W. Neil, and Katharine Vincent. 2005. Uncertainty in Adaptive Capacity. *Comptes Rendus Geoscience* 337 (4): 399–410.

Adopt a Negotiator. 2014. "The Ultimate Guide for Lima's Final ADP."

African Group proposal. 2009. "Submission on the Outcome of the Ad Hoc Working Group on Long Term Cooperative Action under the Convention under Item 3." December 12 Available at http://unfccc.int/files/kyoto_protocol/application/pdf/algeriaafrican111209.pdf. Agarwal, Anil, and Sunita Narain. 1991. *Global Warming in an Unequal World: A Case of Environmental Colonialism*. New Delhi: Center for Science and Environment.

Agency France Presse. 2009. "Waste Pickers of the World Unite at Climate Talks." December 8.

Alber, Gotelind. 2009. "The Women and Gender Constituency in the Negotiations." *Talking Points*, no. 2.

Alessi, Christopher. 2012. "Expanding China-Africa Oil Ties." Council on Foreign Relations, February 8.

Alliance of Small Island States. 2009. "Proposal by the Alliance of Small Island Developing States (AOSIS) for the Survival of the Kyoto Protocol." December Available at http://germanwatch.org/klima/c15aosis.pdf.

Anderson, John W. 1998. *The Kyoto Protocol on Climate Change: Background, Unresolved Issues and Next Steps*. Washington, DC: Resources for the Future.

Anderson, Kevin. 2012. The Inconvenient Truth of Carbon Offsets. *Nature* 484:7.

Anderson, Kevin, and Alice Bows. 1934. 2011. "Beyond 'Dangerous' Climate Change: Emission Scenarios for a New World. *Philosophical Transactions of the Royal Society A: Mathematical, Physical and Engineering Sciences* 369: 20–44.

Andonova, Liliana B., Michele M. Betsill, and Harriet Bulkeley. 2009. Transnational Climate Governance. *Global Environmental Politics* 9 (2): 52–73.

Andrée, Peter. 2005. The Genetic Engineering Revolution in Agriculture and Food: Strategies of the 'Biotech Bloc. In *The Business of Global Environmental Governance*, ed. David Levy and Peter Newell, 135–166. Cambridge, MA: MIT Press.

Angus, Ian, ed. 2010. *The Global Fight for Climate Justice: Anticapitalist Responses to Global Warming and Environmental Destruction*. Black Point, NS: Fernwood.

Advisors, Arabella. n.d. "Measuring the Global Fossil Fuel Divestment Movement."Availableathttp://www.arabellaadvisors.com/research/measuring-the -global-fossil-fuel-divestment-movement/.

Arrighi, Giovanni. 1994. *The Long Twentieth Century*. London: Verso.

Arrighi, Giovanni, and Beverly Silver. 2001. Capitalism and World (Dis)Order. *Review of International Studies* 5 (27): 257–279.

Arrighi, Giovanni, Beverly J. Silver, and Benjamin D. Brewer. 2003. Industrial Convergence, Globalization, and the Persistence of the North-South Divide. *Studies in Comparative International Development* 38:1.

Athanasiou, Tom. 2012. "Global Warming's Terrifying New Math: Bill McKibben's Call for a Carbon Divestment Movement." EcoEquity.org website blog.

Ayers, Jessica, and David Dodman. 2010. Climate Change Adaptation and Development: The State of the Debate. *Progress in Development Studies* 10 (2): 161–168.

Ayers, Jessica M., and Saleemul Huq. 2009. Supporting Adaptation to Climate Change: What Role for Official Development Assistance? *Development Policy Review* 27:675–692.

Baer, Paul, T. Athanasiou, and S. Kartha. 2013. "The Three Salient Global Mitigation Pathways Assessed in Light of the IPCC Carbon Budgets." Saveourselvesnow.net.

Baer, Paul, Glenn Fieldaman, Tom Athanasiou, and Sivan Kartha. 2008. Greenhouse Development Rights: Towards an Equitable Framework for Global Climate Policy. *Cambridge Review of International Affairs* 21:649–669.

Barnett, M. N., and R. Duvall. 2005. *Power in Global Governance*. Cambridge: Cambridge University Press.

*BBC News*. 2009. "Copenhagen Climate Change Negotiations 'Suspended.'" December 14.

*BBC News*. 2003. "Millions join global anti-war protests." February 17.

Beck, Ulrich. 1994. *Reflexive Modernization: Politics, Tradition and Aesthetics in the Modern Social Order*. Stanford, CA: Stanford University Press.

Bee, Beth, Maureen Biermann, and Petra Tschakert. 2013. Gender, Development, and Rights-Based Approaches: Lessons for Climate Change Adaptation and Adaptive Social Protection. In *Research, Action and Policy: Addressing the Gendered Impacts of Climate Change*, ed. Margaret Alston and Kerri Whittenbury, 95–108. New York: Springer.

Bernauer, Thomas, Tobias Böhmelt, and Vally Koubi. 2013. Is There a Democracy–Civil Society Paradox in Global Environmental Governance? *Global Environmental Politics* 13 (1): 88–107.

Bernauer, Thomas, and Robert Gampfer. 2013. Effects of Civil Society Involvement on Popular Legitimacy of Global Environmental Governance. *Global Environmental Change* 23 (2): 439–449.

Berners-Lee, Mike, and Duncan Clark. 2013. *The Burning Question: We Can't Burn Half the World's Oil, Coal and Gas. So How Do We Quit?* London: Profile Books.

Betsill, Michele. 2006. Transnational Actors in International Environmental Politics. In *Advances in Environmental Politics*, ed. Michele M. Betsill, Kathryn Hochstetler and Dimitris Stevis, 172–202. London: Palgrave Macmillan.

Betsill, Michele M., and Elisabeth Corell. 2001. NGO Influence in International Environmental Negotiations: A Framework for Analysis. *Global Environmental Politics* 1 (4): 65–85.

Betsill, Michele Merrill, and Elisabeth Corell, eds. 2008. *NGO Diplomacy: The Influence of Nongovernmental Organizations in International Environmental Negotiations*. Cambridge, MA: MIT Press.

Betzold, Carola. 2010. 'Borrowing' Power to Influence International Negotiations: AOSIS in the Climate Change Regime, 1990–1997. *Politic* 30 (3): 131–148.

Bhattacharya, D., and S. Dasgupta. 2012. Global Financial and Economic Crisis: Exploring the Resilience of the Least Developed Countries. *Journal of International Development* 24 (6): 673–685.

Biermann, F. 2007. Earth System Governance as a Crosscutting Theme of Global Change Research. *Global Environmental Change* 17 (3): 326–337.

Biermann, Frank, Michele M. Betsill, Joyeeta Gupta, Norichika Kanie, Louis Lebel, Diana Liverman, Heike Schroeder, Bernd Siebenhüner, and Ruben Zondervan. 2010. Earth System Governance: A Research Framework. *International Environmental Agreement: Politics, Law and Economics* 10 (4): 277–298.

Biermann, Frank, and Ingrid Boas. 2010. Preparing for a Warmer World: Towards a Global Governance System to Protect Climate Refugees. *Global Environmental Politics* 10 (1): 60–88.

Biermann, Frank, Philipp Pattberg, Harro Van Asselt, and Fariborz Zelli. 2009. The Fragmentation of Global Governance Architectures: A Framework for Analysis. *Global Environmental Politics* 9 (4): 14–40.

Björkdahl, Annika. 2008. Norm Advocacy: A Small State Strategy to Influence the EU. *Journal of European Public Policy* 15 (1): 135–154.

*Bloomberg Business.* 2012. "Fossil-Fuel Subsidies of Rich Nations Five Times Climate Aid." December 3.

*Bloomberg Business.* 2013. "Carbon Market Dropped 36% in 2012 as Permits Declined." January 3.

*Bloomberg Business.* 2014. "China's Treasury Holdings Climate to Record in Government Data." January 15.

*Bloomberg New Energy Finance.* 2014. "Global Trends in Renewable Energy Investment 2014: Key Findings."

Block, Ben. N.d. "U.S. Environmental Groups Divided on 'Clean Coal.'" Washington, DC: Worldwatch Institute.

*Blue and Green Tomorrow.* 2014. "Germany Pledges $1bn to UN Green Climate Fund." July 16.

Bodansky, Daniel. 2001. The History of the Global Climate Change Regime. In *International Relations and Global Climate Change*, ed. Urs Luterbacher and Detlef F. Sprinz, 23–40. Cambridge, MA: MIT Press.

Bodansky, Daniel. 2009. Introduction: Climate Change and Human Rights: Unpacking the Issues. *Georgia Journal of International and Comparative Law* 38:511.

Bolivia. 2009. "Bolivia Proposal 13 Dec 2009." December 13. Available at https://unfccc.int/files/kyoto_protocol/application/pdf/bolivia141209.pdf.

Bond, P. 2012. *The Politics of Climate Justice.* London: University of KwaZulu-Natal Press.

Boyer, R. 2000. Is the Finance-Led Growth Regime a Viable Alternative to Fordism? A Preliminary Analysis. *Economy and Society* 29 (1): 111–145.

Bradbury, James, and Michael Obeiter. 2013. "A Close Look at Fugitive Methane Emissions from Natural Gas." Washington, DC: World Resources Institute. April 2.

Brauch, H. G. 2009. Securitizing Global Environmental Change. *Facing Global Environmental Change* 4:65–102.

Brulle, Robert J. 2000. *Agency, Democracy and Nature: The U.S. Environmental Movement from a Critical Theory Perspective.* Cambridge, MA: MIT Press.

Brulle, Robert J., and J. Craig Jenkins. 2005. Foundations and the Environmental Movement: Priorities, Strategies, and Impact. In *Foundations for Social Change: Critical Perspectives on Philanthropy and Popular Movements*, ed. Daniel Faber and Debra McCarthy. Lanham, NJ: Rowman & Littlefield.

Brulle, Robert J. 2014. Institutionalizing Delay: Foundation Funding and the Creation of US Climate Change Counter-Movement Organizations. *Climatic Change* 122:681–694.

Buchner, Barbara, Angela Falconer, Morgan Hervé-Mignucci, and Chiara Trabacchi. 2012. *The Landscape of Climate Finance.* Venice, Italy: Climate Policy Initiative.

Buchner, B., M. Stadelmann, J. Wilkinson, F. Mazza, A. Rosenberg, and D. Abramskiehn. 2014. *The Global Landscape of Climate Finance 2014*. Venice, Italy: Climate Policy Initiative.

Betsill, Michele M., and Harriet Bulkeley. 2006. "Cities and the Multilevel Governance of Global Climate Change." *Global Governance: A Review of Multilateralism and International Organizations* 12 (2): 141–159.

Bulkeley, H., and P. Newell. 2009. *Governing Climate Change*. London: Taylor & Francis.

Bullard, Robert Doyle. 2000. *Dumping in Dixie: Race, Class, and Environmental Quality*. vol. 3. Boulder, CO: Westview Press.

Bunker, Stephen G. 1985. *Underdeveloping the Amazon*. Champaign-Urbana: University of Illinois Press.

Burgiel, Stanley W. 2008. Non State Actors and the Cartagena Protocol on Biosafety. In *NGO Diplomacy: The Influence of Nongovernmental Organizations in International Environmental Negotiations*, ed. Michele Merrill Betsill and Elisabeth Corell, 67–100. Cambridge, MA: MIT Press.

Burton, Ian. 1996. The Growth of Adaptive Capacity: Practice and Policy. In *Adapting to Climate Change: An International Perspective*, ed. Joel B. Smith and Neelo Bhatti. New York, NY: Springer-Verlag New York Inc.

Cahill, Damien. 2011. Beyond Neoliberalism? Crisis and Prospects for Progressive Alternatives. *New Political Science* 33 (4): 479–492.

Caney, Simon. 2010. Climate Change and the Duties of the Advantaged. *Critical Review of International Social and Political Philosophy* 13 (1): 203–228.

Caney, Simon, and Cameron Hepburn. 2011. Carbon Trading: Unethical, Unjust and Ineffective? *Royal Institute of Philosophy Supplement* 69:201–234.

Carbon Market Watch. 2011. "CDM Waste Methodologies in the Spotlight: Guest Article by GAIA."

Carbon Trade Watch. 2013. "It Is Time the EU Scraps Its Carbon Emissions Trading System." February 18.

Cardoso, Fernando Henrique, and Ernesto Faletto. 1979. *Dependency and Development in Latin America*. Berkeley: University of California Press.

Carr, Edward. 2008. Between Structure and Agency: Livelihoods and Adaptation in Ghana's Central Region. *Global Environmental Change* 18 (4): 688–699.

Castells, Manuel. 2011. *The Rise of the Network Society: The Information Age: Economy, Society, and Culture*. Hoboken, NJ: Wiley.

Cathles, Lawrence M., III, Larry Brown, Milton Taam, and Andrew Hunter. 2012. A Commentary on "The Greenhouse-Gas Footprint of Natural Gas in Shale Formations." *Climatic Change* 113 (2): 525–535.

Center for American Progress. 2014. "With Only $93 Billion in Profits, the Big Five Oil Companies Demand to Keep Tax Breaks." February 10.

Center for Public Integrity. 2009. "The Climate Lobby from Soup to Nuts." December 27.

Center for Public Integrity. 2011. "BINGOs and the Global Lobbyist: Industry Climate Change Reps 'Loitering' instead of Lobbying. August 3.

*Climate Progress*. 2012. "Big Oil's Banner Year: High Profits, Record Profits, Less Oil." February 8.

CDM EB 55th meeting. Available at http://unfccc2.meta-fusion.com/kongresse/cdm55/templ/play.php?id_kongresssession=2920.

CDM EB 58th meeting. Available at http://unfccc2.meta-fusion.com/kongresse/cdm58/templ/play.php?id_kongresssession=3206.

CDM EB 59th meeting. Available at http://unfccc2.meta-fusion.com/kongresse/cdm59/templ/play.php?id_kongresssession=3324.

CDM. 2009. "Question and Answer Meeting." Available at http://cop15.meta-fusion.com/kongresse/cop15/templ/play.php?id_kongresssession=2367&theme=unfccc.

CDM. 2012. Approved Baseline and Monitoring Methodology AM0025 version 14.0. Alternative Waste Treatment Processes, 4.

CDM. 2012. "Small-Scale Methodology for Recycling Solid Waste." Available at http://www.no-burn.org/downloads/CDM%20comment_AMS-III-AJ.pdf.

Chandhoke, Neera. 2005. How Global Is Global Civil Society? *Journal of World-systems Research* 11 (2): 354–371.

Chase-Dunn, C. K. 1998. *Global Formation: Structures of the World-Economy*. Lanham, MD: Rowman & Littlefield.

Chayes, Abram, and Charlotte Kim. 1998. China and the United Nations Framework Convention on Climate Change. In *Energizing China: Reconciling Environmental Protection and Economic Growth*, ed. M. B. Elroy, C. P. Nielson and P. Lydon, 503–540. Cambridge, MA: Harvard University Press.

Aid, Christian. 1999. "Who Owes Who: Climate Change, Debt, Equity and Survival." Available at http://www.gci.org.uk/Documents/Who_Owes_Who_a.pdf.

*Christian Science Monitor*. 2011. "Food Inflation, Land Grabs Spur Latin America to Restrict Foreign Ownership." May 6.

Christoff, Peter. 2010. Cold Climate in Copenhagen: China and the United States at COP15. *Environmental Politics* 19 (4): 637–656.

Ciplet, David. 2014. Contesting Climate Injustice: Transnational Advocacy Network Struggles for Rights in UN Climate Change Politics. *Global Environmental Politics* 14 (4): 75–96.

Ciplet, David. 2015. Rethinking Cooperation: Consent and Inequality in International Climate Change Politics. *Global Governance* 21 (2): 247–274.

Ciplet, David, Spencer Fields, Keith Madden, Mizan Khan, and Timmons Roberts. 2012. *The Eight Unmet Promises of Fast-Start Climate Finance*. London: International Institute for Environment and Development.

Ciplet, David, Timmons Roberts, and Mizan Khan. 2013. The Politics of International Climate Adaptation Funding: Divisions in the Greenhouse. *Global Environmental Politics* 13 (1): 49–68.

Ciplet, David, Timmons Roberts, and Mizan Khan. N.d. "A New World Climate Order." In *Edward Elgar Research Handbook on Climate Governance.* Forthcoming.

Ciplet, David, Timmons Roberts, Mizan Khan, Spencer Fields, and Keith Madden. 2013. *Least Developed, Most Vulnerable: Have Climate Finance Promises Been Fulfilled for the LDCs?* Oxford: European Capacity Building Initiative.

Ciplet, David, J. Timmons Roberts, Mizan Khan, Linlang He, and Spencer Fields. 2011. *Adaptation Finance: How Can Durban Deliver on Past Promises?* London: International Institute for Environment and Development.

Ciplet, David, Timmons Roberts, Pa Ousman, Achala Abeysinghe, Alexis Durand, Daniel Kopin, Olivia Santiago, Keith Madden, and Sophie Purdom. 2013. *A Burden to Share? Addressing Unequal Climate Impacts in the Least Developed Countries.* London: International Institute for Environment and Development.

Ciplet, David, J. Timmons Roberts, Martin Stadelmann, Saleemul Huq, and Achala Chandani. 2011. *Scoring Fast-Start Climate Finance: Leaders and Laggards in Transparency.* London: International Institute for Environment and Development.

Clabough, Andrea. 2013. "Beyond Oil: The New Energy Politics of the Middle East." *Georgetown Security Studies Review* 1 (3).

Clapp, J. 1998. The Privatization of Global Environmental Governance: ISO 14000 and the Developing World. *Global Governance* 4:290–316.

Clapp, Jennifer, and Peter Dauvergne. 2005. *Paths to a Green World.* Cambridge, MA: MIT Press.

Clapp, Jennifer, and Linda Swanston. 2009. Doing Away with Plastic Shopping Bags: International Patterns of Norm Emergence and Policy Implementation. *Environmental Politics* 18 (3): 315–332.

Clark, Ann Marie, Elisabeth Friedman, and Kathryn Hochstetler. 1998. The Sovereign Limits of Global Civil Society. *World Politics* 51 (1): 1–39.

Clémençon, Raymond. 2006. What Future for the Global Environment Facility? *Journal of Environment & Development* 15 (1): 50–74.

*Climate Connections.* 2011. "Showdown at the Durban Disaster: Challenging the 'Big Green' Patriarchy." December 16. Available at http://climate-connections. org/2011/12/16/showdown-at-the-durban-disaster-challenging-the-big-green -patriarchy/.

Climate Funds Update. 2014. *Adaptation Fund Project Data 2014.* United Framework Convention on Climate Change.

Climate Interactive. 2013. The Climate Scoreboard. *Update.* 19 (April): Available at http://climateinteractive.org/scoreboard.

Climate Justice Now. 2007. "What's Missing from the Climate Talks: Justice!" December 14. Available at http://www.climate-justice-now.org/about-cjn/history/ bali-cjn-founding-presse-release/.

Climate Progress 2013. "What You Need to Know about the Biggest Free Trade Agreement Ever and How It Affects Climate Change." October 3.

*ClimateWire.* 2013. "China Is a Major Investor in Overseas Renewable Energy Projects." June 11.

*ClimateWire.* 2014. "Geoengineering Proposals to Slow Climate Change Get More Serious Attention." March 6.

CNN. 2009. "Activists Arrested in Copenhagen Protests." December 16

CNN. 2012. "Supreme Court Won't Consider Blocking $18B Judgment against Chevron." October 24.

Commoner, B. 1971. *The Closing Circle.* New York: Knopf.

Cook, John, Dana Nuccitelli, Sarah A. Green, Mark Richardson, Bärbel Winkler, Rob Painting, Robert Way, Peter Jacobs, and Andrew Skuce. 2013. Quantifying the Consensus on Anthropogenic Global Warming in the Scientific Literature. *Environmental Research Letters* 8 (2): 024024.

Corell, Elisabeth, and Michele M. Betsill. 2001. A Comparative Look at NGO Influence in International Environmental Negotiations: Desertification and Climate Change. *Global Environmental Politics* 1 (4): 86–107.

CorpWatch. 2002. "Bali Principles of Climate Justice." Available at http://www.corpwatch.org/article.php?id=3748

Cox, R. W. 1987. *Production, Power, and World Order: Social Forces in the Making of History.* vol. 1. New York: Columbia University Press.

Cox, R. W. 1992. Multilateralism and World Order. *Review of International Studies* 18 (2): 161–180.

Cox, R. W. 1993. Gramsci, Hegemony and International Relations: An Essay in Method. *Cambridge Studies in International Relations* 26:49–66.

*CSR Hub.* 2013. "Could the Wind Industry End Fossil Fuel Subsidies?" January 22. Available at http://www.csrhub.com/blog/2013/01/could-the-wind-industry-end-fossil-fuel-subsidies.html.

*Daily Mail.* 2009. "Tear Gas Fired at Climate Talks: Police Make 230 Arrests as Protesters Storm Copenhagen Summit." December 16.

DARA. 2012. *Climate Vulnerability Monitor 2012.*

Deffeyes, K. S. 2011. *Hubbert's Peak: The Impending World Oil Shortage.* New ed. Princeton, NJ: Princeton University Press.

Deitelhoff, Nicole, and Linda Wallbott. 2012. Beyond Soft Balancing: Small States and Coalition-Building in the ICC and Climate Negotiations. *Cambridge Review of International Affairs* 25 (3): 345–366.

Dellas, Eleni, Philipp Pattberg, and Michele Betsill. 2011. Agency in Earth System Governance: Refining a Research Agenda. *International Environmental Agreement: Politics, Law and Economics* 11 (1): 85–98.

Dellink, Rob, Michel den Elzen, Harry Aiking, Emmy Bergsma, Frans Berkhout, Thijs Dekker, and Joyeeta Gupta. 2009. Sharing the Burden of Financing Adaptation to Climate Change. *Global Environmental Change* 19 (4): 411–421.

*Democracy Now!* 2010. "US Cancels Climate Aid to Bolivia, Ecuador over Copenhagen Opposition." April 12.

*Democracy Now!* 2009. "Voices from Small Island States." December 17.

Den Elzen, Michel, and Niklas Höhne. 2008. Reductions of Greenhouse Gas Emissions in Annex I and Non-Annex I Countries for Meeting Concentration Stabilisation Targets. *Climatic Change* 91 (3–4): 249–274.

Depledge, Joanna. 2005. Against the Grain: The United States and the Global Climate Change Regime. *Global Change, Peace & Security* 17 (1): 11–27.

Depledge, Joanna. 2008. Striving for No: Saudi Arabia in the Climate Change Regime. *Global Environmental Politics* 8 (4): 9–35.

DeSombre, Elizabeth R. 2000. *Domestic Sources of International Environmental Policy: Industry, Environmentalists, and US Power.* Cambridge, MA: MIT Press.

DiMuzio, Tim. 2012. Capitalizing a Future Unsustainable: Finance, Energy and the Fate of Market Civilization. *Review of International Political Economy* 19 (3): 363–388.

Dombrowski, Kathrin. 2010. Filling the Gap? An Analysis of Non-Governmental Organizations Responses to Participation and Representation Deficits in Global Climate Governance. *International Environmental Agreement: Politics, Law and Economics* 10 (4): 397–416.

Doolittle, Amity A. 2010. The Politics of Indigeneity: Indigenous Strategies for Inclusion in Climate Change Negotiations. *Conservation & Society* 8 (4): 286.

Dubash, Navroz K., and Bronwen Morgan. 2012. Understanding the Rise of the Regulatory State of the South. *Regulation and Governance* 6 (3): 261–281.

Dunlap, Riley E., and Angela G. Mertig, eds. 2014. *American Environmentalism: The US Environmental Movement, 1970–1990.* London: Taylor & Francis.

Earth Day Network. N.d. "Earth Day: The History of a Movement." Available at http://www.earthday.org/earth-day-history-movement.

Earth Negotiations Bulletin. 1995. "Working Group 1." December 11. Available at http://www.iisd.ca/vol12/1211011e.html.

East Asian Forum. 2012. "The Trans-Pacific Partnership, the Environment and Climate Change." September 22.

ECO. 2012. "En-Gender-Ing the Process." December 1.

EcoEquity. 2009. "After Copenhagen: On Being Sadder But Wiser, China, and Justice as the Way Forward." February 19.

Ecofys. 2013. "Mapping Carbon Pricing Initiatives: Development and Prospects." May 29. Available at http://www.ecofys.com/en/publication/mapping-carbon-pricing-initiatives-developments-and-prospects/.

Economic Coalition for Latin America and the Caribbean. 2011. *People's Republic of China and Latin America and the Caribbean: Ushering in a New Era in the Economic and Trade Relationship.* Santiago de Chile: ECLAC.

*Economist.* 2009. "Cap and Trade, with Handouts and Loopholes." May 21.

*Economist.* 2010. "Climate Change: A Surprising Success." December 11.

Economy, E. 1997. Chinese Policy-Making and Global Climate Change: Two-Front Diplomacy and the Internationality Community. In *The Internationalization of Environmental Protection*, ed. M. A. Scheurs and E. Economy, 19–41. Cambridge: Cambridge University Press.

EcoWatch. 2014. "McKibben to Obama: Fracking May Be Worse Than Burning Coal." September 8.

Edwards, Guy, and J. Timmons Roberts. 2012. "Three Hungry Giants? China, the U.S. and the E.U.'s Battle over Latin America's Natural Resources, and Its Implications for Climate Change and Resource Scarcity." Paper presented at Beyond Competition? China, Climate Change, Security and the Developing World, Brown University, April 6.

Edwards, Guy, and J. Timmons Roberts. 2015. "A Fragmented Continent: Latin America and the Global Politics of Climate Change." MIT Press.

Ehrlich, P., and J. Holdren. 1970. The People Problem. *Saturday Review* 4 (42): 42–43.

Environmental Defense Fund. N.d. "Carbon Capture and Sequestration: Storing Carbon to Reduce Emissions."

Environmental Justice Organizations, Liabilities and Trade. 2013. "The Colombian Mining Locomotive Has Halted." February 14.

 E. T. C. Group (Pat Mooney, Kathy Jo Wetter and Diana Bronson). 2012. "Darken the Sky and Whiten The Earth—the Dangers of Geoengineering." In *What Next Volume III: Climate, Development and Equity*, edited by Niclas Hallstrom, 210–237. What Next Forum. http://www.whatnext.org/Publications/Volume_3/Volume_3_main.html.

Ethiotube 2009. "Joint Press Conference by Ethiopian PM Meles Zenawi and French President Nicolas Zarkozy." December 15.

Evans, Peter. 2008. Is an Alternative Globalization Possible? *Politics & Society* 36 (2): 271–305.

Falkner, Robert. 2000. *The Role of Firms in Global Environmental Politics: The Case of Ozone Layer Protection.* New York: Oxford University Press.

Fankhauser, Samuel, and Nat Martin. 2010. The Economics of the CDM Levy: Revenue Potential, Tax Incidence and Distortionary Effects. *Energy Policy* 38 (1): 357–363.

*Financial Express.* 2010. "Government Faces Opposition's Fire for Cancun Statement." December 11.

*Financial Times.* 2013. "EU Carbon Prices Crash to Record Low." January 24.

Finnemore, Martha, and Kathryn Sikkink. 1998. International Norm Dynamics and Political Change. *International Organization* 52 (4): 887–917.

Fisher, Dana R. 2010. COP-15 in Copenhagen: How the Merging of Movements Left Civil Society Out in the Cold. *Global Environmental Politics* 10 (2): 11–17.

Fisher, Dana R., and Jessica F. Green. 2004. Understanding Disenfranchisement: Civil Society and Developing Countries' Influence and Participation in Global Governance for Sustainable Development. *Global Environmental Politics* 4 (3): 65–84.

Ford, James D. 2012. Indigenous Health and Climate Change. *American Journal of Public Health* 102 (7): 1260–1266.

Ford, L. H. 2003. Challenging Global Environmental Governance: Social Movement Agency and Global Civil Society. *Global Environmental Politics* 3 (2): 120–134.

Forgacs, D. 2000. *The Antonio Gramsci Reader*. London: Lawrence and Wishart.

Foucault, Michel. 2003. *"Society Must be Defended." Lectures Given at the College de France 1975–1976*. New York: Picador Press. ( 1976).

Franck, T. M. 1990. *The Power of Legitimacy among Nations*. New York: Oxford University Press.

Friedman, Thomas L. 2006. *The World Is Flat : A Brief History of the Twenty-First Century (updated and expanded edition)*. New York: Macmillan.

Friedman, Thomas L. 2013. "It's Lose-Lose vs. Win-Win-Win-Win-Win." *New York Times,* March 17, 1.

Friends of the Earth. 2010. "US Export-Import Bank to Vote on 'Carbon Increase Plan.'" March 9.

*Frontline*. 2012. "Climate of Denial." October 23.

Füssel, H. M. 2009. An Updated Assessment of the Risks from Climate Change Based on Research Published since the IPCC Fourth Assessment Report. *Climatic Change* 97 (3): 469–482.

Gardiner, Stephen M. 2004. Ethics and Global Climate Change. *Ethics* 114 (3): 555–600.

Gardiner, Stephen M. 2011. Some Early Ethics of Geoengineering the Climate: A Commentary on the Values of the Royal Society Report. *Environmental Values* 20 (2): 163–188.

Gareau, Brian. 2013. *From Precaution to Profit: Contemporary Challenges to Environmental Protection in the Montreal Protocol*. New Haven, CT: Yale University Press.

Garnaut, R. 2008. *The Garnaut Climate Change Review*. Cambridge: Cambridge University Press.

Gender and Climate Change Network. 2007. "Gender: Women from All Regions of the World Strongly Oppose including Nuclear Energy into the CDM." Available at http://www.wrm.org.uy/actors/CCC/Bali/Nuclear_Energy.pdf.

German Advisory Council on Global Change. 2009. *Solving the Climate Dilemma: The Budget Approach*. Berlin, Germany: WBGU.

German Advisory Council on Global Change. 2010. "Climate Policy Post-Copenhagen." Policy paper 6.

Giddens, Anthony. 1994. "Risk, Trust, Reflexivity." In *Reflexive Modernization*, edited by Ulrich Beck, Anthony Giddebs, and Scott Lash, 184–197. Stanford, CA: Stanford University Press.

Gill, Stephen. 2002. Constitutionalizing Inequality and the Clash of Globalizations. *International Studies Review* 4 (2): 47–65.

Global Alliance for Incinerator Alternatives. 2011. "The Green Climate Fund: Effective Community Ally or Corporate Giveaway."

Global Environment Facility. 2010. *System for Transparent Allocation of Resources*. STAR.

Facility, Global Environmental. 2011. "Report of the Global Environmental Facility to the Conference of Parties Seventeenth Session." December 19.

Global Environment Facility. 2014a. "Global Environment Facility Least Developed Countries Fund Progress Report."

Global Environment Facility. 2014b. "Global Environment Facility Special Climate Change Fund Financial Progress Report."

Goldman, Michael. 2007. How 'Water for All!' Policy Became Hegemonic: The Power of the World Bank and Its Transnational Policy Networks. *Geoforum* 38 (5): 786–800.

Gramsci, A. 2012. *1971. Selections from the Prison Notebooks*. New York: International Publishers.

Grasso, Marco. 2010. An Ethical Approach to Climate Adaptation Finance. *Global Environmental Change* 20 (1): 74–81.

Grasso, Marco, and J. Timmons Roberts. 2013. A Fair Compromise to Break the Climate Impasse: A Major Economies Forum Approach to Emissions Reductions Budgeting. In *Policy Paper 2013-02*. Washington, DC: Brookings Institution.

Grasso, Marco, and J. Timmons Roberts. 2014. A Compromise to Break the Climate Impasse. *Nature Climate Change* 4 (7): 543–549.

Green Climate Fund. 2013. "Members of the Board." Available at http://gcfund.net/?id=10.

Gulbrandsen, Lars H., and Steinar Andresen. 2004. NGO Influence in the Implementation of the Kyoto Protocol: Compliance, Flexibility Mechanisms, and Sinks. *Global Environmental Politics* 4 (4): 54–75.

Gupta, Joyeeta. 1997. *The Climate Change Convention and Developing Countries: From Conflict to Consensus?* Dordrecht: Kluwer Academic.

Gupta, J. 2000. North-South Aspects of the Climate Change Issue: Towards a Negotiating Theory and Strategy for Developing Countries. *International Journal of Sustainable Development* 3 (2): 115–135.

*Guardian*. 2009a. "China and India Agree to Cooperate on Climate Change Policy." October 22.

*Guardian*. 2009b. "Copenhagen Police Release Hundreds of Detained Activists." December 13.

*Guardian*. 2010a. "US Embassy Cables: US Urges Ethiopia to Back Copenhagen Climate Accord." December 3.

*Guardian*. 2010b. "US Embassy Cables: Maldives Tout $50 Million Climate Projects to US." December 3.

*Guardian*. 2010c. "US Denies Climate Aid to Countries Opposing Copenhagen Accord." April 9.

*Guardian*. 2010d. "Climate aid threat to countries that refuse to back Copenhagen accord,", April 10.

*Guardian*. 2010e. "WikiLeaks Cables Reveal How US Manipulate Climate Accord," *The Guardian,* December 3.

*Guardian*. 2010f. "Climate Funds Recycled from Existing Aid Budget, UK Government Admits." January 15.

*Guardian*. 2010g. "Global Work Party: 10/10/10 Day of Climate Action." October 11.

*Guardian*. 2010h. "Grassroot Summit Calls for International Climate Court." April 23.

*Guardian*. 2010i. "The Powerful Coalition that Wants to Engineer the World's Climate." September 13.

*Guardian*. 2011a. "Geo-Engineering: Greed vs. Green in the Race to Cool the Planet." July 9.

*Guardian*. 2011b. "Durban Climate Talks—Friday as It Happened." December 9.

*Guardian*. 2012a. "Philippines Negotiator Makes Emotional Plea at Doha Climate Talks—Video." Available at http://www.theguardian.com/environment/video/2012/dec/06/philippines-negotiator-emotional-plea-doha-climate-talks-video.

*Guardian*. 2012b. "Phasing Out Fossil Fuel Subsidies 'Could Provide Half of Global Carbon Target.'" January 19.

*Guardian*. 2012c. "Global Carbon Trading System Has 'Essentially Collapsed.'" September 10.

*Guardian*. 2012d. "Fossil Fuel Subsidies: A Tour of the Data." January 19.

*Guardian*. 2013a. "World Carbon Dioxide Emissions by Country." July 15.

*Guardian*. 2013b. "Why Are Carbon Markets Failing?" April 12.

*Guardian*. 2013c. "Canadian and US Aboriginal Groups Vow to Block Oil Pipelines." March 21.

*Guardian*. 2013d. "Keystone XL Protestors Pressure Obama on Climate Change Promise." February 17.

*Guardian*. 2014a. "World Carbon Dioxide Emissions Data by Country: China Speeds Ahead of the Rest." January 31.

*Guardian*. 2014b. "Snowden Revelations of NSA Spying on Copenhagen." January 30.

*Guardian*. 2014c. "Fossil Fuel Subsidies Growing Despite Concerns." April 29.

*Guardian.* 2014d. "World Council of Churches Pulls Fossil Fuels Investments." July 11.

Gupta, Joyeeta. 2010. A History of International Climate Change Policy. *Climatic Change* 1 (5): 636–653.

Gustafsod, Per E. 1998. Gender Differences in Risk Perception: Theoretical and Methodological Perspectives. *Risk Analysis* 18 (6): 805–811.

Haas, Peter M. 1992. Banning Chlorofluorocarbons: Epistemic Community Efforts to Protect Stratospheric Ozone. *International Organization* 46 (1): 187–224.

Habermas, Jürgen. 2001. *The Postnational Constellation: Political Essays.* Ed. Max Pensky. Cambridge, MA: MIT Press.

Hamilton, Clive. 2011a. "Ethical Anxieties about Geoengineering: Moral Hazard, Slippery Slope and Playing God." Paper presented in *Australian Academy of Science Conference: Geoengineering the Climate.*

Hamilton, Clive. 2011b. "The Powerful Coalition that Wants to Engineer the World's Climate." *The Guardian.*

Hamilton, Clive. 2013. *Earthmasters: Playing God with the Climate.* London: Allen & Unwin.

Hansen, Sarah. 2013. *Cultivating the Grassroots: A Winning Approach for Environment and Climate Funders.* Washington, DC: National Committee for Responsive Philanthropy. February.

Harmeling, Sven. 2010. *Global Climate Risk Index 2010: Who Is Most Vulnerable? Weather-Related Loss Events since 1990 and How Copenhagen Needs to Respond.* Bonn: Germanwatch.

Harrison, K., and L. M. I. Sundstrom, eds. 2010. *Global Commons, Domestic Decisions: The Comparative Politics of Climate Change.* Cambridge, MA: MIT Press.

Harvey, David. 1982. *Limits to Capital.* Oxford: Basil Blackwell.

Harvey, David. 1987. Flexible Accumulation through Urbanization: Reflections on "Post-Modernism" in the American City. *Antipode* 19 (3): 260–286.

*Havas Worldwide.* 2009. "Tck Tck Tck." Available at http://www.havasworldwide.com/our-work/tcktcktck.

Hayes, Christopher. 2014. The New Abolitionism. *Nation* (April): 22.

Hecht, Susanna B., Anthony B. Anderson, and Peter May. 1988. The Subsidy from Nature: Shifting Cultivation, Successional Palm Forests, and Rural Development. *Human Organization* 47 (1): 25–35.

Heede, Richard. 2014. Tracing Anthropogenic Carbon Dioxide and Methane Emissions to Fossil Fuel and Cement Producers, 1854–2010. *Climatic Change* 122 (1–2): 229–241.

Heggelund, G. 2007. China's Climate Change Policy: Domestic and International Developments. *Asian Perspective* 31:155–191.

Heinrich Böll Foundation. 2008. "Towards Climate Justice in Asia." July.

Held, David. 1999. The Transformation of Political Community: Rethinking Democracy in the Context of Globalization. In *Democracy's Edges*, ed. Ian Shapiro and Casiano Hacker-Cordón, 84–111. Cambridge: Cambridge University Press.

Held, David. 2007. *Globalization/Anti-Globalization*. Cambridge: Polity Press.

Held, David. 2010. *Cosmopolitanism: Ideals and Realities*. Cambridge: Polity Press.

Held, David, and Anthony G. McGrew, eds. 2007. *Globalization Theory: Approaches and Controversies*. Cambridge: Polity Press.

Helleiner, Eric, Stephano Pagliari, and Hubert Zimmerman. 2009. *Global Finance in Crisis: The Politics of International Regulatory Change*. New York: Routledge.

Hemmati, Minu, and Ulrike Röhr. 2009. Engendering the Climate-Change Negotiations: Experiences, Challenges, and Steps Forward. *Gender and Development* 17 (1): 19–32.

Hicks, Robert L., Bradley C. Parks, J. Timmons Roberts, and Michael J. Tierney. 2008. *Greening Aid? Understanding the Environmental Impact of Development Assistance*. New York: Oxford University Press.

*Hindustan Times*. 2009. "Presenting *HT*'s Stars of the Summit." December 17. Available at http://georgewbush-whitehouse.archives.gov/news/releases/2007/09/20070927.html.

Howarth, Robert W., Renee Santoro, and Anthony Ingraffea. 2012. Venting and Leaking of Methane from Shale Gas Development: Response to Cathles et al. *Climatic Change* 113 (2): 537–549.

Howarth, Robert W., Renee Santoro, and Anthony Ingraffea. 2011. Methane and the Greenhouse-Gas Footprint of Natural Gas from Shale Formations. *Climatic Change* 106 (4): 679–690.

*Huffington Post*. 2012a. "Why the Environmental Movement Is Not Winning." February 29.

*Huffington Post*. 2012b. "Problems with the Math: Is 350's Carbon Divestment Campaign Complete?" November 29.

*Huffington Post*. 2012c. "Why Is the TPP Such a Big Secret?" May 4.

*Huffington Post*. 2013. "Obama Golfed with Oil Men as Climate Protesters Descended on White House." February 20.

*Huffington Post* 2014. "U.S. Challenges India's Solar Industry, Again." February 18.

Hulme, M. 2010. *Why We Disagree about Climate Change*. Cambridge: Cambridge University Press.

Hultman, Nathan E., Simone Pulver, Leticia Guimarães, Ranjit Deshmukh, and Jennifer Kane. 2012. Carbon Market Risks and Rewards: Firm Perceptions of CDM Investment Decisions in Brazil and India. *Energy Policy* 40:90–102.

Hultman, Nathan E., Simone Pulver, Sergio Pacca, Samir Saran, Lydia Powell, Viviane Romeiro, and Tabitha Benney. 2011. Carbon Markets and Low-Carbon

Investment in Emerging Economies: A Synthesis of Parallel Workshops in Brazil and India. *Energy Policy* 39 (10): 6698–6700.

Humphreys, David. 2004. Redefining the Issues: NGO Influence on International Forest Negotiations. *Global Environmental Politics* 4 (2): 51–74.

Hurd, I. 1999. Legitimacy and Authority in International Politics. *International Organization* 53 (2): 379–408.

Hurrell, Andrew, and Sandeep Sengupta. 2012. Emerging Powers, North–South Relations and Global Climate Politics. *International Affairs* 88 (3): 463–484.

Intergovernmental Panel on Climate Change. 2001. *Synthesis Report 2001: Climate Change 2001: The Scientific Basis*. Ed. J. T. Houghton, Y. Ding, D. J. Griggs, M. Noguer, P. J. Van der Linden, X. Dai, K. Maskell, and C. A. Johnson. Cambridge: Cambridge University Press.

Intergovernmental Panel on Climate Change. 2007a. *Synthesis Report: Climate Change 2007: Impacts, Adaptation and Vulnerability. Edited by O. F. Canziani M. L. Parry, J. P. Palutikof, P. J. van der Linden, and C. E. Hanson.* Cambridge: Cambridge University Press.

Intergovernmental Panel on Climate Change. 2007b. *The Physical Science Basis.* Edited by S. Solomon, Dahe Qin, Martin Manning, Z. Chen, M. Marquis, K. B. Avery M. Tignor, and H. L. Miller. Cambridge: Cambridge University Press.

Intergovernmental Panel on Climate Change. 2012. *Managing the Risks of Extreme Events and Disasters to Advance Climate Change Adaptation: A Special Report of Working Groups I and II of the Intergovernmental Panel on Climate Change.* Edited by C.B Field., V. Barros, T. F. Stocker, D. Qin, D. J. Dokken, K. L. Ebi, M. D. Mastrandrea, K. J. Mach, G.-K. Plattner, S. K. Allen, M. Tignor, and P. M. Midgley. Cambridge: Cambridge University Press.

Intergovernmental Panel on Climate Change. 2013. *Climate Change 2013: The Physical Science Basis.* Ed. T. F. Stocker, D. Qin, G. K. Plattner, M. Tignor, S. K. Allen, J. Boschung, A. Nauels, Y. Xia, B. Bex, and B. M. Midgley. Cambridge: Cambridge University Press.

*International Business Times*. 2014. "China Economy Surpasses US In Purchasing Power, But Americans Don't Need to Worry." October 8.

International Energy Agency. 2007. "World Energy Outlook 2007: China and India Insights: Executive Summary."

International Energy Agency. 2012. "World Energy Outlook 2012: Executive Summary."

International Energy Agency. 2013. "Redrawing the Energy-Climate Map: World Energy Outlook Special Report." June 10.

International Energy Agency. 2014a. "World Energy Outlook 2014: Executive Summary."

International Energy Agency. 2014b. "Coal Medium-Term Market Report 2014."

International Indigenous Peoples Forum on Climate Change. 2011. "IIPFCC Opening Statement on AWG-KP 16.

International Indigenous Peoples Forum on Climate Change. 2012a. "Open Statement on the Green Climate Fund.

International Indigenous Peoples Forum on Climate Change. 2012b. "Statement to the UNFCCC Subsidiary Body for Implementation.

Jackson, R. H. 1993. *Quasi-States: Sovereignty, International Relations and the Third World*. Cambridge: Cambridge University Press.

Jacobsen, S. F. 1997. "The Determinants of the National Position of Brazil on Climate Change: Empirical Reflections." CDR WP 97.1. Available at http://www .library@cdr.dk.

Jagers, Sverker C., and Goran Duus-Ottestrom. 2008. Dual Climate Change Responsibility: On Moral Divergence between Mitigation and Adaptation. *Environmental Politics* 17 (4): 576–591.

Jansen, M. 2009. *Income Volatility in Small and Developing Economies: Export Concentration Matters*. World Trade Organization.

Jasanoff, Sheila. 2013. *States of Knowledge: The Co-Production of Science and the Social Order*. New York: Routledge.

Joint Statement of Indigenous Peoples Networks. 2000. "The Quito Declaration on Climate Change Negotiations." Quito, Ecuador.

Joint Statement of Indigenous Peoples Networks. 2007. "Statement by the International Forum of Indigenous Peoples on Climate Change." Bali, Indonesia.

Joint Statement of Indigenous Peoples Networks. 2008. "Indigenous Peoples, Local Communities and NGOs Outraged at the Removal of Rights from UNFCCC Decision on REDD. Joint Statement. The Quito Declaration." Quito, Ecuador. Available at http://www.indigenousclimate.org/index.php?option=com _content&view=article&id=58&Itemid=58.

Joint Statement of Indigenous Peoples Networks. 2000–2004. "The Hague Declaration" (2000), The Hague, Netherlands; "the Bonn Statement" (2001), Bonn, Germany; "the Marrakesh Statement" (2001), Marrakesh, Morocco; "Indigenous Peoples' Caucus Statement on Climate Change" (2002), New Delhi, India; "the Milan Declaration" (2003), Milan, Italy; and "the Declaration of COP 10" (2004), Buenos Aires, Argentina. Available at http://www.indigenousclimate.org/ index.php?option=com_content&view=article&id=58&Itemid=58.

Jones, Van. 2009. *The Green Collar Economy*. New York: HarperCollins.

Jorgenson, Andrew K. 2014. Economic Development and the Carbon Intensity of Human Well-Being. *Nature Climate Change* 4:186–189.

Jorgenson, Andrew K., and Brett Clark. 2009. The Economy, Military, and Ecologically Unequal Exchange Relationships in Comparative Perspective: A Panel Study of the Ecological Footprints of Nations, 1975—2000. *Social Problems* 56 (4): 621–646.

Kaldor, Mary. 2003. The Idea of Global Civil Society. *International Affairs* 79 (3): 583–593.

Karl, T. L. 1997. *The Paradox of Plenty: Oil Booms and Petro-States*. Berkeley: University of California Press.

Karliner, Joshua. 1997. *The Corporate Planet: Ecology and Politics in the Age of Globalization*. Berkeley: University of California Press.

Kartha, S., and P. Erickson. 2011. *Comparison of Annex 1 and Non-Annex 1 Pledges under the Cancun Agreements*. Somerville, MA: Stockholm Environment Institute.

Kasa, Sjur, Anne T. Gullberg, and Gørild Heggelund. 2008. The Group of 77 in the International Climate Negotiations: Recent Developments and Future Directions. *International Environmental Agreement: Politics, Law and Economics* 8 (2): 113–127.

Kasperson, Jeanne X., and Roger E. Kasperson. 2001. *Global Environmental Risk*. Tokyo, London: United Nations University Press and Earthscan.

Kates, Robert W. 2000. Cautionary Tales: Adaptation and the Global Poor. In *Societal Adaptation to Climate Variability and Change*, ed. Sally M. Kane and Gary W. Yohe, 5–17. New York: Springer.

Keck, M. E., and K. Sikkink. 1998. *Activists beyond Borders: Advocacy Networks in International Politics*. Cambridge: Cambridge University Press.

Keohane, R. O. 1980. The Theory of Hegemonic Stability and Changes in International Economic Regimes, 1967–1977. In *Change in the International System*, ed. Ole R. Holsti, Randolph M. Siverson and Alexander L. George, 131–162. Boulder, CO: Westview Press.

Keohane, Robert O. 1982. The Demand for International Regimes. *International Organization* 36 (2): 325–355.

Keohane, Robert O., Stephen Macedo, and Andrew Moravcsik. 2009. Democracy-Enhancing Multilateralism. *International Organization* 63 (1): 1–31.

Khan, Mizan. 2013. *Towards a Binding Climate Change Adaptation Regime: A Proposed Framework*. London: Routledge.

Khan, Mizan R., and J. Timmons Roberts. 2013. Adaptation and International Climate Policy. *Climatic Change* 4 (3): 171–189.

Khastagir, N. 2002. *The Human Face of Climate Change: Thousands Gather in India to Demand Climate Justice*. London: CorporateWatch.

Kinchy, Abby J. 2012. *Seeds, Science, and Struggle: The Global Politics of Transgenic Crops*. Cambridge, MA: MIT Press.

Klein, Naomi. (May/June 2001). Reclaiming the Commons. *New Left Review* 9:81–89.

Klein, Naomi. 2007. *The Shock Doctrine: The Rise of Disaster Capitalism*. New York: Macmillan.

Klein, Naomi. November 9 2011. "Capitalism vs. the Climate. *The Nation*. Available at http://www.thenation.com/article/164497/capitalism-vs-climate.

Klein, Naomi. 2014. *This Changes Everything*. New York: Simon and Schuster.

Klein, Richard. 2010. *Which Countries Are Particularly Vulnerable? Science Doesn't Have the Answer!* Somerville, MA: Stockholm Environmental Institute.

Koopman, Jeanne. 2012. Land Grabs, Government, Peasant and Civil Society Activism in the Senegal River Valley. *Review of African Political Economy* 39 (134): 655–664.

Kotin, Adam Charles. 2012. "Farm Disaster Policy, Crop Insurance, and Climate Change: Investigating Support Mechanisms for Agricultural Adaptation." Master's thesis, Brown University.

Kozloff, Nikolas. 2012. *Durban's Legacy: A More Complex, Unstable Geopolitical Climate Order Emerges—Part 1.* Citizen Action Monitor.

Kraska, James, ed. 2011. *Arctic Security in an Age of Climate Change.* Cambridge: Cambridge University Press.

Krasner, Stephen D. 1982. Structural Causes and Regime Consequences: Regimes as Intervening Variables. *International Organization* 36 (2): 185–205.

Krasner, S. D. 1985. *Structural Conflict: The Third World against Global Liberalism.* Berkeley: University of California Press.

Kuik, Onno, Jereon Aerts, Franciscus Berkhout, Frank Biermann, Jos Bruggink, Joyeeta Gupta, and Richard S. J. Tol. 2008. Post-2012 Climate Policy Dilemmas: A Review of Proposals. *Climate Policy* 8:317–336.

Lamb, William F., J. K. Steinberger, A. Bows-Larkin, G. P. Peters, J. T. Roberts, and F. R. Wood. 2014. Transitions in Pathways of Human Development and Carbon Emissions. *Environmental Research Letters* 9 (1): 014011.

Lang, Graeme, and Ying Xu. 2013. Anti-Incinerator Campaigns and the Evolution of Protest Politics in China. *Environmental Politics* 22 (5): 832–848.

Larson, Mary Jo. 2003. Low-Power Contributions in Multilateral Negotiations: A Framework Analysis. *Negotiation Journal* 19 (2): 133–149.

Launder, Brian Edward, and Michael T. Thompson, eds. 2010. *Geo-Engineering Climate Change: Environmental Necessity or Pandora's Box?* Cambridge: Cambridge University Press.

Le Quéré, C., M. R. Raupach, J. G. Canadell, G. Marland, 2009. Trends in the Sources and Sinks of Carbon Dioxide. *Nature Geoscience* 2 (12): 831–836.

Leggett, Jeremy. 1999. *The Carbon War: Global Warming and the End of the Oil Era.* New York: Routledge.

Leggett, L. M. W., and D. A. Ball. 2011. The Implication for Climate Change and Peak Fossil Fuel of the Continuation of the Current Trend in Wind and Solar Energy Production. *Energy Policy* 41:610–617.

Levy, D. L., and D. Egan. 1998. Capital Contests: National and Transnational Channels of Corporate Influence on the Climate Change Negotiations. *Politics & Society* 26:337–362.

Levy, D. L., and D. Egan. 2003. A Neo-Gramscian Approach to Corporate Political Strategy: Conflict and Accommodation in the Climate Change Negotiations. *Journal of Management Studies* 40 (4): 803–829.

Levy, D. L., and P. J. Newell. 2002. Business Strategy and International Environmental Governance: Toward a Neo-Gramscian Synthesis. *Global Environmental Politics* 2 (4): 84–101.

Levy, D. L., and P. J. Newell, eds. 2005. *The Business of Global Environmental Governance*. Cambridge, MA: MIT Press.

Lieberman, Ben, and Brett D. Schaefer. 2007. "The Major Economies Meeting on Energy Security and Climate Change: A Badly Needed Alternative to the Kyoto Protocol." Washington, DC: Heritage Foundation.

Lindblom, Charles. 1965. *The Intelligence of Democracy*. New York: Free Press.

Lipschutz, Ronnie. 2005. Global Civil Society and Global Governmentality: or, the Search for Politics and the State amidst the Capillaries of Social Power. In *Power in Global Governance*, ed. Michael Barnett and Raymond Duvall, 229–248 Cambridge: Cambridge University Books.

Litfin, Karen. 1994. *Ozone Discourses: Science and Politics in Global Environmental Cooperation*. New York: Columbia University Press.

Liverman, Diana, and Emily Boyd. 2008. The CDM, Ethics and Development. In *A Reformed CDM—including New Mechanisms for Sustainable Development*, ed. Karen Holm Olsen and Jørgen Villy Fenhann. Roskilde, Denmark: UNEP Risø Centre.

Logan, J. R., and H. L. Molotch. 1987. *Urban Fortunes: The Political Economy of Place*. Berkeley: University of California Press.

Lohmann, Larry. 2006. *Carbon Trading: A Critical Conversation on Climate Change, Privatisation and Power*. Uppsala: Dag Hammarskjold Centre.

Lohmann, Larry. 2008. Carbon Trading, Climate Justice and the Production of Ignorance: Ten Examples. *Development* 51 (3): 359–365.

Lohmann, Larry, Niclas Hällström, Robert Österbergh, and Olle Nordberg. 2006. *Carbon Trading: A Critical Conversation on Climate Change, Privatisation and Power*. Uppsala: Dag Hammarskjöld Foundation.

*Los Angeles Times*. 2008. "At the U.N., Leaders Call for New Rules for Global Markets." September 24.

Lovins, L. Hunter, and Boyd Cohen. 2011. *Climate Capitalism: Capitalism in the Age of Climate Change*. London: Macmillan.

Lumumba Di-Aping, C. O. P. Briefing. December 11, 2009. Available at https://www.youtube.com/watch?v=s0_wvZw0fOU.

Lund, Emma. 2013. Environmental Diplomacy: Comparing the Influence of Business and Environmental NGOs in Negotiations on Reform of the Clean Development Mechanism. *Environmental Politics* 22 (5): 739–759.

Lynas, M. 2008. *Six Degrees: Our Future on a Hotter Planet*. Washington, DC: National Geographic Books.

Marceau, Gabrielle, and Peter N. Pedersen. 1999. Is the WTO Open and Transparent? A Discussion of the Relationship of the WTO with Nongovernmental

Organisations and Civil Society's Claims for More Transparency and Public Participation. *Journal of World Trade* 33 (1): 5–49.

Martinez-Alier, J. 2003. *The Environmentalism of the Poor: A Study of Ecological Conflicts and Valuation.* Cheltenham: Edward Elgar.

Martinez-Alier. 2013. "The Colombian Mining Locomotive Has Halted." Environmental Justice Organizations, Liabilities and Trade. February 14. Available at http://www.ejolt.org/2013/02/the-colombian-mining-locomotive-has-halted.

Martone, Francesco, and Tom Griffiths. 2013. "Safeguards in REDD+ Financing Schemes." *FPP E-Newsletter.* April 29.

McCarthy, James. 2005. Commons as Counterhegemonic Projects. *Capitalism, Nature, Socialism* 16 (1): 9–24.

McCormick, Sabrina. 2007. Democratizing Science Movements: A New Framework for Mobilization and Contestation. *Social Studies of Science* 37 (4): 609–623.

Meckling, J. 2011. *Carbon Coalitions: Business, Climate Politics, and the Rise of Emissions Trading.* Cambridge, MA: MIT Press.

Mehra, Melini. 2009. *India Starts to Take On Climate Change.* Washington, DC: Worldwatch Institute.

Meinshausen, Malte, Nicolai Meinshausen, William Hare, Sarah C. B. Raper, Katja Frieler, Reto Knutti, David J. Frame, and Myles R. Allen. 2009. Greenhouse-Gas Emission Targets for Limiting Global Warming to 2° C. *Nature* 458 (7242): 1158–1162.

Messner, Dirk, John Schellnhuber, Stefan Rahmstorf, and Daniel Klingenfeld. 2010. The Budget Approach: A Framework for a Global Transformation toward a Low-Carbon Economy. *Journal of Renewable and Sustainable Energy* 2 (3): 031003.

Michaelowa, Axel. 2010. Copenhagen and the Consequences. *Inter Economics* 45 (1): 2–3.

Michaelowa, Axel, and Pallav Purohit. 2007. *Additionality Determination of Indian CDM Projects.* London: Climate Strategies.

McKibben, Bill. 2012. "Global Warming's Terrifying New Math." *Rolling Stone.* July 19.

Mitchell, Ronald B. 2011. Transparency for Governance: The Mechanisms and Effectiveness of Disclosure-Based and Education-Based Transparency Policies. *Ecological Economics* 70 (11): 1882–1890.

Mol, Arthur P. J. 1997. Ecological Modernization: Industrial Transformations and Environmental Reform. In *The International Handbook of Environmental Sociology,* ed. Michael Redclift and Graham Woodgate, 138–149. Cheltenham, UK: Edward Elgar.

Moore, Frances C. 2012. Negotiating Adaptation: Norm Selection and Hybridization in International Climate Negotiations. *Global Environmental Politics* 12 (4): 30–48.

Morris, L., ed. 2006. *Rights: Sociological Perspectives*. London: Routledge.

Mouffe, Chantal. 2013. *Hegemony, Radical Democracy and the Political*. London: Routledge.

Moyo, Dambisa. 2012. *Winner Take All: China's Race for Resources and What It Means for Us*. New York: Basic Books.

Muetzelfeldt, Michael, and Gary Smith. 2002. Civil Society and Global Governance: The Possibilities for Global Citizenship. *Citizenship Studies* 6 (1): 55–75.

Müller, Benito. 1999. *Justice in Global Warming Negotiations: How to Obtain a Procedurally Fair Compromise*. Oxford: Oxford Institute for Energy Studies.

Müller, Benito. 2013. *'Enhanced Direct Access' through 'National Funding Entities': Etymology and Examples*. Oxford Institute for Energy Studies.

Murphy, C. N. 1998. Understanding IR: Understanding Gramsci. *Review of International Studies* 24 (3): 417–425.

Myers, Norman. 2002. Environmental Refugees: A Growing Phenomenon of the 21st Century. *Philosophical Transactions of the Royal Society of London. Series B, Biological Sciences* 357 (1420): 609–613.

Najam, Adil. 2005. Developing Countries and Global Environmental Governance: From Contestation to Participation to Engagement. *International Environmental Agreement: Politics, Law and Economics* 5 (3): 303–321.

Najam, Adil, Saleemul Huq, and Youba Sokona. 2003. Climate Negotiations beyond Kyoto: Developing Countries' Concerns and Interests. *Climate Policy* 3 (3): 221–231.

Neumayer, Eric, and Thomas Plümper. 2007. The Gendered Nature of Natural Disasters: The Impact of Catastrophic Events on the Gender Gap in Life Expectancy, 1981–2002. *Annals of the Association of American Geographers* 97 (3): 551–566.

Newell, P. 2000. *Climate for Change: Non-State Actors and the Global Politics of the Greenhouse*. Cambridge: Cambridge University Press.

Newell, P. 2005. Climate for Change? Civil Society and the Politics of Global Warming. In *Global Civil Society 2005/6*, ed. Marlies Glasius, Mary Kaldor and Helmut Anheier, 90–120. London: Sage.

Newell, Peter. 2008. Civil Society, Corporate Accountability and the Politics of Climate Change. *Global Environmental Politics* 8 (3): 122–153.

Newell, Peter J., and Matthew Paterson. 1998. A Climate for Business: Global Warming, the State and Capital. *Review of International Political Economy* 5 (4): 679–703.

Newell, P., and M. Paterson. 2010. *Climate Capitalism: Global Warming and the Transformation of the Global Economy*. Cambridge: Cambridge University Press.

Newell, Peter, Diana Tussie, and Phyllida Cox. 2006. *Civil Society Participation in Trade Policy-Making in Latin America: Reflections and Lessons*. Brighton: Institute of Development Studies, University of Sussex.

*New York Times*. 2010. "Leaked Cables Show US Pressured Saudis to Accept Copenhagen Accord." November 30.

*New York Times*. 2012. "A Rogue Climate Experiment Enrages Scientists." October 18.

*New York Times*. 2014a. "U.S. and China Reach Climate Accord After Months of Talks." November 11.

*New York Times*. 2014b. "Rockefellers, Heirs to an Oil Fortune, Will Divest Charity of Fossil Fuels." September 21.

Nordaus, Ted, and Michael Shellenberger. 2013. "Against Technology Tribalism: Why We Need Innovation to Make Energy Clean, Cheap and Reliable." Breakthrough Institute. Available at http://thebreakthrough.org/index.php/voices/michael-shellenberger-and-ted-nordhaus/against-technology-tribalism.

Norgaard, Kari Marie. 2009. "Cognitive and Behavioral Challenges in Responding to Climate Change." Policy Research working paper 4940.

No REDD+. 2012. "Indigenous Leaders Rejecting California REDD Hold Governor Responsible for Their Safety."

Nussbaum, Martha C. 2001. *Women and Human Development: The Capabilities Approach*. Cambridge: Cambridge University Press.

Nye, J. S., Jr. 2010. American and Chinese Power after the Financial Crisis. *Washington Quarterly* 33 (4): 143–153.

O'Brien, Karen, and Jon Barnett. 2013. Global Environmental Change and Human Security. *Annual Review of Environment and Resources* 38:373–391.

OECD Development Assistance Committee. 2014. "Climate-Related Development Finance in 2013—Improving the Statistical Picture."

Okereke, Chukwumerije. 2009. The Politics of Interstate Climate Negotiations. In *The Politics of Climate Change: A Survey*, ed. Max Boykoff, 42–61. London: Routledge.

Okereke, Chukwumerije, H. Bulkeley, and H. Schroeder. 2009. Conceptualizing Climate Governance beyond the International Regime. *Global Environmental Politics* 9 (1): 58–78.

Okereke, Chukwumerije, Philip Mann, Henny Osbahr, B. Muller, and Johannes Ebeling. 2007. "Assessment of Key Negotiating Issues at Nairobi Climate COP/MOP and What It Means for the Future of the Climate Regime." Tyndall Centre for Climate Research.

Oil Change International. 2012. "No Time to Waste: The Urgent Need for Transparency in Fossil Fuel Subsidies."

Olivier, Jos G.J., Greet Janssens-Maenhout, Marilena Muntean, and Jeroen A.H.W. Peters. 2014. *Trends in Global CO2 Emissions: 2014 Report*. The Hague, Netherlands: PBL Netherlands Environmental Assessment Agency.

O'Neill, Kate, Jörg Balsiger, and Stacy D. VanDeveer. 2004. Actors, Norms, and Impact: Recent International Cooperation Theory and the Influence of the Agent-Structure Debate. *Annual Review of Political Science* 7:149–175.

Önis, Z., and A. B. Güven. 2011. The Global Economic Crisis and the Future of Neoliberal Globalization: Rupture versus Continuity. *Global Governance* 17 (4): 469–488.

Ott, H. 2003. Warning Signs from Delhi: Troubled Waters Ahead for Global Climate Policy. *Yearbook of International Environmental Law* 13:261–270.

Ott, Hermann, E. W. Sterk, and R. Watanabe. 2008. The Bali Roadmap: New Horizons for Global Climate Policy. *Climate Policy* 8:91–95.

Overholt, W. H. 2010. China in the Global Financial Crisis: Rising Influence, Rising Challenges. *Washington Quarterly* 33 (1): 21–34.

Oxfam America. 2012. "The Climate Finance Cliff: An Evaluation of Fast Start Finance and Lessons for the Future." November 25.

Oxfam International. 2010. "Call for a Fair Global Climate Fund at Cancun Climate Conference COP-16.

Pacala, S., and R. Socolow. 2004. Stabilization Wedges: Solving the Climate Problem for the Next 50 Years with Current Technologies. *Science* 305:968–971.

Panke, Diana. 2012. Dwarfs in International Negotiations: How Small States Make Their Voices Heard. *Cambridge Review of International Affairs* 25 (3): 313–328.

Parliament.com. 2012. "MEPs Threaten to 'Withdraw Support' for UN Agency." July.

Parry, Martin L. 2009. *Assessing the Costs of Adaptation to Climate Change: A Review of the UNFCCC and Other Recent Estimates*. London: International Institute for Environment and Development.

Parson, Edward, and David Keith. 2013. End the Deadlock on Governance of Geoengineering Research. *Science* 229 (6125): 1278–1279.

Paterson, Matthew. 2001. Principles of Justice in the Context of Global Climate Negotiations. In *International Relations and Global Climate Change*, ed. U. Luterbacher and D. Sprinz, 119–126. Cambridge, MA: MIT Press.

Patzek, T. W., and G. D. Croft. 2010. A Global Coal Production Forecast with Multi-Hubbert Cycle Analysis. *Energy* 35 (8): 3109–3122.

Pearson, Ben. 2007. "Market Failure: Why the Clean Development Mechanism Won't Promote Clean Development." *Journal of Cleaner Production* 15 (2): 247–252.

*People and Planet*. 2009. "Blame Canada—Fossil of the Day at Copenhagen." December 10.

People's Climate March website. September 2014. "The People's Climate March Lineup." Available at http://peoplesclimate.org/lineup/.

Pepinsky, Thomas. 2012. The Global Economic Crisis and the Politics of Non-Transitions. *Government and Opposition* 47 (2): 135–161.

Pieterse, J. N. 2011. Global Rebalancing: Crisis and the East–South Turn. *Development and Change* 42 (1): 22–48.

Piewitt, Martina, Meike Rodekamp, and Jens Steffek. 2010. Civil Society in World Politics: How Accountable Are Transnational CSOs? *Journal of Civil Society* 6 (3): 237–258.

Pilifosova, Olga. 2000. "Where Is Adaptation Going in the UNFCCC?" In *Proceedings of SURVAS Expert Workshop on European Vulnerability and Adaptation to Impacts of Accelerated Sea-Level Rise*, 19–21. Hamburg.

Piven, Frances Fox, and Richard A. Cloward. 1979. *Poor People's Movements: Why They Succeed, How They Fail.* New York: Vintage.

Podesta, John, and Peter Ogden. 2008. The Security Implications of Climate Change. *Washington Quarterly* 31 (1): 115–138.

Polanyi, Karl. 1944. *The Great Transformation: Economic and Political Origins of Our Time.* New York: Rinehart.

Prebisch, Raul. 1950. *The Economic Development of Latin America and Its Principal Problems.* New York: United Nations.

Preston, Christopher J. 2013. Ethics and Geoengineering: Reviewing the Moral Issues Raised by Solar Radiation Management and Carbon Dioxide Removal. *Climatic Change* 4 (1): 23–37.

Princen, Thomas, Matthias Finger, M. L. Clark, and J. P. Manno. 1994. *Environmental NGOs in World Politics: Linking the Local and the Global.* London: Routledge.

Princen, Thomas, Michael Maniates, and Ken Conca, eds. 2002. *Confronting Consumption.* Cambridge, MA: MIT Press.

PriceWaterhouseCooper. 2014. "Low Carbon Economy Index 2014 I 2 Degrees of Separation: Ambition and Reality."

Pulver, Simone. 2005. Organising Business: Industry NGOs in the Climate Debates. In *The Business of Climate Change: Corporate Responses to Kyoto*, ed. Kathryn Begg, Frans van der Woerd and David Levy, 47–60. Sheffield, UK: Greenleaf Publishing.

Putnam, Robert D. 1988. Diplomacy and Domestic Politics: The Logic of Two-Level Games. *International Organization* 42 (3): 427–460.

Rainforest Action Network, Sierra Club and BankTrack. 2014. "Extreme Investments, Extreme Consequences: Coal Finance Report Card 2014."

Rainforest Action Network. 2013. "Experts Urge Bank of America to Phase Out Coal Investments." March 20.

Rawls, John. 1999. *A Theory of Justice.* Rev. ed. Oxford: Oxford University Press.

Reus-Smit, C. 2007. International Crises of Legitimacy. *International Politics* 44 (2): 157–174.

*Reuters.* 2012. "Climate Talks Adopt Gender Balance Goal." December 9.

*Reuters.* 2010. "Climate Talks End with Modest Deal and Standing Ovation." December 11.

RGGI. 2013a. "About the Regional Greenhouse Gas Initiative (RGGI)." Fact sheet. Available at http://www.rggi.org/docs/documents/RGGI_Fact_Sheet.pdf.

RGGI. 2013b. First U.S. Carbon Market Begins Sixth Year of CO2 Auctions." Press release. Available at http://www.rggi.org/docs/Auctions/19/PR031513 _Auction19.pdf.

Rice, James. 2007. Ecological Unequal Exchange: International Trade and Uneven Utilization of Environmental Space in the World System. *Social Forces* 85 (3): 1369–1392.

Richardson, K., W. Steffen, and D. Liverman, eds. 2011. *Climate Change: Global Risks, Challenges and Decisions*. Cambridge: Cambridge University Press.

Rising Tide. 2000. "Dissenting Voices COP 6 Climate Talks."

Roberts, J. Timmons, and Bradley C. Parks. 2009. Ecologically Unequal Exchange, Ecological Debt, and Climate Justice the History and Implications of Three Related Ideas for a New Social Movement. *International Journal of Comparative Sociology* 50 (3–4): 385–409.

Roberts, J. Timmons. 2011. Multipolarity and the New World Dis(Order): US Hegemonic Decline and the Fragmentation of the Global Climate Regime. *Global Environmental Change* 21 (3).

Roberts, J. Timmons, and Amy Hite. 2007. *The Globalization and Development Reader: Perspectives on Development and Social Change*. 2nd ed. London: Blackwell.

Roberts, Timmons J., and Bradley C. Parks. 2007. *A Climate of Injustice*. Cambridge, MA: MIT Press.

Roberts, Timmons, and Bradley Parks. 2009. Ecologically Unequal Exchange, Ecological Debt, and Climate Justice: The History and Implications of Three Related Ideas for a New Social Movement. *International Journal of Comparative Sociology* 50 (3–4): 385–409.

Roberts, J. Timmons, and Nikki Demetria Thanos. 2003. Trouble in Paradise. In *Globalization and Environmental Crises in Latin America*. New York, London: Routledge.

Roberts, J. Timmons, and Melissa M. Toffolon-Weiss. 2001. *Chronicles from the Environmental Justice Frontline*. Cambridge: Cambridge University Press.

Robinson, Kaleigh. 2010. "Brazil's Global Warming Agenda." World Resources Institute. March 1.

Robinson, W. I. 1996. *Promoting Polyarchy: Globalization, US Intervention, and Hegemony*. Cambridge: Cambridge University Press.

Rogelj, Joeri, Julia Nabel, Claudine Chen, William Hare, Kathleen Markmann, Malte Meinshausen, Michiel Schaeffer, Kirsten Macey, and Niklas Höhne. 2010. Copenhagen Accord Pledges Are Paltry. *Nature* 464 (7292): 1126–1128.

Royal Society. (2009). "Geoengineering the Climate: Science, Governance, and Uncertainty." London: Science Policy Centre. https://royalsociety.org/policy/ publications/2009/geoengineering-climate.

Rübbelke, Dirk T. G. 2011. International Support of Climate Change Policies in Developing Countries: Strategic, Moral and Fairness Aspects. *Ecological Economics* 70 (8): 1470–1480.

Rudel, Thomas. 2013. *Defensive Environmentalists and the Dynamics of Global Reform.* Cambridge: Cambridge University Press.

Shandra, John M., Christopher Leckband, Laura A. McKinney, and Bruce London. 2009. Ecologically Unequal Exchange, World Polity, and Biodiversity Loss: A Cross-National Analysis of Threatened Mammals. *International Journal of Comparative Sociology* 50 (3–4): 285–310.

Schipper, E., and F. Lisa. 2006. Conceptual History of Adaptation in the UNFCCC Process. *Review of European Community & International Environmental Law* 15 (1): 82–92.

Schnaiberg, Allan. 1980. *Environment: From Surplus to Scarcity.* New York: Oxford University Press.

Scholte, Jan Aart. 2002. Civil Society and Democracy in Global Governance. *Global Governance* 8:281.

Schroeder, Heike. 2010. Agency in International Climate Negotiations: The Case of Indigenous Peoples and Avoided Deforestation. *International Environmental Agreement: Politics, Law and Economics* 10 (4): 317–332.

Scientists' Letter. 2014. "Re: Recommendation to Accurately Account for Warming Effects of Methane. Letter to Obama Administration." Letter from F. Stuart Chapin III, Ph.D., Institute of Arctic Biology, University of Alaska Fairbanks, and two dozen other scientists. July 29. Available at http://www.eenews.net/assets/2014/07/30/document_gw_02.pdf.

Sen, Amartya. 2001. *Development as Freedom.* New York: Oxford University Press.

Sen, Amartya. 2010. *The Idea of Justice.* London: Penguin.

Sending, Ole Jacob, and Iver B. Neumann. 2006. Governance to governmentality: analyzing NGOs, states, and power. *International Studies Quarterly* 50 (3): 651–672.

Sethi, Nitin. 2014. "Lima Climate Talks Avert Disaster." *Business Standard.* December 15.

Shadlen, K. C. 2004. Patents and Pills, Power and Procedure: The North-South Politics of Public Health in the WTO. *Studies in Comparative International Development* 39 (3): 76–108.

Shepherd, J. G. 2009. *Geoengineering the Climate: Science, Governance and Uncertainty.* London: Royal Society.

Shue, Henry. 1992. The Unavoidability of Justice. In *The International Politics of the Environment,* ed. Hurrell Andrew and Barbara Kingsbury, 373–397. Oxford: Clarendon Press.

Shue, Henry. 1999. Global Environment and International Inequality. *International Affairs* 75 (3): 531–545.

Shuhan, Debra, and Christopher Marcoux. 2010. "Assessing Arab Aid: Trends, Explanations, and Unreported Transfers." Unpublished manuscript from *Aid-Data Conference at Oxford University,* March 22–25.

Simms, Andrew. 2001. *Ecological Debt: Balancing the Environmental Budget and Compensating Developing Countries.* London: International Institute for Environment and Development.

Singer, Peter. 2002. One Atmosphere. In *One World: The Ethics of Globalization,* ed. Peter Singer, 14–50. New Haven, CT: Yale University Press.

*Sixty Minutes.* 2012. "David Nilsson: Carbon Cowboy."

Sklair, Leslie. 2009. The Transnational Capitalist Class and the Politics of Capitalist Globalization. In *Politics of Globalization,* ed. Samir Dasgupta and Jan Nederveen Pieterse, 82–97. London: Sage.

Smaller, Carin, and Howard Mann. 2009. *A Thirst for Distant Lands: Foreign Investment in Agricultural Land and Water.* Winnipeg: International Institute for Sustainable Development, May.

Smil, Vaclav. 2005. *Energy at a Crossroads: Global Perspectives and Uncertainties.* Cambridge, MA: MIT Press.

Smit, Barry, and Olga Pilifosova. 2003. Adaptation to Climate Change in the Context of Sustainable Development and Equity. *Sustainable Development* 8 (9): 9.

Smith, Christopher. 2010. The Bali Firewall and Member States' Future Obligations within the Climate Change Regime. *Law, Environment & Development Journal* 6:284.

Smith, Paul J. 2007. Climate Change, Mass Migration and the Military Response. *Orbis* 51 (4): 617–633.

Somers, Margaret R. 2008. *Genealogies of Citizenship: Markets, Statelessness, and the Right to Have Rights.* Cambridge: Cambridge University Press.

Somers, Margaret, and Christopher Roberts. 2008. Towards a New Sociology of Rights: A Genealogy of `Buried Bodies' of Citizenship and Human Rights. *Annual Review of Law and Society* 4:385–425.

Spaargaren, Gert, and Arthur P. J. Mol. 1992. Sociology, Environment, and Modernity: Ecological Modernization as a Theory of Social Change. *Society & Natural Resources* 5 (4): 323–344.

Stadelmann, Martin, Axel Michaelowa, and J. Timmons Roberts. 2011. Difficulties in Accounting for Private Finance in International Climate Policy. *Climate Policy* 13 (6): 718–737.

Stadelman, Martin J., J. Timmons Roberts, and Saleemul Huq. 2010. *Baseline for Trust: Defining 'New and Additional' Climate Funding.* London: International Institute for Environment and Development.

Stanford Report. 2014. "Stanford to Divest for Coal Companies." May 6.

Steinberger, J. K., and J. T. Roberts. 2010. From Constraint to Sufficiency: The Decoupling of Energy and Carbon from Human Needs, 1975–2005. *Ecological Economics* 70 (2): 425–433.

Steinberger, Julia K., J. Timmons Roberts, Glen P. Peters, and Giovanni Baiocchi. 2012. Pathways of Human Development and Carbon Emissions Embodied in Trade. *Nature Climate Change* 2 (2): 81–85.

Stern, N. 2007. *The Economics of Climate Change: The Stern Review.* Cambridge: Cambridge University Press.

Stockholm International Peace and Research Institution. 2014. "SIPRI Military Expenditure Database." Available at http://www.sipri.org/research/armaments/milex/milex_database/milex_database

Svoboda, Toby, Klaus Keller, Marlos Goes, and Nancy Tuana. 2011. Sulfate Aerosol Geoengineering: The Question of Justice. *Public Affairs Quarterly* 25 (3): 157–180.

Tellus Institute with Sound Resource Management. 2011. "More Jobs, Less Pollution: Growing the Recycling Economy in the U.S." November 14. Available at http://www.bluegreenalliance.org/news/publications/more-jobs-less-pollution.

*TerraViva.* 2009. "Zenawi Out on His Own in Africa. December 17.

Terry, Geraldine. 2009. No Climate Justice without Gender Justice: An Overview of the Issues. *Gender and Development* 17 (1): 5–18.

White House. 2014. "United States and Japan Announce $4.5 Billion in Pledges to Green Climate Fund." November 15. Available at http://www.whitehouse.gov/the-press-office/2014/11/15/united-states-and-japan-announce-45-billion-pledges-green-climate-fund-g.

Third World Network. 2013. "NGOs Call on EU to Abolish Its Emissions Trading System." February.

Thompson, A. 2006. Management under Anarchy: The International Politics of Climate Change. *Climatic Change* 78 (1): 7–29.

*Times of India.* 2009. "Jairam for Major Shift at Climate Talks." October 19.

Traxler, Martino. 2002. Fair Chore Division for Climate Change. *Social Theory and Practice* 28 (1): 101–134.

Tucker, Aviezer. 2013. The New Power Map: World Politics after the Boom in Unconventional Energy. *Foreign Affairs* (December): 19.

UN and Climate Change. 2014a. "Lima Conference Paves the Way Toward a Climate Agreement in Paris." December 14.

UN and Climate Change. 2014b. "Green Climate Fund Exceeds Initial Capitalization Target of $10 billion." December 10.

UNfairplay. 2011. "Leveling the Playing Field: A Report to the UNFCCC on Negotiating Capacity and Access to Information." April.

United Nations. 2008. "United Nations Declaration on the Rights of Indigenous Peoples."

United Nations Conference on Trade and Development. 2012. "Enabling the Graduation of LDCs: Enhancing the Role of Commodities and Improving Agricultural Productivity."

United Nations Development Programme. 2006. *Human Development Report.* New York: UNDP.

United Nations Development Programme. 2007. *Human Development Report 2007/2008.* New York: UNDP.

United Nations Environment Programme. 2010. *The Emissions Gap Report: Are the Copenhagen Accord Pledges Sufficient to Limit Global Warming to 2°C or 1.5°C?* Nairobi, Kenya: UNEP.

United Nations Environment Programme. 2013. "The Impact of Corruption on Climate Change: Threatening Emissions Trading Mechanisms?" *UNEP Global Environmental Alert Service Bulletin,* March.

United Nations Environment Programme. 2014. *The Adaptation Gap Report: A Preliminary Assessment.* Nairobi.

United Nations Framework Convention on Climate Change. 1992. Available at http://unfccc.int/resource/docs/convkp/conveng.pdf.

United Nations Framework Convention on Climate Change. 2009. "The Copenhagen Accord."

United Nations Framework Convention on Climate Change. 2011. "Establishment of an Ad Hoc Working Group on the Durban Platform for Enhanced Action." Available at http://unfccc.int/resource/docs/2011/cop17/eng/09a01.pdf.

United Nations Framework Convention on Climate Change. 2013. Decision1/CP.19, Further advancing the Durban Platform."

United Nations Framework Convention on Climate Change. 2014a. "ADP Text December 13."

United Nations Framework Convention on Climate Change. 2014b. "Lima Call for Climate Action—Decision/CP20." Available at https://unfccc.int/files/meetings/lima_dec_2014/application/pdf/auv_cop20_lima_call_for_climate_action.pdf.

United Nations Framework Convention on Climate Change. 2015. "Ad Hoc Working Group on the Durban Platform for Enhanced Action: Work of the Contact Group on Item 3."

United Nations Framework Convention on Climate Change. 2011. "Non-Governmental Organizations Constituencies." Available at http://unfccc.int/files/parties_and_observers/ngo/application/pdf/constituency_2011_english.pdf.

United Nations Department of Economic and Social Affairs. Population Division. 2013. *World Population Prospects: The 2012 Revision.* Vol. 1: Comprehensive Tables ST/ESA/SER.A/336. New York: United Nations.

Unmüssig, Barbara. 2011. NGOs and Climate Crisis: Fragmentation, Lines of Conflict and Strategic Approaches. Heinrich Böll Stiftung. Available at http://www.boell.de/de/ecology/ecology-society-ngos-climate-crisis-12261.html.

USCAN. 2010. "Who's On Board with the Copenhagen Accord?" Available at http://www.usclimatenetwork.org/policy/copenhagen-accord-commitments.

US Energy Information Administration. 2012. "China Consumes Nearly as Much Coal as the Rest of the World Combined." Available at http://www.eia.gov/todayinenergy/detail.cfm?id=9751#.

US Federal Advisory Committee. 2014. "National Climate Assessment."

Victor, David G. 2011. *Global Warming Gridlock: Creating More Effective Strategies for Protecting the Planet*. Cambridge: Cambridge University Press.

Vidal, John. 2011. "Geo-engineering: Greed versus green in the race to save the planet. The Guardian.

Vihma, A., Y. Mulugetta, and S. Karlsson-Vinkhuyzen. 2011. Negotiating Solidarity? The G77 through the Prism of Climate Change Negotiations. *Global Change, Peace & Security* 23 (3): 315–334.

Vihma, Antto. 2010. "Elephant in the Room–The New G77 and China Dynamics in Climate Talks." Briefing paper 6.

Viola, Eduardo. 1998. "Globalisation, Environmentalism and New Transnational Social Forces." In *Globalisation and the Environment: Perspectives from OECD and Dynamic Non-Member Economies*, 39–52. Paris: OECD.

Voorhar, Ria, and Lauri Myllyvirta. 2013. *Point of No Return: The Massive Climate Threats We Must Avoid*. Greenpeace International.

Vormedal, Irja. 2008. The Influence of Business and Industry NGOs in the Negotiation of the Kyoto Mechanisms: The Case of Carbon Capture and Storage in the CDM. *Global Environmental Politics* 8 (4): 36–65.

Wade, R. H. 2011. Emerging World Order? From Multipolarity to Multilateralism in the G20, the World Bank, and the IMF. *Politics & Society* 39 (3): 347–378.

Wallerstein, Immanuel. 1988. World-Systems Analysis. In *Social Theory Today*, ed. Anthony Giddens and Jonathan H. Turner, 309–324. Palo Alto, CA: Stanford University Press.

Wallerstein, I. 2011. *The Modern World-System I: Capitalist Agriculture and the Origins of the European World-Economy in the Sixteenth Century, with a New Prologue*. Berkeley: University of California Press.

Walker, Richard, and Michael Storper. 1991. *The Capitalist Imperative: Territory, Technology and Industrial Growth*. New York: Wiley-Blackwell.

Walter, Kerstin. 2012. "Mind the Gap: Exposing the Protection Gaps in International Law for Environmentally Displaced Citizens of Small Island States." Ph.D. dissertation, University of British Columbia.

*Washington Post*. 2010. "U.S. Cuts Aid to Colombia, But They're Still Drug War Partners." February 11.

*Washington Post*. 2013. "Within Mainstream Environmental Groups, Diversity Is Lacking." March 24.

Weather Channel. 2013. "Interview with Sheldon Whitehouse at the Anti-Keystone Pipeline/Climate Change Rally." February 17.

Weinberg, Adam S., David N. Pellow, and Allan Schnaiberg. 2000. *Urban Recycling and the Search for Sustainable Community Development*. Princeton, NJ: Princeton University Press.

Welz, Adam. 2009. "Emotional Scenes at Copenhagen: Lumumba Di-Aping @ Africa Civil Society Meeting." Adam Welz's Weblog. December 8.

Wheeler, David, and Dan Hammer. 2010. "The Economics of Population Policy for Carbon Emissions Reduction in Developing Countries." Center for Global Development working paper 229.

White House 2007. "Fact Sheet: Major Economies Meeting on Energy Security and Climate Change U.S. Takes the Lead to Forge Consensus on Energy Security and Climate Change." September 27.

Winkler, Harald, and Judy Beaumont. 2010. Fair and Effective Multilateralism in the Post-Copenhagen Climate Negotiations. *Climate Policy* 10 (6): 638–654.

Women and Gender Constituency. 2009. "Call for a Gender/Women Paragraph in the Shared Vision Document." Available at http://www.gendercc.net/fileadmin/inhalte/Dokumente/UNFCCC_conferences/Road_to_COP15/women_and _gender-input-shared-vision.pdf.

Women's Environment and Development Organization. 2010. "Key Principles for Incorporating a Gender Dimension into the Green Climate Fund."

Women's Environment and Development Organization. 2012. "Women's Participation in UN Climate Negotiation 2008–2012."

World Bank. 2000. "Small States: Meeting Challenges in the Global Economy." Report of the Commonwealth Secretariat/World Bank Joint Task Force on Small States.

World Bank. 2010. *World Development Report 2010: Development and Climate Change*. Washington, DC: World Bank.

World Bank. 2012. "Turn Down the Heat: Why a 4°C Warmer World Must Be Avoided."

World Bank. 2014a. "World Bank Climate Lending Group Grows to over $11 Billion." September 9.

World Bank. 2014b. "Switching On Power Sector Reform in India." June 24.

World Bank and UN-REDD Programme. 2011. "A Review of Three REDD+ Safeguard Initiatives."

World Commission on Environment and Development. 1987. *Our Common Future*. Oxford: Oxford University Press.

Worldwatch Institute. 2010. "Interview with Tuvalu Climate Negotiator Ian Fry."

World Wildlife Fund. N.d. "The Gold Standard." Available at http://www .goldstandard.org/about-us.

Yamin, Farhana. 2001. NGOs and International Environmental Law: A Critical Evaluation of Their Roles and Responsibilities. *Review of European Community & International Environmental Law* 10 (2): 149–162.

Yamin, Farhana. 2011. Pathways and Partnerships for Progress for Durban and Beyond." In *A Future for International Climate Politics–Durban and Beyond*, edited by Lili Fuhr, Barbara Unmuessig, Hans J. H. Verolme and Farhana Yamin, Berlin: Heinrich Böll Foundation.

Yohe, Gary, Elizabeth Malone, Antoinette Brenkert, Michael Schlesinger, Henk Meij, and Xiaoshi Xing. 2006. Global Distributions of Vulnerability to Climate Change. *Integrated Assessment Journal* 6 (3): 35–44.

York, Richard. 2010. The Paradox at the Heart of Modernity. *International Journal of Sociology* 40 (2): 6–22.

York, Richard. 2012. Do Alternative Energy Sources Displace Fossil Fuels? *Nature Climate Change* 2 (6): 441–443.

York, Richard, and Eugene A. Rosa. 2003. Key Challenges to Ecological Modernization Theory Institutional Efficacy, Case Study Evidence, Units of Analysis, and the Pace of Eco-Efficiency. *Organization & Environment* 16 (3): 273–288.

York, Richard, Eugene A. Rosa, and Thomas Dietz. 2003. STIRPAT, IPAT and ImPACT: Analytic Tools for Unpacking the Driving Forces of Environmental Impacts. *Ecological Economics* 46 (3): 351–365.

Young, I. M. 1990. *Justice and the Politics of Difference*. Princeton, NJ: Princeton Univ. Press.

YouTube. 2009. "Venezuela: "Claudia Salerno Won't Be Sidelined." December 10.

YouTube. 2013. "Weather Channel Interview of Sheldon Whitehouse at the Anti-Keystone Pipeline/Climate Change Rally." February 17.

Zehner, Ozzie. 2012. *Green Illusions: The Dirty Secrets of Clean Energy and the Future of Environmentalism*. Lincoln: University of Nebraska Press.

# Index

Abrahams, Yvette, 189
Acción Ecológica, 81
Active inaction, 3, 47
Adaptation
apartheid, 60
carbon dioxide reduction and solar
  radiation management technology
  and, 128–129
dodge issue, 120–124
early scientific uncertainty about
  impacts of climate change and, 103
finance, 101–102, 108–109,
  268–269n62
finance justice, 109–111, 112–113
fund creation, 121–122
gap, 114–117
IPCC Third Assessment Report on,
  105
politics, 101–103, 129–131
public protest regarding, 106
reluctance to address, 104
runaway warming and, 124–128
wedge issue of allocation, 117–120
Adaptation Framework Committee,
  107
Adaptation Fund, 157
*Adaptation Gap Report,* 114
Adger, W. Neil, 5
Agreement on Subsidies and
  Countervailing Measures, 178
Agreement on Trade-Related
  Investment Measures, 178
Alliance of Bolivarian States (ALBA),
  54, 214–215

Alliance of Small Island States
  (AOSIS), 31, 54, 57, 58–60, 80,
  214–215, 241
adaptation and, 105, 110
runaway warming and, 127
vulnerability identity and, 61–62
Amin, Idi, 247
Anderson, Anthony, 104
Anderson, Kevin, 147
Annan, Kofi, 107
Arabella Advisors, 164
Arctic Council, 128
Arrighi, Giovanni, 41, 42, 43, 44, 247
Asia-Pacific Partnership for Clean
  Development and Climate,
  261–262n35
Avaaz, 172, 249

Bali Action Plan, 53, 62–63, 64, 81,
  157, 170
adaptation and, 107, 118
"Bali Principles of Climate Justice,"
  169
Ban Ki Moon, 156
Bank of America, 163
BASIC countries. *See* Brazil, South
  Africa, India, and China (BASIC)
  countries
Berners-Lee, Mike, 148
Betsill, Michele M., 166
Bierman, Frank, 31
Big green advocacy, 161–163, 176
BINGOs. *See* Nongovernmental
  organizations (NGOs)

BMW, 175
Boomerang effect, 186, 244
Bottom-up bargaining, 28
Boyd, Emily, 15
Brazil, South Africa, India, and China
  (BASIC) countries, 54–55, 241
 adaptation and, 108
 splintering of, 67–72
 vulnerability identity and, 61–62
Bread for the World, 169
British Petroleum, 141, 144, 161
Brundtland Commission, 138–139
*Burning Question, The,* 147–148
Bush, George W., 14, 62, 213
Business. *See* Fossil fuels industry
Business boomerang, 186, 244

Calderon, Felipe, 89
California cap-and-trade program,
  149
Campbell, Nick, 150
Cancun Agreements, 2, 75–77, 96–97,
  242
 adaptation and, 107, 110, 114–117
 concessions in, 89–90
 gender equality and, 188
 rethinking cooperation and, 77–79
Capability, 185–186
Carbon
 budget approach, 10
 capture and storage (CCS)
   technology, 128–129, 216–220
 trading and offsets, 146–147,
   148–149, 151–152, 170, 244
Carbon dioxide reduction (CDR),
  128–129, 216
Care International, 146
Castells, Manuel, 49
Center for Public Integrity, 150
Central American Integration System,
  54
Chemical industry, 139
Chevron, 164
China, 10–11, 207–208
 natural resource needs of, 40
 rise of, 40–45, 69, 239
Chrétien, Jean, 206

Christian Aid, 81, 169
Civil society, 34–35, 153. *See also*
  Climate injustice/justice
 access and sway on negotiations,
   166–167
 activism's early days, 167–169
 activist protests and, 155–156
 big green advocacy and, 161–163,
   176
 Copenhagen conflicts and, 171–173
 corporate social responsibility (CSR)
   programs and, 163–166
 emergence of global climate justice
   organizations in, 169–171
 environmental justice movement and,
   158–161
 ineffectiveness of, 175–179, 244–246
 new historic bloc and, 247–252
 role in international negotiations,
   156–157, 177–178
 transnational, 48–51, 179–180
 waning of access and sway by,
   173–175
Clark, Duncan, 148
Clean Development Mechanism
  (CDM), 141–142, 145–146, 149,
  157, 181
 gender equality and, 188
 indigenous rights and, 193–194
 right to livelihood and, 196–197,
   199
Climate Action Network (CAN), 58,
  62, 90, 92, 137, 141, 167–168, 171
Climate change politics. *See also* Civil
  society; Climate injustice/justice;
  Future of climate politics; Global
  power
 active inaction in, 3, 47
 activism, 155–156
 civil society role in, 156–157,
   177–178
 climate governance and, 31–34,
   179–180
 community and organizational efforts
   regarding, 14–17
 concessions in, 89–90
 conflict in, 80–89, 98

cooperation in, 77–79, 90–92,
  99–100, 149
in Doha, Warsaw, and Lima, 92–95
dominance of fossil fuel industry in,
  148–152, 243–244
Durban Platform for Enhanced
  Action and, 90–92
governmentality in, 30, 179
Gramscian lens on, 23–24
historical shifts relevant to, 35–51
historic bloc in, 27–28, 246–252
Intended Nationally Determined
  Contributions in, 94
international negotiations, 1–3,
  13–14
natural disasters prompting action
  on, 235–238
reasons for concessions made by
  low-income and developing states,
  95–99, 242
regime rights in, 183–186, *200–201*
runaway warming and, 124–128,
  247–248
shortcomings in, 240–243
treaty criteria, 11–12
trends since 2009, 238–240
Climate injustice/justice, 6–12,
  254n27
adaptation finance and, 109–111,
  112–113
civil society and, 158–161
to climate justice, 6–12
defined, 5–6
future scenario with global,
  229–232
gender equality and, 185, 186–190
global power and, 3–5
historic bloc and, 27–28, 246–248
indigenous rights and, 185, 190–195
marginal groups and, 181–183,
  199–203
regime rights and, 183–186, *200–201*
right to livelihood and, 181,
  195–198, 199
theory of justice and, 12–13, 110
Climate Justice Alliance, 249
Climate Justice Now! (CJN!), 170

*Climate of Injustice, A,* 36, 55
Climate refugees, 126
Clinton Initiative, 198
Coal industry, 46–47, 137, 142,
  144–145. *See also* Fossil fuels
  industry
  civil society and, 163–164
Coalition of Rainforest Nations, 54,
  190–191
Cochabamba People's Conference,
  227
Coercive forces, 29
Cohen, Boyd, 138
Community-led approach, 226–229
Conference of Parties (COP), 53, 58,
  88, 103
Conflict and climate change politics,
  80–89
Conrad, Kevin, 63
Consent, negotiated, *80,* 99–100
in Cancun, 89–90
conflict in Copenhagen and, 80–89
in Durban, 90–92
Intended Nationally Determined
  Contributions in, 94
low-income and developing states'
  reasons for, 95–99
Conservation International, 162
Constructive ambiguity, 96
Contemporary globalization, 48–49
Cooperation, 77–79, 99–100
  among NGO actors, 149
  in Durban, 90–92
  new historic bloc and, 246–252
Copenhagen Accord. *See* United
  Nations Copenhagen Accord
Corell, Elisabeth, 166
Corporate social responsibility (CSR)
  programs, 139, 163–166
Corpuz, Victoria Tauli, 192
Counterhegemony, 250
Cox, Robert, 34, 35, 48, 51, 52, 238,
  242

De Alba, Luis Alfonso, 89, 96
Decisive nucleus of economic activity,
  34–35

Deepwater Wind, 224
Delhi Ministerial Declaration on
  Climate Change and Sustainable
  Development, 106
Desombre, Elizabeth, 249
Desperate technofixes scenario,
  216–220
Development space, 45–48
Di-Aping, Lumumba, 66, 85,
  101
DiMuzio, Tim, 47
Disenfranchisement, 185
Dobransky, Paula, 63
Doha negotiations, 92–93, 97, 157
  gender equality and, 189
  indigenous rights and, 193
Domestic content rules, 36
Dow Chemical Company, 161
Duke Energy, 161
DuPont, 141
Durban Declaration on Carbon
  Trading, 170
  gender equality and, 188
  indigenous rights and, 192–193
Durban Platform for Enhanced
  Action, 90–92, 253n4

Earth Day, 156
EcoEquity, 10
Ecological collapse, 45–48
Ecosocialist perspective, 25, 26
E3G, 168
Egan, Daniel, 26, 28, 100
Emissions gap, 10
ENGOs. *See* Nongovernmental
  organizations (NGOs)
Environmental Defense Fund, 50, 141,
  161, 176, 219
Environmentalism of the poor,
  158–161
Environmental justice movements,
  158–161. *See also* Climate justice
Environmental Protection Agency,
  45
Espinosa, Patricia, 75, 89
European Bank for Reconstruction
  and Development, 16, 46, 144

European Capacity Building Initiative,
  264n26
European Investment Bank, 46, 144
European Union, 58–59
  Emissions Trading System, 16, 145,
    149, 163
  Green Group, 106
Exclusive action scenario, 213–216
Exclusive economic zones (EEZs), 127
Exclusive inaction scenario, 209–213
Extended rights, 186, 188
Exxon Mobil, 150

Fascism, 247
Figueres, Christiana, 75, 189
Finnemore, Martha, 185
Fisher, Dana, 172
Flannery, Brian, 150
Flexible accumulation, 250–251
Focus on the Global South, 170–171
Fossil fuels industry
  business groups, 136–138,
    260–261n13
  civil society activism and, 164, 179
  in climate negotiations, 133–136
  dominance in international climate
    policy, 148–152, 243–244
  early efforts to reduce emissions,
    138–140
  fracturing among groups in, 141–142
  lobbying, 149–151
  marketing diversion by, 142–148
  obstruction by, 140, 151
  profits, 144
  renewable energy and, 143–144, 151
  unconventional, 45–46
Foucault, Michel, 30, 179
Fragmented climate governance,
  31–34
Franco, Francisco, 247
Friedman, Thomas, 205–206
Friends of the Earth, 88, 143, 162,
  164, 172
Fry, Ian, 1, 167, 242
Future of climate politics
  desperate technofixes scenario,
    216–220

examining scenarios in, 205–209
exclusive action scenario, 213–216
exclusive inaction scenario, 209–213
global climate justice scenario, 229–232
going local scenario, 226–229
riding renewables scenario, 221–226
wagering on warming worlds and, 232–234

G77. *See* Group of 77
Gareau, Brian, 4, 166
Gender Climate Change, 187
Gender equality, 185, 186–190
General Agreement on Tariffs and Trade (GATT), 178–179
General Motors, 175
Geoengineering, 129, 220
Germanwatch, 168
Global Alliance of Waste Pickers and Allies (GAWA), 195–196, 199
Global circulation models (GCMs), 2
Global Cities Covenant on Climate, 15
Global Climate Coalition, 140, 142–143
Global Environment Facility (GEF), 104, 119
Global governance architecture, 31
Globalization, contemporary, 48–49
Global power, 3–5, 51–52. *See also* Climate change politics; Transnational advocacy networks (TANs)
decline of US and rise of China, 40–45, 69, 239
ecological collapse and development space and, 45–48
four perspectives on, 24–27
fragmented climate governance and, 31–34
shifting world order and, 36–41
states, markets, ecosystems, and civil society in, 34–35
transnational coalitions and, 27–31
Global Warming Solutions Act (California), 16

Governance, climate, 31–34, 179–180
Governmentality, 30, 179
Gramsci, Antonio, 4, 23–24, 100, 236–237
on counterhegemony, 250
on decisive nucleus of economic activity, 34–35
on hegemony, 27–28, 78
ideology and, 79
on new historic bloc, 247
on political domination, 29
on unstable equilibria, 79
on war of position, 79, 95–96
Grasso, Marco, 109, 214
Grassroots activism, 14, 226–229
Great Recession of 2008, 36, 39, 51, 238–239, 243
Green Climate Fund (GCF), 107, 122, 124, 157
gender equality and, 188–189
right to livelihood and, 196
Green Group, 106
Greenpeace, 50, 143, 162
Group of 77, 56–58
adaptation and, 104–105, 108–109, 118
conflict among, 80–89, 98
emerging climate order and, 72–74
splintering of, 67–72
vulnerability identity and, 61
Group of Twenty, 3
*Guardian, The,* 222

Harper, Stephen, 206
Harvey, David, 49, 250
Hedegaard, Connie, 88
Heede, Richard, 147
Hegemony, 27–28, 78
counter-, 250
crises, 41–42
Historic bloc, 27–28
new, 246–252
Home Depot, 163
Hone, David, 150
Horner, Kate, 88
Hu Jintao, 52
Humphreys, David, 168

Idle No More movement, 159, 165, 228, 252

Independent Association of Latin America and the Caribbean, 93, 241

India, 69–72
 protests in, 106, 108

Indigenous peoples' networks, 185, 190–195

INGOs, 171

Institutionalist perspective, 24, 25–26

Intellectual property rules, 36

Intended Nationally Determined Contributions (INDCs), 94

Inter-American Development Bank, 191, 225

Intergovernmental Panel on Climate Change (IPCC), 2, 8–9, 57, 101, 220
 adaptation and, 105, 108
 on fossil fuel use and climate change, 144
 runaway warming and, 124–125

International Chamber of Commerce, 140, 150

International Climate Change Partnership, 140, 141

International Emissions Trading Association, 137, 142

International Energy Agency, 46, 143, 147, 211

International Indigenous Peoples Forum on Climate Change (IIPFCC), 190–191, 194

International Monetary Fund, 38, 143

International Petroleum Environmental Conservation Association, 140

International Policy Studies, 163

International Whaling Commission, 42

Jones, Van, 162

JP Morgan Chase, 47

Kaldor, Mary, 48

Kasperson, Jeanne, 5–6

Kasperson, Roger, 5–6

Keck, M. E., 186

Keohane, Robert, 41

Keystone XL pipeline, 155, 165

Kim, Jim Yong, 225

Klein, Naomi, 236

Kyoto Protocol, 1–2, 42, 51, 53, 54, 59, 81, 85–86, 98, 157, 241
 adaptation funding and, 108, 122–124
 business interests and, 141–142
 expiration of, 99
 indigenous rights and, 190
 post period, 142
 second commitment period, 68, 90–91, 253n4
 top down approach, 62

League of Nations, 206

Least Developed Countries (LDCs), 7, 31, 40, 54, 214–215, 241, 264n26
 adaptation and, 105, 110, 118, 122–123
 concessions in Cancun and, 89–90
 conflict among, 80–89
 Doha, Warsaw, and Lima negotiations, 92–95
 Durban discussions and, 90–92
 reasons for commitments made by, 95–99, 242
 trade dependency of, 97–98
 vulnerability identity and, 61, 96

Legitimacy, 30

Levy, David, 26, 28, 100

Lima Call for Climate Action, 93–95, 96, 157
 gender equality and, 189–190
 indigenous rights and, 193

Liverman, Diana, 15

*Long Twentieth Century, The,* 42

Lovins, Amory, 221

Lovins, Hunter, 138

Mad Max scenario, 209–213

Major Economies Forum on Climate and Energy (MEF), 213–216

March on Washington, 2013, 155

Market pragmatist perspective, 25, 26

McCarthy, James, 229
McKibben, Bill, 46, 147, 155
Meckling, Jonas, 50, 141
Mercy Corps, 146
Mexico City Pact, 15
Miliband, Ed, 87
Mitigation. *See* Adaptation
Mohamed, Abdul Ghafoor, 87
Mol, Arthur, 140
Monbiot, George, 222
Montreal Protocol, 166, 179, 186
Morales, Evo, 155–156
Müller, Benito, 142
Multilateralism, 34
Mussolini, Benito, 247

Naidoo, Kumi, 86–87
Nairobi Work Program, 107
Najam, Adil, 37
Nasheed, Mohamed, 54
National Adaptation Programmes of
    Action, 105
Natural disasters, 235–238
Natural gas, 47
Natural resource extraction, 37, 40
Natural Resources Defense Council
    (NRDC), 161, 162, 171, 219
*Nature,* 147
Nature Conservancy, 161
Negotiated consent, *80*
Neoliberalism, 38–41
Neumann, Iver B., 30
Newell, Peter, 28, 185
*New York Times,* 205
NIABY (Not In Anyone's Back Yard)
    perspective, 160
NIMBY (Not In My Back Yard)
    perspective, 160, 221
Nongovernmental organizations
    (NGOs), 50, 81, 136
  adaptation and, 106–107
  business, 134, 137, 166, 171
  emergence of global justice, 169–171
  environmental, 137, 166, 171
  indigenous, 171
  influence of, 166
  professionalized, 157

recent trends, 239–240
research institution, 171
youth, 171
NOPE (Not on Planet Earth)
    perspective, 160
Norgaard, Kari, 222
North-South politics
  adaptation finance and, 102
  Bali action plan and, 53, 62–63, *64*
  cooperation and, 77–79
  Copenhagen Accord and, 65–67
  Durban negotiations and, 91
  emerging climate order and, 72–74
  new identity of vulnerability and,
    60–62
  old world order and, 55–60
  recent trends, 239–240
  splintering of the South in, 67–72
  trade dependency and, 97–98

Obama, Barack, 45, 52, 53, 65, 155,
    206, 213
Occupy movement, 49
Offsets and trading, carbon, 146–147,
    148–149, 151–152, 170, 244
Organization of the Petroleum
    Exporting Countries (OPEC),
    42–43, 58–59, 80–81, 260–261n13
  adaptation and, 105
  vulnerability identity and, 61–62
*Our Common Future,* 138
Overseas Private Investment
    Corporation, 163
Oxfam, 169
Oxfam International, 168
Oxford Institute for Energy Studies,
    142

Pacific Environment, 163
Parks, Bradley, 36, 55
Patterson, Matthew, 43
People's Climate March, 231, 249,
    252, 282–283n24
Pershing, Jonathan, 88
Petromarket civilization, 47
Pew Center, 141, 176
Polanyi, Karl, 247

Polish Energy Group, 175
Political domination, 29
Poor, environmentalism of the, 158–161
"Post-Hegemonic Climate Politics?." 43
Power, global. *See* Global power
PriceWaterhouseCooper, 10
Princeton Review, 15
Project identity movements, 49
Pulver, Simone, 137
Putnam, Robert, 228

Rainforest Action Network (RAN), 163
Ramesh, Jairam, 70
Rasmussen, Lars Løkke, 1, 53
Realization-focused comparison, 110
REDD+ (reducing emissions from deforestation and land degradation in developing countries), 149, 191–194, 199
Refugees, climate, 126
Regime rights, 183–186, *200–201*
Regional Greenhouse Gas Initiative, 16, 149
Renewable energy, 143–144, 151, 221–226
Rights of Mother Earth, 155
Right to development, 36
Right to livelihood, 181, 195–198, 199
RINGOs, 171
Rio Earth Summit, 7, 56, 139, 168
Rising Tide, 49, 169
Roberts, J. Timmons, 36, 55, 214
Robinson, Mary, 189
Rockefeller, John D., 164
Rockefeller Brothers Fund, 164
Royal Dutch Shell, 150–151

Salerno, Claudia, 87
Schwarzenegger, Arnold, 16
Sen, Amartya, 12–13, 102, 110
Sending, Ole Jacob, 30
Sethi, Nitin, 96
Shell Oil, 161

Shifting world order, 36–41
*Shock Doctrine, The,* 236
Sierra Club, 14–15, 50, 143, 162, 168, 249
Sikkink, Kathryn, 185, 186
Silver, Beverly, 41, 42, 43, 44, 247
*Sixty Minutes,* 164
Skocpol, Theda, 161–162
Small island developing states (SIDS), 81, 86, 118
Solar radiation management (SLM), 128–129, 216
Solón, Pablo, 66, 211
Spaargaren, Gert, 140
Strategic analytical framework, *32, 35*
Strong, Maurice, 139
Structuralist perspective, 24, 25
Superstorm Sandy, 235–236
Sustainable Endowments Institute, 15

Tacit power, 244
Tangri, Neil, 197–198
Tck Tck Tck, 172
Tea Party, 162
Technology, carbon, 128–129, 216–220
Theory of justice, 12–13, 110
Third World Network, 171
350.org movement, 46, 155–156, 249
Translocal approach, 226–229
Transnational advocacy networks (TANs), 181–183
  assessing gains by, 198–199
  gender equality and, 185, 186–190
  indigenous rights and, 185, 190–195
  new historic bloc and, 246–252
  regime rights and, 183–186, *200–201*
  right to livelihood and, 181, 195–198, 199
Transnational civil society, 48–51, 153, 169–173, 179–180
Transnational coalitions, 27–31
Trans Pacific Partnership, 178
Tutu, Desmond, 60

Tyndall Centre for Climate Research, 15, 147

Unconventional fossil fuels, 45–46
Union Carbide, 139
Union of Concerned Scientists, 168
United Nations
Conference of Parties, 53, 58, 88, 103, 118, 189
Conference on the Human Environment, 56
Conventions on Statelessness, 126
Declaration for the Rights of Indigenous Peoples (UNDRIP), 191–192
Nuclear Non-Proliferation Treaty, 220
United Nations Copenhagen Accord, 1–3, 11, 42, 53–55, 65–67, 95–96
adaptation and, 107, 108, 114–117
Clean Development Mechanism and, 181
conflict over, 80–89, 171–173
emerging climate order from, 72–74
faith lost in UN system after, 75
integrated into Cancun Agreements, 90
United Nations Framework Convention on Climate Change (UNFCCC), 13, 15, 17, 28, 33, 56, 102–103, 241
adaptation finance justice and, 110–111, 112–113, 114–116
Cancun Agreements compared to, 89–90
civil society and, 156–157, 177–178
ENGOs and, 166
founding of, 136
gender equality and, 187
indigenous rights and, 190–192
scaled-up funding, 116
United States decline relative to China, 40–45, 239
Unstable equilibria, 79
Uribe, Alvaro, 87–88

US Climate Action Partnership (USCAP), 149, 161–162
US Export-Import Bank, 163, 164

*Vanity Fair,* 164
Via Campesina, La, 49, 159, 165, 227
Victoria's Secret, 163
Vihma, Antto, 57
Viola, Eduardo, 207
Vormedal, Irja, 136, 150
Vulnerability, 60–62, 96

Wade, R. H., 44
War of position, 79, 95–96
Warsaw negotiations, 92–93, 157
indigenous rights and, 193
Washington Consensus, 38, 40
*Washington Post,* 162
Waste pickers, 181, 195–198
Wells Fargo, 47
Wen Jiabao, 65
Whitehouse, Sheldon, 155
White House Report on Climate Change, 125
WikiLeaks, 42, 87, 88, 145
Williamson, John, 38
Wilson, Woodrow, 206
Women and Gender Constituency, 187, 188
Women in Europe for a Common Future, 187
Women's Environment and Development Organization (WEDO), 187, 188
World Bank, 16, 38, 46, 125, 163, 179, 198, 225, 227
fossil fuel industry and, 143, 144–145
Global Environment Facility (GEF), 104
Global Facility for Disaster Risk Reduction, 127
indigenous rights and, 193
World Council of Churches, 164
World Health Organization, 33
World People's Conference on Climate Change, 155, 211

World Social Forum, 252
World Systems Theory, 37
World Trade Organization, 16, 36,
   169, 178–179
World Wildlife Fund, 50, 146, 162,
   168, 171, 272n61

Xie Zhenhua, 71–72

York, Richard, 144
Young, Iris, 183
YUNGOS, 171

Zenawi, Meles, 87
Zuma, Jacob, 72